Reinhard Selten (Ed.)

Rational Interaction

Essays in Honor of John C. Harsanyi

With 25 Figures

Springer-Verlag Berlin Heidelberg New York London
Paris Tokyo Hong Kong Barcelona Budapest

Prof. Dr. Reinhard Selten
University of Bonn
Department of Economics
Adenauerallee 24-42
D-5300 Bonn 1, FRG

ISBN 3-540-55067-4 Springer-Verlag Berlin Heidelberg New York Tokyo
ISBN 0-387-55067-4 Springer-Verlag New York Berlin Heidelberg Tokyo

2142/7130-543210 - Printed on acid - free paper

Preface

This book is a collection of essays in honor of John C. Harsanyi. Originally, we hoped that it would be ready in the year of his 70th birthday, but we did not quite succeed. It sometimes takes longer than anticipated to do things as well as one wants. I think that John Harsanyi will understand this.

It is my desire to express our thanks to all those who helped to make the book possible, especially to the publishing house, the Springer Verlag. I am grateful to Hildegard Grober who assisted us in the organizational tasks connected to our editorial effort.

I am deeply indebted to Horace W. Brock. His active support was extremely valuable. The results of my editorial efforts should partly be attributed to him.

Reinhard Selten

Table of Contents

IV. Game Models

V. Utilitarianism and Related Topics

VI. Reflections

Introduction

This book contains 23 essays in honor of John C. Harsanyi. The title "Rational Interaction" points to a common theme. Most of the contributions to this volume concern game theory or make use of it. Game theory can be described as the theory of rational interaction of self–interested individuals. Harsanyi is not only a leading game theorist, but also the founder of modern utilitarianism, which is a theory of rational interaction in an ethical context. Some of the contributions relate to this aspect of his work. Rational interaction in the search for truth is a venerable theme within the philosophy of science. Harsanyi's contributions to the development of a Bayesian epistemological position have also found an echo in this volume.

The volume has six parts:

I	Cooperative game theory
II	Mechanism design
III	Basic questions of non–cooperative game theory
IV	Game models
V	Utilitarianism and related topics
VI	Reflections

Part I contains 6 papers. Maschler and Owen present a generalization of the Shapley value to characteristic function games without side–payments. Myerson discusses fictitious transfer solutions in various contexts of cooperative game theory. Owen defines the "finagle point" which comes nearest to dominating all other imputations in a distance sense. Peleg looks at the rule of voting by count and account which has been used historically by Jewish communities. Rosenmüller investigates the set of incentive compatible mechanisms for a special class of cooperative games with incomplete information. His paper would also fit in part II on mechanism design. Sergiu Hart and Andreu Mas–Collel provide a new interpretation for the Shapley value of characteristic function games with side–payments. The Shapley value is shown to be the optimal strategy of one of the players in an auxiliary 2–person zero–sum game.

Part II contains 3 papers. Kalai and Samet establish the equivalence of Nash–implementability and perfect implementability for the case where the type probabilities of different players are stochastically independent. Van Damme investigates mechanisms of fair division under asymmetric information. D'Aspremont and Gérard–Varet explore incentive compatible mechanisms in problems without transferable utility.

Part III concerns basic questions of non–cooperative game theory. Güth proposes a criterion for bilateral comparisons between equilibrium points which may be used in equilibrium selection theories. The papers by Hardin and Sugden attack the logic of backward induction in games like the iterated Prisoner's Dilemma.

Under the heading "Game Models", part IV presents 3 papers. Friedman analyses a price–quantity oligopoly with adjustment costs. Selten describes a finite non–cooperative model of characteristic function bargaining. Shubik discusses questions of game theoretic modelling arising in an attempt to reconcile micro– and macroeconomics.

Part V contains 5 contributions on the ethics of rational interaction, especially utilitarianism, and related topics. Hammond presents a new and simpler proof for Harsanyi's basic theorem on linear welfare functions and discusses the problem of utilitarian judgments under incomplete information. Binmore defines the social contract as a convention used in the selection of equilibria. He describes an evolutionary process in a population of Bayesian individuals converging to the Nash solution as an equilibrium selection convention in a two–person Nash demand game. (This paper could have been included in part III.) Arrow shows that the avoidance of sex differentiation in annuities cannot be based on egalitarian arguments; however, it is required by utilitarian ethics. Roemer applies Greenberg's theory of social situations under the veil of ignorance in the context of a simple economic environment. Broome considers the problem of intertemporal personal welfare judgments and axiomatizes the "temporal good" of a rational individual as measured by a cardinal utility function.

Part VI is called "Reflections", since it presents 3 contributions "on science" rather than "within science". Samuelson examines an analogy between thermodynamics and economics suggested by von Neumann in connection with the von Neumann multisector growth model. Brock discusses the role of symmetry and invariance in theory construction, especially in the work of John Harsanyi, but also in general with examples from various disciplines. Finally Selten tries to give an appraisal of the significance of John C. Harsanyi as a system builder and conceptual innovator.

Reinhard Selten

I. Cooperative Game Theory

The Consistent Shapley Value for Games without Side Payments

MICHAEL MASCHLER

Department of Mathematics
The Hebrew University
Jerusalem 91904, Israel

GUILLERMO OWEN

Department of Mathematics
Naval Post Graduate School
Monterey, California 93940

Abstract. In [Maschler and Owen, 1989], a new generalization of the Shapley value for a class of NTU games was introduced. The motivation was a desire to preserve as much as possible the consistency property of the Shapley value for TU games, in the sense of [Hart and Mas-Colell, 1989]. It turned out that the new value resulted from an intuitive dynamic process which was interesting also for the class of TU games. Unfortunately, the class of NTU games was quite narrow; namely, the class of hyperplane games. The purpose of this paper is to extend the definition to the general class of NTU games, whose coalition functions satisfy (essentially) the usual requirements.

1. Introduction

In [Maschler and Owen, 1989] we introduced an extension of the Shapley value, which we called "*the consistent Shapley value*", to the class of *hyperplane games*. This class consists of non-transferable-utility (NTU) games whose coalitional [characteristic] function is a half space of $\Re^S$ for each coalition S.

The consistent value was justified by several nice properties:

(1) It generalized in some sense the [Hart and Mas-Colell] *reduced game property* of the Shapley value for transferable utility (TU) games (see Hart and Mas-Colell [1989]).
(2) It was an obvious generalization of the random order procedure to get the Shapley value of TU games. (We could therefore call this solution concept also "a random order Shapley value".)
(3) An intuitively justifiable dynamic procedure—interesting also in the TU case—leads the players from any arbitrary imputation to the value.

Unfortunately, the random order procedure of our [1989] paper did not produce satisfactory results for other NTU games. Usually it yielded outcomes which were not even Pareto-optimal. The reason is that, unlike the case of a hyperplane game, a convex combination of Pareto-optimal points need not be Pareto-optimal. It became necessary, therefore, to get a more satisfactory solution for NTU games—one that will coincide with the consistent value when the games happen to be hyperplane games. That is the purpose of this chapter.

The paper is organized as follows: In Section 2 we set the notation and recall the definition of the consistent Shapley value for hyperplane games. In Section 3 we introduce the desired extension and prove that it is never an empty set. In Section 4 we present an example which shows how the solution can sometimes be computed.

2. Preliminaries

Let $N := \{1, 2, \ldots, n\}$ be a finite set of *players*. Non-empty subsets of N are called *coalitions*. We denote by $\Re^S$ the real $|S|$-dimensional space with coordinates named after the members of S.

DEFINITION 2.1. *A game over N is a function V which assigns to each coalition S a non-empty closed subset $V(S)$ of $\Re^S$. It is denoted $(N; V)$.*

It is customary to require

(A1) $V(S)$ is *comprehensive*, in the sense that if $x \in V(S)$ and $y \leq x$, and $y \in \Re^S$, then $y \in V(S)$.

(A2) $V(S)$ is *bounded above*, in the sense that for each x in $V(S)$, the set $\{y \in V(S)\colon y \geq x\}$ is compact.

In the simplest case, each $V(S)$ consists of points in $\Re^S$ satisfying $\sum_{i \in S} x_i \leq r^S$ for some numbers r^S. Such games are known as the transferable-utility (TU) games. A somewhat more general case occurs when each $V(S)$ is given by a linear inequality:

$$V(S) = \{x \in \Re^S : \sum p_i^S x_i \leq r^S\}, \tag{2.1}$$

where p_i^S, $i \in S$, are positive numbers and r^S is a number. Such games are known as *hyperplane games*.

Let π be a permutation of N. We define recursively the *marginal worth vector* $y(V; \pi)$ to be an n-tuple y satisfying

$$y_{\pi(1)} = \max\{x_{\pi(1)} : x \in V(\{\pi(1)\})\} \tag{2.2}$$

and if $y_{\pi(j)}$ is already defined for $j = 1, 2, \ldots, s-1$, then

$$y_{\pi(s)} = \max\{x_{\pi(s)} : x \in V(\{\pi(1), \ldots, \pi(s)\}) \text{ and } x_{\pi(j)} = y_{\pi(j)} \text{ for } j = 1, 2, \ldots, s-1\}. \tag{2.3}$$

Thus, for a given order $\pi(N)$, each player successively obtains the maximum he can get if he joins the previous players who have already gotten their payments.

The consistent Shapley value φ for hyperplane games is the arithmetic mean of these vectors:

$$\varphi(N; V) = \sum_{\pi} y(V; \pi)/n! \,. \tag{2.4}$$

We refer the reader to [Maschler and Owen, 1989] for the study of the properties of this solution concept.

3. The extension to general NTU games

One way to extend the hyperplane solution to the general class of NTU games with convex $V(S)$'s, suggested by Moshe Yarom,[1] is to pass arbitrary supporting hyperplanes to the

[1] Oral communication.

various sets $V(S)$. These hyperplanes determine a hyperplane game for which we know the solution. If this solution happens to lie on the Pareto-optimal surface of $V(N)$, we call it a solution of the given game $(N; V)$. One particular case of this solution is the classical NTU value given in [Shapley, 1988], which is obtained if we require that all the hyperplanes be parallel to each other when we properly imbed all the $V(S)$'s in the $\Re^N$ space.[2]

Yarom's idea yields, in our opinion, too many solution points when the game is not a hyperplane game and played by more than two players. We therefore propose to restrict the supporting hyperplanes by requiring for each proper subcoalition S that the associated hyperplane passes through the solution to the game $(S; V|S)$, where $V|S$ is the restriction of V to subsets of S. This leads to a recursive definition.

We also wish to have a definition for cases in which the $V(S)$'s are not always convex, and supporting hyperplanes through each Pareto-optimal point do not always exist. We want to replace them by tangent hyperplanes. We therefore prefer to impose the following restrictions on the coalition function:

(A3) The Pareto-optimal surface $\bar{V}(S)$ of each $V(S)$ is smooth; namely, this surface has, at each point, a unique normal vector, $p^S = \{p_i^S : i \in S\}$ which is normalized such that the sum of its coordinates in the outward direction is 1.

(A4) These p_i^S are continuous functions on $\bar{V}(S)$.

(A5) There exists a positive number δ, such that for each S and every i in S, $p_i^S > \delta$.

(A6) The origin 0^S lies in $V(S)$ for each S.

With these requirements we can proceed now to define the value:

DEFINITION 3.1. *Let $(N; V)$ be an NTU game whose coalition function satisfies (A1)–(A6). A consistent Shapley value for this game is a collection of points $\{x^S : \emptyset \neq S \subseteq N\}$ such that*

(a) *Each point x^S lies on the Pareto-optimal surface $\bar{V}(S)$,*

(b) *For each coalition S, the set $\{x^R : \emptyset \neq R \subseteq S\}$ is a consistent Shapley value for the subgame $(S; V|S)$ obtained from $(N; V)$ by restriction to the players in S.*

(c) *x^N is a solution to the hyperplane game $(N; \tilde{V})$ determined by the hyperplanes $H(x^S; S)$ in $\Re^S$, $\emptyset \neq S \subseteq N$, that pass through x^S and are tangent to $\bar{V}(S)$.*

REMARK 3.2. *One can avoid the recursive character of the above definition by replacing (b) and (c) by the following:*

(b′) *For each coalition S, x^S is the solution in the sense of Section 2, of the hyperplane game $(S; \tilde{V}|S)$.*

Note that instead of defining a solution as a point in $\bar{V}(N)$, we now define a solution as a collection of points—one for each coalition. This approach is similar to [Hart, 1985] approach of axiomatizing (essentially) the Harsanyi NTU value.

[2] Note that this solution is meaningless when the game is a hyperplane game because, in this case, there do not exist in general parallel supporting hyperplanes to all the various values of V. We are indebted to Lloyd Shapley for this observation which shows that the classical NTU value and the value defined in Section 2 are defined over different classes of games whose intersection is the class of TU games.

Note also that $\{\varphi(S): \emptyset \neq S \subseteq N\}$, where $\varphi(S)$ is the consistent Shapley value, as defined in Section 2, is the unique Shapley value in the sense of Definition 3.1 of $(S; \tilde{V}|S)$ (see Remark 3.2). In this sense, Definition 3.1 is an extension of the definition to hyperplane games.

The question is whether such a value exists. We answer this in the affirmative:

THEOREM 3.3. *Every NTU game satisfying (A1)–(A6) has a consistent solution.*

PROOF: We shall provide a proof by induction on the number of players n. Actually we prove the following stronger claim: *For any choice* $\{x^R: \emptyset \neq R \subsetneq N\}$, *such that for any coalition* S *other than* N, *the collection* $\{x^R: \emptyset \neq R \subseteq S\}$ *is a consistent value of the game* $(S; V|S)$, *there exists a point* x^N *on* $\bar{V}(N)$ *which complements the above collection to a solution of* $(N; V)$.

For $n = 1$, the claim is obviously true.

For $n = 2$, if $V(\{1,2\})$ is convex and contains the point $(\bar{V}(\{1\}), \bar{V}(\{2\}))$ then the solution is $\{\bar{V}(\{1\}), \bar{V}(\{2\}), \bar{x}^{\{1,2\}}\}$, where $\bar{x}^{\{1,2\}}$ is the Nash solution for the unanimity game $V(\{1,2\})$ with constant threatpoint $(\bar{V}(\{1\}), \bar{V}(\{2\}))$.

If the above conditions do not hold, there may be several values. Each collection $\{\bar{V}(\{1\}), \bar{V}(\{2\}), x^{\{1,2\}}\}$, where $x^{\{1,2\}}$ is a point in the *Nash set* for the above unanimity game, is a consistent Shapley value of the 2-person game, and there are no other values. The Nash set is defined to be the set of points $x^{\{1,2\}}$ on $\bar{V}(\{1,2\})$ for which the Nash product

$$(x_1 - \bar{V}(\{1\}))(x_2 - \bar{V}(\{2\})) \tag{3.1}$$

is stationary. We refer the reader to [Maschler, Owen and Peleg, 1988], where this set was introduced and studied.

Suppose now that the claim holds for fewer than n players, $n \geq 3$. Then, there exists a collection $\{x^R: \emptyset \neq R \subsetneq N\}$ such that for each coalition S other than N, $\{x^R: \emptyset \neq R \subseteq S\}$ is a consistent value to the game $(S; V|S)$.

Let

$$A = \max\{|x_i^R|: \emptyset \neq R \subsetneq N,\ i \in R\}. \tag{3.2}$$

For any Pareto-optimal point z in $\bar{V}(N)$, let $\varphi(z)$ be the consistent value, in the sense of Section 2, for the hyperplane game determined by the hyperplanes $H(x^R, R)$ for $\emptyset \neq R \subsetneq N$ and $H(z, N)$. We want to show that there exists a point z^N in $\bar{V}(N)$ for which $\varphi(z^N) = z^N$. If that is the case, adding it to the above collection yields a consistent value to the original game.

Now, $\varphi(z) = (\varphi_1(z), \varphi_2(z), \ldots, \varphi_n(z))$ is given by (2.4) (see also (2.2) and (2.3)). For each coordinate i, let us partition the sum in (2.4) into n sums of $(n-1)!$ terms each, depending on which player is the last in the ordering π:

$$\varphi_i(z) = \frac{1}{n!} \sum_{\{\pi:\ \pi(n)=i\}} y_i(V;\pi) + \frac{1}{n} \sum_{j \in N\setminus\{i\}} \frac{1}{(n-1)!} \sum_{\{\pi:\ \pi(n)=j\}} y_i(V;\pi), \tag{3.3}$$

$i = 1, 2, \ldots, n$. Now, for a fixed j, $j \neq i$,

$$\frac{1}{(n-1)!} \sum_{\{\pi\,:\ \pi(n)=j\}} y_i(V;\pi) = x_i^{N\setminus\{j\}}. \tag{3.4}$$

Indeed, the left side is the average of the marginal payments to player i for all orders in which he arrives before player j. During the processing of these orders he receives exactly the payment he would have received had he participated in the game restricted to the players $N \setminus \{j\}$; namely, $x^{N\setminus\{j\}}$.

As to the first sum in (3.4) we obtain

$$\frac{1}{(n-1)!} \sum_{\{\pi\,:\ \pi(n)=i\}} y_i(V;\pi) = \frac{r^N(z) - \sum_{j\in N\setminus\{i\}} p_j^N(z) x_j^{N\setminus\{i\}}}{p_i^N(z)} \tag{3.5}$$

Indeed, the left side is the average of marginal payments to player i for all orders in which he comes last. In these orders he receives whatever is left to him after all other players got their share, which is precisely $x_i^{N\setminus\{j\}}$ for each j, $j \neq i$. Since the equation of the hyperplane $H(z;N)$ is $\sum_N p_j^N(z)x_i = r^N(z)$, it follows that what is left for player i is the right side of (3.5).

Combining (3.3), (3.4) and (3.5), we obtain

$$\varphi_i(z) = \frac{1}{n} \sum_{j\in N\setminus\{i\}} x_i^{N\setminus\{j\}} + \frac{r^N(z) - \sum_{j\in N\setminus\{i\}} p_j^N(z) x_j^{N\setminus\{i\}}}{np_i^N(z)}. \tag{3.6}$$

By (A3)–(A5), the p_j^N's are a strictly positive and continuous function of z; therefore, $\varphi(z)$ is also a continuous function of z.

Choose M so that $M > A/\delta$, where δ is given by ($A5$), then, by ($A6$), $r^N(z) \geq 0$ and so, from (3.6),

$$n\varphi_i(z) \geq \sum_{j\in N\setminus\{i\}} x_i^{N\setminus\{i\}} - \frac{\sum_{j\in N\setminus\{i\}} p_j^N(z) x_j^{N\setminus\{i\}}}{p_i^N(z)}. \tag{3.7}$$

Consider the second sum in (3.7). By (A3), the sum of the $p_j^N(z)$'s is not more than 1, and so, the sum is not greater than a weighted average of the $x_j^{N\setminus\{i\}}$; i.e., not greater than $M\delta$ (see (3.2)). Also, each term in the first sum is not smaller than $-M\delta$. There are $n-1$ terms in this sum; hence, by (A5),

$$n\varphi_i(z) \geq -(n-1)M\delta - M\delta/p_i^N(z) > -(n-1)M\delta - M, \tag{3.8}$$

and since $\delta < 1$, we finally have:

$$\varphi_i(z) > -M. \tag{3.9}$$

Consider the simplex

$$\Omega = \{x\colon \sum_{i\in N} x_i = 0,\ x_i \geq -M,\ \text{all } i \text{ in } N\} \tag{3.10}$$

and let F be the projection function centered at $\sigma\colon = (-M, -M, \ldots, -M)$ from $D\colon = \sigma + \Re_+^N$ to Ω. Thus, $F(x)$ is the intersection of the line σx with Ω.

Consider also the restriction of F to $\bar{V}(N) \cap D$. Call this function G, then G is a $1-1$ map from $\bar{V}(N) \cap D$ to Ω. Both F and G^{-1} are continuous functions.

The function $f\colon \Omega \to \Omega$ defined by

$$f(x) = F\varphi(G^{-1}(x)) \tag{3.11}$$

is well defined because of (3.9) and continuous. Therefore, it has a fixed point $\bar{x}$. Denote by z^N the point $G^{-1}(\bar{x})$ and apply G^{-1} to both sides of (3.11), we obtain

$$z^N = G^{-1}F\varphi(z^N) = \varphi(z^N) \tag{3.12}$$

and this proves that $\varphi(z^N) = z^N$. ∎

4. An example

Consider a 3-person game where players 1,2 and 3 are to decide on the division of \$1, to be done by majority voting (2 out of 3 being a majority). The players' utility function for money in positive[3] amounts is the same; namely:

$$u_1(x) = x, \tag{4.1}$$

$$u_2(y) = 1 - e^{-y}, \tag{4.2}$$

$$u_3(z) = \ln(1 + z). \tag{4.3}$$

The coalition function of this game is:

$$V(\{i\}) = 0 - \Re_+^{\{i\}},\ i = 1,2,3, \tag{4.4}$$

$$V(\{1,2\}) = \{(u_1, 1 - e^{u_1 - 1}) \in \Re^{\{1,2\}} : 0 \le u_1 \le 1\} - \Re_+^{\{1,2\}}, \tag{4.5}$$

$$V(\{1,3\}) = \{(u_1, \ln(2 - u_1)) \in \Re^{\{1,3\}} : 0 \le u_1, \le 1\} - \Re_+^{\{1,3\}}, \tag{4.6}$$

$$V(\{2,3\}) = \{(1 - e^{-y}, \ln(2 - y)) \in \Re^{\{2,3\}} : 0 \le y \le 1\} - \Re_+^{\{1,3\}}, \tag{4.7}$$

$$V(\{1,2,3\}) = \{(u_1, 1 - e^{-y}, \ln(2 - u_1 - y)) \in \Re^{\{1,2,,3\}} : u_1 + y \le 1, u_1 \ge 0, y \ge 0\} - \Re_+^{\{1,2,3\}}. \tag{4.8}$$

The points on the Pareto optimal surfaces for the various coalitions satisfy the following equations:[4]

$$u_1 - \ln(1 - u_2) = 1 \qquad \text{for } \bar{V}(\{1,2\}), \tag{4.9}$$

$$u_1 + e^{u_3} = 2 \qquad \text{for } \bar{V}(\{1,3\}), \tag{4.10}$$

$$-\ln(1 - u_2) + e^{u_3} = 2 \qquad \text{for } \bar{V}(\{2,3\}), \tag{4.11}$$

$$u_1 - \ln(1 - u_2) + e^{u_3} = 2 \qquad \text{for } \bar{V}(\{1,2,3\}). \tag{4.12}$$

Note that this game does *not* satisfy the requirements (A3) and (A4) of Section 3, but in the interior of the ranges that interest us it does, as can be verified from (4.9)–(4.12).

---

[3]We do not specify utilities for negative amounts, because we do not wish to be engaged in sharing the dollar in a way that some players are becoming worse off.

[4]Note that in the domain which interests us, $0 \le u_2 \le 1 - e^{-1}$.

The consistent Shapley value of this game for the single-person coalitions is, of course, $\bar{x}^{\{i\}} = 0$. For the two-person coalition it must be the Nash point for the corresponding subgame. Calculating it, we find:

$$\bar{x}^{\{1,2\}} = (.5572, .3578) \tag{4.13}$$

$$\bar{x}^{\{1,3\}} = (.5450, .3750) \tag{4.14}$$

$$\bar{x}^{\{2,3\}} = (.3852, .4143) \tag{4.15}$$

To find $\bar{x}^{\{1,2,3\}}$, we pass supporting hyperplanes through the points (4.13)–(4.15) to the corresponding values of the coalition functions. We also pass a supporting hyperplane to $V(\{1,2,3\})$ through an arbitrary point $u = (u_1, u_2, u_3)$ in $\bar{V}(N)$ and compute the value, in the sense of Section 2, for the resulting hyperplane game. We search for that point (u_1, u_2, u_3) that coincides with the value.

The equation of the hyperplane is

$$p_1 z_1 + p_2 z_2 + p_3 z_3 = r, \tag{4.16}$$

where $p_1 = 1$, $p_2 = 1/(1 - u_2)$, $p_3 = e^{u_3}$ and $r = p_1 u_1 + p_2 u_2 + p_3 u_3$. The equations for the fixed point are:

$$\begin{cases} u_1 &= [.5572 + .5450 + (r - .3854 p_2 - .4143 p_3)/p_1]/3, \\ u_2 &= [.3578 + .3854 + (r - .5450 p_1 - .3750 p_3)/p_2]/3, \\ u_3 &= [.3750 + .4143 + (r - .5572 p_1 - .3578 p_2)/p_3]/3 \end{cases} \tag{4.17}$$

These equations are now easy to solve numerically in order to obtain the value

$$\bar{x}^N = (.3770, .2619, .2771). \tag{4.18}$$

From the last relation we infer that the dollar should be shared in accordance with the following split:

$$x = \$.3770,\ y = \$.3037,\ z = \$.3193. \tag{4.19}$$

5. References

Hart, S., *An Axiomatization of Harsanyi's Non-Transferable Utility Solution*, Econometrica **53** (1985), 1295–1313.

Hart, S., and A. Mas-Colell, *Potential, Value and Consistency*, Econometrica **57** (1989), 589–614.

Maschler, M., and G. Owen, *The Consistent Shapley Value for Hyperplane Games*, International Journal of Game Theory **18** (1989), 389–407.

Maschler, M., G. Owen, and B. Peleg, *Paths Leading to the Nash Set*, in "The Shapley Value, Essays in Honor of Lloyd S. Shapley," A. E. Roth, ed., Cambridge University Press, Cambridge–New York–New Rochelle–Melbourne–Sydney, 1988, pp. 321–330.

Shapley, L. S., *Utility Comparison and the Theory of Games*, in "La Décision: Aggrégation et Dynamique des Ordres de Preférénce," Edition du CNRS, Paris, 1969, pp. 251–263;

also, in "The Shapley Value, Essays in Honor of Lloyd S. Shapley," A. E. Roth. ed., Cambridge University Press, Cambridge–New York–New Rochelle–Melbourne–Sydney, 1988, pp. 307–319.

Keywords. Game Theory, Cooperative Games, Games Without Side Payments, NTU Games, Shapley Value, Consistency, Reduced Game Property

Fictitious-Transfer Solutions in Cooperative Game Theory

Roger B. Myerson
J. L. Kellogg Graduate School of Management, Northwestern University
Evanston, Illinois 60208, U.S.A.

It is easy to see that games with transferable utility (TU) really are just a special case of games with nontransferable utility (NTU), because transfer activities can be remodeled as strategic options in a game "without transferable utility." Thus, it is natural to ask why game theorists should have devoted substantial efforts to developing solution concepts for games with transferable utility. When I teach cooperative game theory now, more than 45 years after von Neumann and Morgenstern [1944], I try to motivate the old emphasis on coalitional games with transferable utility by two propositions. First, because coalitional interactions can be very complicated, we may initially want to simplify our analysis by assuming transferable utility, so that the set of feasible utility allocations for each coalition can be described by a single number. Second, by the method of fictitious transfers of weighted utility (or λ-transfers), we can easily generalize any solution concept for transferable-utility games to the case of games without transferable utility. Thus, from the perspective of 1990, the method of fictitious transfers appears to justify the original decision of von Neumann and Morgenstern [1944] to concentrate on games with transferable utility. However, the method of fictitious transfers was recognized only after the pivotal breakthrough of Harsanyi [1963]. My purpose in this paper is to reexamine the importance of this method, in the context of the history of its development and some recent results.

The assumption of transferable utility seems to have been adopted almost without discussion in the seminal work of von Neumann [1928] and von Neumann and Morgenstern [1944]. In the index of von Neumann and Morgenstern [1944], there are only two references to transferability of utility. Remarkably, in the book where it is first proven that cardinal utility is measurable from the risk preferences of a rational decision maker, the main justification of transferability appears to be to avoid measurement questions. Von

Neumann and Morgenstern [1944, page 8] assert:

> We wish to concentrate on one problem -- which is not that of measurement of utilities and of preferences -- and we shall therefore attempt to simplify all other characteristics as far as reasonably possible. We shall therefore assume that the aim of all participants in the economic system, consumers as well as entrepreneurs, is money, or equivalently a single monetary commodity. This is supposed to be unrestictedly divisible and substitutable, freely transferable and identical, even in the quantitative sense, with whatever "satisfaction" or "utility" is desired by each participant.

But from von Neumann and Morgenstern's own theory of utility, we now understand that, with great generality, the options available to a coalition can be represented by a convex set of expected utility allocations. This result was used by Nash [1950]; but it was not until his second paper on bargaining (Nash [1953]) that Nash emphasized that, by studying games without transferable utility, he was dropping one of the restrictive assumptions of the previous work in game theory.

In the same volume of Econometrica, Shapley and Shubik [1953] published a one-page abstract that describes how von Neumann and Morgenstern's solution theory can be generalized to games without transferable utility, and they found that nontransferable utility did not seem to create major new conceptual problems. However, if Shapley and Shubik had also considered the problem of generalizing the Shapley [1953] value, no doubt they would have had more difficulties to report. By the late 1950s, published work in game theory reflected a much greater frustration with the prevailing assumption of transferable utility. Luce and Raiffa [1957, pp. 233-4] remark:

> The assumption ... that there exists a transferable utility in which sidepayments are effected is exceedingly restrictive -- for many purposes it renders n-person theory next to useless.

It is at this point that John Harsanyi began the search for a one point solution concept for n-person cooperative games without transferable utility. He had two partial solutions to build on: the Shapley value for n-person games with transferable utility and the Nash bargaining solution for two-person games without transferable utility. Each of these solution concepts had been compellingly derived as the unique solution satisfying a natural set of axioms, and each selects a unique payoff allocation for each game in its respective domain. In the fourth Princeton volume on game theory, Harsanyi [1959] published a general solution concept that determines a payoff allocation for any n-person cooperative game with nontransferable utility. He showed that this bargaining solution coincides with the Nash bargaining solution in the two-person case, and coincides with the Shapley value in the transferable utility case. Thus, for the first time, it was suggested that the Nash bargaining solution and the Shapley value might be understood as special cases of a more general solution concept.

There is a easy way to generalize the Nash bargaining solution to any number of players, by simply maximizing the product of all players' utility gains over the disagreement point. However, this simple n-person Nash bargaining solution is unacceptable because it neglects the power of subcoalitions, and thus it does not coincide with the Shapley value in the TU case. In particular, this simple multiplayer Nash bargaining solution would generally give positive payoff to a dummy player with no strategic options. In his first general solution concept, Harsanyi [1959] brought all coalitions into the analysis by assuming that every coalition (or nonempty set of players) would negotiate a vector of "dividends," such that the sum of all coalitions' dividend vectors would be a feasible allocation for the grand coalition of all players. Within each coalition S, Harsanyi [1959] tried to determine the dividend vector by applying the simple multiplayer Nash bargaining solution, where the disagreement point would be the sum of all dividend vectors for subsets of the coalition S. A difficulty with this approach arises because, for games with three or more players (such as the three-person majority game), the disagreement point for some coalitions may be outside of the feasible set, so that negative dividends may be required. Harsanyi's [1959] attempt to deal with this negative-dividend case seemed unsatisfactory, and Isbell [1960] showed that the solution can violate the basic individual-rationality constraints (that each player must get at least what he could earn on his own). Isbell [1960] suggested an alternative solution concept that always satisfies individual rationality and generalizes the Shapley value, but does not generalize the Nash bargaining solution. (From a modern perspective, Isbell's solution looks more like a generalization of the Kalai-Smorodinsky [1975] solution.)

In response to such objections, Harsanyi re-examined the Nash bargaining solution, to identify other ways of extending it to the multiplayer case. The key insight appears as Theorem 1 in Harsanyi [1963]. We may state it here as follows. In a two-person given game, let $V(\{1,2\})$ denote the set of feasible utility allocations for players 1 and 2 when they cooperate. Let $\Delta(C_i)$ denote the set of randomized strategies for player i, and let $u_i(\sigma_1,\sigma_2)$ denote the expected utility payoff for player i when players 1 and 2 use the randomized strategies σ_1 and σ_2 respectively. Then (x_1,x_2) is the Nash bargaining solution and (τ_1,τ_2) are the rational threats for the game with variable threats iff there exist nonnegative numbers λ_1 and λ_2, not both zero, such that

(1) $$(x_1,x_2) \in V(\{1,2\}),$$

(2) $$\lambda_1 x_1 + \lambda_2 x_2 = \max_{w \in V(\{1,2\})} (\lambda_1 w_1 + \lambda_2 w_2),$$

(3) $$\lambda_1(x_1 - u_1(\tau_1,\tau_2)) = \lambda_2(x_2 - u_2(\tau_1,\tau_2)),$$

(4) $$\tau_1 \in \operatorname{argmax}_{\sigma_1 \in \Delta(C_1)} (\lambda_1 u_1(\sigma_1,\tau_2) - \lambda_2 u_2(\sigma_1,\tau_2)),$$

(5) $$\tau_2 \in \operatorname{argmax}_{\sigma_2 \in \Delta(C_2)} (\lambda_2 u_2(\tau_1,\sigma_2) - \lambda_1 u_1(\tau_1,\sigma_2)).$$

These conditions have a natural interpretation. With nontransferable utility, we have no grounds for interpersonal comparison of utility, so we may feel free to rescale either player's utility separately by a positive scaling factor or utility weight λ_i. Now, in a rescaled version of the game, pretend that the weighted-utility payoffs are transferable. Suppose that each player chooses a threat that maximizes the difference between his own and the other player's weighted-utility payoff (4,5), and they then divide among themselves the maximal transferable weighted-utility worth that they can earn together (2), in such a way that each player gets the same weighted-utility gain over the threat point (3). If the resulting utility allocation would actually be feasible even without these fictitious transfers (1), then it is a Nash bargaining solution.

Given a game in strategic (or normal) form, Von Neumann and Morgenstern [1944] defined the worth of a coalition S to be the maximum total payoff that the members of S can guarantee themselves against the most offensive threat from the complementary coalition N\S (where N denotes the set of all players). Conditions (4) and (5), derived from Nash's [1953] rational threats criterion, pose a challenge to von Neumann and Morgenstern's definition of the characteristic function. For a game with transferable utility, condition (2) just implies that $\lambda_1 = \lambda_2$, and then (4) and (5) assert that the worth of each one-person coalition should be defined in terms of a threat game where each player seeks to maximize the difference between the worth of his (one-person) coalition and the complementary coalition. Harsanyi [1963] generalized this criterion to create a new way to derive coalitional-form games from games in strategic form, for both the TU and NTU cases. Around the same time, Aumann and Peleg [1960] and Jentzsch [1964] developed other ways of deriving NTU coalitional-form games from strategic-form games, but Aumann and Peleg and Jentzsch's derivations coincide with von Neumann and Morgenstern's minimax definition in the TU case.

For our present purposes, however, the most important consequence of these conditions (1)-(5) is that they revealed new ways to extend the Nash bargaining solution to games with more than two players. We may present these ideas most simply by assuming that we are given a game in NTU coalitional form. When N denotes the finite set of players, an <u>NTU game in coalitional form</u> is a function V such that, for each coalition S that is a nonempty subset of N, V(S) is a convex subset of $\mathbb{R}^S$ (the set of all vectors of real numbers indexed on the members of S). We interpret V(S) as the set of expected utility allocations that the members of S could cooperatively guarantee themselves. We assume throughout this paper that V has the following properties, for all coalitions S and T:

(6) V(S) is a nonempty closed convex subset of $\mathbb{R}^S$;

(7) if $x \in \mathbb{R}^S$, $y \in V(S)$, and $x_i \le y_i$, $\forall i \in S$, then $x \in V(S)$;

(8) if $S \cap T = \emptyset$ then
$$V(S \cup T) \supseteq \{x \in \mathbb{R}^{S \cup T} | (x_i)_{i \in S} \in V(S), (x_j)_{j \in T} \in V(T)\}.$$

(Here x_i and y_i respectively denote the i-components of the vectors x and y.) Condition (7) asserts that each set V(S) is <u>comprehensive</u> and (8) asserts that V is <u>superadditive</u>. As an additional technical condition, we may say that V is <u>finitely generated</u> or <u>polyhedral</u> iff, for every coalition S, there exist a finite set of points W(S) such that V(S) is the smallest convex comprehensive set containing W(S); that is,

(9) $x \in V(S)$ iff there exists a function $q: W(S) \rightarrow \mathbb{R}$ such that $q(y) \geq 0$ for every y in W(S), $\Sigma_{z \in W(S)} q(z) = 1$, and $x_i \leq \Sigma_{z \in W(S)} q(z) z_i$, $\forall i \in S$.

Let $\partial V(S)$ denote the boundary of V(S), so that $x \in \partial V(S)$ iff $x \in V(S)$ and there does not exist any y in V(S) such that $y_i > x_i$ for all i in S.

A feasible payoff allocation $\bar{x}$ is a <u>Harsanyi NTU value</u>, in the sense of Harsanyi's [1963] simplified bargaining solution, if there exists a nonnegative vector $\lambda = (\lambda_i)_{i \in N}$ and, for each coalition S, there exists a coalitional allocation $x^S = (x_i^S)_{i \in S}$, such that:

(10) $$\bar{x} = x^N \in \operatorname{argmax}_{w \in V(N)} \Sigma_{i \in N} \lambda_i w_i;$$

(11) $$x^S \in \partial V(S), \quad \forall S \subseteq N;$$

(12) $$\lambda_i (x_i^S - x_i^{S-j}) = \lambda_j (x_j^S - x_j^{S-i}), \quad \forall S \subseteq N, \forall i \in S, \forall j \in S-i.$$

Condition (10) generalizes condition (2) of the Nash bargaining solution. Condition (12), called <u>balanced contributions</u> by Myerson [1980], asserts that each pair of players make equal weighted-utility contributions to each other in each coalition, and generalizes the λ-equity condition (3) of the Nash bargaining solution. In the domain of two-person games, the Harsanyi NTU value coincides with the Nash bargaining solution. In the domain of coalitional games with transferable utility (that is, where there exists some TU coalitional game v such that

$$V(S) = \{x \in \mathbb{R}^S \mid \Sigma_{i\in S}\, x_i \leq v(S)\}$$

for each coalition S), the Harsanyi NTU value coincides with the Shapley value.

Shapley [1964] (see also Aumann [1967] and Shapley [1969]) found Harsanyi's NTU value to be hard to work with in large games, and so he proposed another NTU value that is also naturally derived from Harsanyi's conditions (1)-(3) for the Nash bargaining solution. Let $\mathbb{R}^N_{++}$ denote the set of all strictly positive vectors indexed on N, so

$$\mathbb{R}^N_{++} = \{\lambda = (\lambda_i)_{i\in N} \mid \lambda_i > 0,\ \forall i \in N\}.$$

For any positive vector λ in $\mathbb{R}^N_{++}$, let $F(V,\lambda)$ be a set of TU coalitional games such that $v \in F(V,\lambda)$ iff

$$(13) \qquad v(N) \geq \max_{y\in V(N)} \Sigma_{i\in N}\, \lambda_i y_i, \quad \text{and}$$

$$v(S) = \max_{y\in V(S)} \Sigma_{i\in S}\, \lambda_i y_i, \quad \forall S \subset N.$$

That is, $F(V,\lambda)$ is the set of TU coalitional game that we would get if we enlarged the feasible set for the grand coalition N, rescaled each player i's utility by the weighting factor λ_i, and then (fictitiously) allowed every coalition to make interpersonal transfers of these weighted-utility payoffs. Now let

$$\phi(v) = (\phi_i(v))_{i\in N}$$

denote the Shapley value of any game v in $F(V,\lambda)$. If we believe in the Shapley value as a predictive solution concept for TU games, then we should predict the weighted-utility allocation $\phi(v)$ in the fictitious game with transferable weighted utility. To convert each player i's payoff $\phi_i(v)$ back to the original utility scale, we must divide by weight λ_i. So let

$$(14) \qquad \Phi(V,\lambda) = \{((\phi_i(v)/\lambda_i)_{i\in N} \mid v \in F(V,\lambda)\}.$$

It can be shown that, for any such positive vector λ, $\bar{x} \in \Phi(V,\lambda)$ if and only if there exist coalitional allocations x^S, for each coalition S, such that

(15) $\bar{x} = x^N, \quad \forall i \in N;$

(16) $\Sigma_{i\in N} \lambda_i x_i^N \geq \max_{y\in V(N)} \Sigma_{i\in N} \lambda_i y_i;$

(17) $\Sigma_{i\in S} \lambda_i x_i^S = \max_{y\in V(S)} \Sigma_{i\in S} \lambda_i y_i, \quad \forall S \subset N;$

(18) $\lambda_i (x_i^S - x_i^{S-j}) = \lambda_j (x_j^S - x_j^{S-i}), \quad \forall S \subseteq N, \ \forall i \in S, \ \forall j \in S-i.$

Thus, the allocations in $\Phi(V,\lambda)$ satisfy an analogue of the conditions (10)-(12) for the TU games that is derived from V by expanding the grand coalition's feasible set and pretending that λ-weighted utility is transferable for all coalitions. An allocation $\bar{x}$ in $\Phi(V,\lambda)$ is a Shapley NTU value of V if it is actually feasible without any such fictitious transfers, that is, if $\bar{x} \in \Phi(V,\lambda) \cap V(N)$.

To be able to prove a general existence theorem, we must define Shapley NTU values more broadly, in a way that allows some λ_i components to go to zero. However, condition (14) is undefined when λ_i is zero, so some limit-condition is needed. There are several ways of constructing such a limit condition; here we consider a technical formulation slightly different from that of Shapley [1969]. Let us say that $\bar{x}$ is a Shapley NTU value of V iff $\bar{x} \in V(N)$ and, for every positive number ε, there exists vector λ in $\mathbb{R}^N_{++}$ such that and a TU coalitional game v in $F(V,\lambda)$ such that

(19) $\bar{x}_i + \varepsilon \geq \phi_i(v)/\lambda_i, \quad \forall i \in N.$

That is, a Shapley NTU value is a feasible allocation $\bar{x}$ such that, by slightly expanding the grand coalition's feasible set and making weighted utility transferable, we could generate a game for which the Shapley value is not substantially better than $\bar{x}$ (in the original utility scales) for any player.

Owen [1972] proposed a third NTU value that also extends by the Nash bargaining solution and the TU Shapley value, but Owen's NTU value requires the solution of a complicated set of differential equations and has not attracted much further interest. Shapley's and Harsanyi's NTU values have been subsequently derived axiomatically by Aumann [1985] and Hart [1985] for games where the boundary of V(N) is differentiable. On the other hand, Roth [1980] has given an example of a game in which the boundary of V(N) is not differentiable and the Shapley and Harsanyi NTU values seem quite unreasonable.

Shapley [1969] remarked that the method of fictitious transfers could be used to extend any TU solution concept to the more general NTU case. To see how, suppose that we are given a solution concept that specifies a set of solutions G(v) for each TU coalitional game v. Then to extend this solution concept G to NTU games by the <u>method of fictitious transfers</u>, we may say that an allocation $\bar{x}$ is a solution for V iff $\bar{x} \in V(N)$ and, for every (small) positive number ε, there exists a positive vector λ in $\mathbb{R}^N_{++}$, a TU coalitional game v in $F(V,\lambda)$, and a vector z in G(v) such that

$$(20) \qquad \bar{x}_i + \varepsilon \geq z_i/\lambda_i, \quad \forall i \in N.$$

That is, a solution of V is a feasible allocation $\bar{x}$ such that, by (slightly) increasing the feasible set for the grand coalition and making weighted utility transferable for all coalitions, we could generate a game for which there exist solutions that are not substantially better than $\bar{x}$ for any player.

When we use the core as our TU solution concept, the NTU solution concept generated by the method of fictitious transfers is called the <u>inner core</u> (see Shubik [1982], page 155). As the name suggests, the inner core is a subset of the core as originally defined for NTU games. The <u>core</u> of V has been defined (see Aumann and Peleg [1960]) to be the set of all allocations x such that $x \in V(N)$ and there does not exist any nonempty coalition S and any allocation y in V(S) such that $y_i > x_i$ for every player i in S. That is, x is in the core iff it is feasible for the grand coalition N, and there does not exist any coalition that could guarantee a higher payoff to all its members.

The comparison between the core and the inner core can provide us with a new test of power of the method of fictitious transfers. As we have seen, fictitious transfers were first used by Shapley in the early 1960s, to define a tractable generalization of the Shapley value for NTU games. At that time, there was not a controversy about the definition of the core for NTU coalitional games (although there was controversy about how to construct a game in NTU coalitional form to represent any given strategic-form game). So now, unasked-for, the method of fictitious transfers gives us an alternative way to define cores for NTU games. If the method of fictitious transfers is really an appropriate way to generate NTU solution concepts, then there should be some fundamental rationale for considering the inner core to be the appropriate extension of the TU core to NTU games.

To pose this question more concretely, let us consider the Banker game analyzed by Owen [1972]. In this three-person game, player 1 is the worker, player 2 is the helper, and player 3 is the banker. By himself, each player can get zero payoff. The pairs {1,3} and {2,3} cannot do any better, because the worker and helper are both needed to create anything of positive value. Players 1 and 2 together can generate a payoff of 100 for player 1, and player 1 can try to transfer any fraction of this payoff to player 2 but, without the banker, three-quarters of the transfered amount will be lost. With the banker, the players can transfer utility among themselves in any way. So we may let

$$V(\{i\}) = \{x_i \mid x_i \le 0\}, \quad \forall i \in N = \{1,2,3\},$$

$$V(\{1,2\}) = \{(x_1,x_2) \mid x_1 \le 100,\ x_2 \le (100 - x_1)/4,\ x_2 \le 25\},$$

$$V(\{1,3\}) = \{(x_1,x_3) \mid x_1 + x_3 \le 0\},$$

$$V(\{2,3\}) = \{(x_2,x_3) \mid x_2 + x_3 \le 0\}, \quad \text{and}$$

$$V(\{1,2,3\}) = \{(x_1,x_2,x_3) \mid x_1 + x_2 + x_3 \le 100\}.$$

The core for this game is

$$\{(x_1,x_2,x_3) \mid x_1 + x_2 + x_3 = 100,\ x_1 + 4x_2 \ge 100,\ x_i \ge 0,\ \forall i\},$$

which is a triangle with corners (100,0,0), (0,100,0), and (0,25,75). To compute the inner core, let v^λ denote the minimal TU game in $F(V,\lambda)$, so

$$v^\lambda(S) = \max_{y\in V(S)} \Sigma_{i\in S}\, \lambda_i y_i .$$

Notice that $v^\lambda(\{1,2,3\})$ would be $+\infty$ unless $\lambda_1 = \lambda_2 = \lambda_3$. When all λ_i are equal and positive,

$$v^{\lambda}(S) = 100\lambda_1 \quad \text{if} \quad S \supseteq \{1,2\},$$
$$= 0 \quad \text{otherwise,}$$

and so the core of the TU game v^{λ} is

$$\{(w_1,w_2,w_3) \mid w_1 + w_2 = 100\lambda_1,\ w_1 \geq 0,\ w_2 \geq 0,\ w_3 = 0\}.$$

Dividing payoffs by λ_1 to translate them back into the original utility scales, the core of v^{λ} then maps to the set

$$\{(x_1,x_2,x_3) \mid x_1 + x_2 = 100,\ x_1 \geq 0,\ x_2 \geq 0,\ x_3 = 0\},$$

which is in V(N) and therefore is the inner core. Increasing the worth of the grand coalition would not generate any other core allocations that are feasible in the original game. Thus, the inner core is the portion of the core where player 3 (the banker) gets zero.

The unique Shapley NTU value for this example is (50,50,0), which is the Shapley value of the TU game v^{λ} when $\lambda_1 = \lambda_2 = \lambda_3 = 1$. The Harsanyi NTU value is (40,40,20), which satisfies conditions (10)-(12) when

$$\lambda_1 = \lambda_2 = \lambda_3 = 1,$$
$$x_1^{\{1,2\}} = 20, \quad x_2^{\{1,2\}} = 20$$
$$x_1^{\{1,2,3\}} = 40, \quad x_2^{\{1,2,3\}} = 40, \quad x_3^{\{1,2,3\}} = 20,$$
$$\text{and} \quad x_i^S = 0 \quad \text{for each } i \text{ in any other coalition } S.$$

The Owen NTU value for this game is (51.88, 47.57, 0.58). Owen [1972] suggests intuitively that a reasonable solution should give a positive payoff to player 3, but the inner core and the Shapley NTU value contradict this intuition. To support the inner core, we now construct an intuitive justification for giving zero to the banker.

To be specific, suppose that the players expect to get the Harsanyi NTU allocation (40,40,20) unless some blocking coalition forms. The allocation (40,40,20) is in the core, as originally defined, so it might seem that no such blocking coalition could be organized. But suppose that a blocking agent (an outside mediator who is not a player) can try to organize a blocking coalition according to a pre-announced randomized plan. That is, suppose that an agent will approach some set of players and ask them to sign over power of attorney to him for all bargaining purposes. If any player who is approached by the agent refuses to give him power of attorney, then the agent will not block the (40,40,20) allocation; but if he gets power of attorney from all the players that he approaches, then the agent will block (40,40,20) and implement instead some allocation that is feasible for the coalition that he controls. Suppose that, when the agent approaches a player, the agent does not have to reveal the identity of the other players whom he has decided to approach, but he does have to describe the randomized plan by which he has made this decision and the expected utility allocation that he would implement for each such possible coalition.

Such a randomized blocking plan could be as follows: with probability .41, the agent will ask players 1 and 2 to join the blocking coalition and implement the allocation $(x_1,x_2) = (100,0)$; and with probability .59, the agent will ask all players to join the blocking coalition and implement the allocation $(x_1,x_2,x_3) = (0,79,21)$. Under this rule, players 1 and 2 are always approached by the blocking agent and get expected payoffs of $.41 \times 100 = 41$ and $.59 \times 79 = 46.61$ respectively. Furthermore, conditionally on being approached by the blocking agent, the expected payoff to player 3 is 21, which is strictly greater than 20. So each player should be willing to accept the agent's invitation to block (40,40,20). Thus, the core allocation of (40,40,20) can be blocked by a randomized blocking rule. Of course, this randomized blocking plan leaves player 3 worse off in expected value. Ex ante, player 3 expects to get $.59 \times 21 + .49 \times 0 = 12.39$ if the agent always succeeds. However, player 3 cannot prevent the blocking plan unless the agent approaches player 3, in which case accepting the plan and giving the agent power of attorney can only help player 3. Thus, unless player 3 can commit himself to reject the blocking coalition before he is invited to join it, the unique equilibrium in undominated strategies is for every player to join the agent's blocking coalition when invited to do so.

Similar tricks can be used to randomly block any other core allocation that gives player 3 a positive payoff. The idea is to randomize between two alternatives: either do not use player 3 and let player 1 keep everything; or use player 3 and transfer everything from player 1 while giving player 3 a slightly higher banker's commission than he would get in the allocation being blocked.

This random-blocking property actually characterizes the inner core quite generally. Let V be any NTU coalitional game on the finite set of players N. Define a randomized blocking plan to be any pair (η,Y) such that η is a probability distribution over the set of all possible coalitions, and Y is a function such that

$$Y(S) \in V(S)$$

for every coalition S. Here $\eta(S)$ represents the probability that the blocking agent will set out to form the coalition S, and Y(S) denotes the expected payoff allocation that he would implement if he succeeded in forming the coalition S.

Let x denote an allocation in $\mathbb{R}^N$ that is interpreted as the status quo, which is to be implemented unless the blocking agent is successful in forming his coalition. Then a randomized blocking plan (η,Y) is viable against x iff

$$\Sigma_{S\supseteq\{i\}}\ \eta(S)(Y_i(S) - x_i) \geq 0, \quad \forall i \in N.$$

That is, (η,Y) is viable against x iff for each player i, conditionally on player i's being invited to join the blocking coalition, his expected payoff in the blocking coalition would be at least x_i. We may say that x is strongly inhibitive iff there do not exist any viable randomized blocking plans against x_i. The following result follows from the work of Myerson [1988, 1991] and by Qin [1990].

Theorem 1. Let V be a finitely generated NTU coalitional game and let $\bar{x}$ be a feasible allocation in V(N). Then $\bar{x}$ is in the inner core of V if and only if there exists a sequence of allocations $(x^k)_{k=1}^{\infty}$ such that $\lim_{k\to\infty} x^k = \bar{x}$, and, for every k, x^k is strongly inhibitive for V.

Proof. Let $\Omega = \{(S,y)\,|\ S$ is a coalition and $y \in W(S)\}$, where

W(S) is the finite set of which V(S) is the comprehensive convex hull. A vector z is strongly inhibitive iff there is no probability distribution μ such that

$$\mu(S,y) \geq 0, \quad \forall (S,y) \in \Omega,$$

$$\Sigma_{S \supseteq \{i\}} \Sigma_{y \in W(S)} \mu(S,y)(z_i - y_i) \leq 0, \quad \forall i \in N,$$

$$\Sigma_{(S,y) \in \Omega} \mu(S,y) = 1.$$

By theorems of the alternative for linear systems, or the duality theorem for linear programming, the nonexistence of such a distribution μ holds iff there exists some vector λ in $\mathbb{R}^N$ such that

$$\lambda_i \geq 0, \quad \forall i \in N,$$

$$\Sigma_{i \in S} \lambda_i (z_i - y_i) > 0, \quad \forall (S,y) \in \Omega.$$

The last of these inequalities when $S = \{i\}$ implies that each λ_i must be strictly positive. Thus, z is strongly inhibitive iff there exists some positive vector λ in $\mathbb{R}^N_{++}$ such that

(21) $$\Sigma_{i \in S} \lambda_i z_i > v^{\lambda}(S), \quad \forall S \subseteq N$$

where

(22) $$v^{\lambda}(S) = \max_{y \in V(S)} \Sigma_{i \in S} \lambda_i y_i, \quad \forall S \subseteq N.$$

Now suppose that $\bar{x}$ is the limit of strongly inhibitive points $(x^k)_{k=1}^{\infty}$. For each k, let λ^k satisfy condition (21) for $z = x^k$, let $w_i^k = \lambda_i^k x_i^k$ for each player i, and let u^k be the TU coalitional game such that

$$u^k(N) = \Sigma_{i \in N} \lambda_i^k x_i^k, \quad \text{and} \quad u^k(S) = v^{\lambda^k}(S), \quad \forall S \subset N.$$

Then $u^k \in F(V,\lambda^k)$, $w^k \in \text{Core}(u^k)$, and $\lim_{k \to \infty} w_i^k/\lambda_i^k = \bar{x}_i$ for each i; and so $\bar{x}$ is in the inner core.

Conversely, suppose that $\bar{x}$ is in the inner core. Then there exists a sequence $(\lambda^k, v^k, w^k)_{k=1}^{\infty}$ such that $\limsup_{k \to \infty} w_i^k/\lambda_i^k \leq \bar{x}_i$ for every player i, and, for each k, $\lambda^k \in \mathbb{R}^N_{++}$, $v^k \in F(V,\lambda^k)$, and $w^k \in \text{Core}(v^k)$. By increasing the components of w^k and increasing the $v^k(N)$ numbers where necessary, we can assume without loss of generality that the $(w_i^k/\lambda_i^k)_{k=1}^\infty$ sequence is decreasing in k and converges to $\bar{x}_i$, for each i. Furthermore, by increasing the components of each w^k allocation slightly, we can assume that the core inequalities are all strictly satisfied for each w^k. Let $x_i^k = w_i^k/\lambda_i^k$ for each i and k. Then

$$\Sigma_{i\in S}\ \lambda_i^k x_i^k > v^{\lambda^k}(S), \quad \forall S \subseteq N.$$

So x^k is strongly inhibitive (by condition (21)). Thus $\bar{x}$ is a limit of strongly inhibitive allocations if it is in the inner core.

Q.E.D.

Harsanyi [1967-8] laid the foundations for the analysis of Bayesian games with incomplete information. Since his paper, perhaps because of rising interest in noncooperative game theory, the extension of cooperative game theory to games with incomplete information has been relatively neglected. (See Harsanyi and Selten [1972], Myerson [1984a, 1984b].)

For any game theorist who wants to develop solution concepts for cooperative games with incomplete information, a natural methodological question is how to extend the method of fictitious transfers to this case. Our technical result about the inner core for complete information games can be generalized and used to give an answer to this question.

Let us consider a simple model in which N is the finite set of players; T_i is the finite set of possible types of player i; C_S denotes the finite set of actions that are jointly feasible for each coalition $S \subseteq N$; $u_i(c_S,t_N)$ denotes the expected utility payoff that player i would get if he joined coalition S, which jointly implemented action c_S, and the profile of all players' types was $t_N = (t_i)_{i\in N}$; and $p_i(t_i)$ denotes positive probability that player i's true type is t_i. We assume here that players' types are independent random variables, but each player learns his own type before the play of the game begins. That is, a player's type is defined to be the state of his private information at the beginning of the game. We also assume that types are <u>unverifiable</u>, in the sense that there is nothing to prevent a player from lying about his type if he is given an incentive to do so.

For any possible coalition S, let $T_S = \times_{i \in S} T_i$ denote the set of all possible combinations of players' types in coalition S, and let

$$p_S(t_S) = \Pi_{j \in S} p_j(t_j), \quad \forall t_S \in T_S. \quad (\text{Here } t_S = (t_j)_{j \in S}.)$$

The status quo that an agent might try to block can be represented by an allocation rule $x:T_N \rightarrow \mathbb{R}^N$ such that, for every possible types profile t_N and every player i, $x_i(t_N)$ is the expected payoff to player i in the status quo if t_N is the profile of all players' types.

Now suppose that an outside agent or mediator attempts to block such a status quo x by forming a blocking coalition according to some random rule. For notational simplicity, we consider here only a very simple (but quite versatile) class of randomized blocking rules. Let us assume that the agent will randomly designate a coalition $S \subseteq N$, a profile of types t_S for this coalition, and a feasible joint action c_S for this coalition. Let $\mu(S,c_S,t_S)$ denote the probability of designating this coalition-types-action triple, under the random blocking plan μ. Then the agent will separately and confidentially approach the members of his designated coalition S and will request that they give him power of attorney in all subsequent bargaining. If any player in the designated coalition rejects the agent's request, then the agent will not block the status quo. Each player who gives the agent power of attorney will then be asked to confidentially tell his type to the agent. If the profile of reported types matches the designated profile t_S, then the agent will implement the designated joint action c_S for the coalition; if the profile of reported types does not match the designated types profile, then the agent will not block the status quo. We assume here that the status quo will be implemented if the agent does not block it. (When x results from the equilibrium of some game, there is a problem about whether the fact that a blocking coalition has not formed might convey information to the players that would change their behavior and thus change the status quo payoffs. One way to get around such considerations is to think of the $\mu(S,c_S,t_S)$ probabilities as infinitesimally small, so that very little information is conveyed by the nonformation of a blocking coalition.)

The agent has two problems in the design of his random blocking rule. He needs that the players in the blocking coalition should expect to do better by joining it than under the status quo, and he needs to give the players in the blocking coalition an incentive to share their type information honestly. So a random blocking plan μ is <u>viable</u> against x iff

(23) $\mu(S,c_S,t_S) \geq 0, \quad \forall S \subseteq N, \quad \forall c_S \in C_S, \quad \forall t_S \in T_S;$

(24) $\Sigma_{S \subseteq N} \Sigma_{c_S \in C_S} \Sigma_{t_S \in T_S} \mu(S,c_S,t_S) > 0;$

(25) $\Sigma_{S \supseteq \{i\}} \Sigma_{t_{N-i} \in T_{N-i}} \Sigma_{c_S \in C_S} \mu(S,c_S,t_S)\, p_{N-i}(t_{N-i})(u_i(c_S,t_N) - x_i(t_N)) \geq 0, \quad \forall i \in N, \quad \forall t_i \in T_i;$

(26) $\Sigma_{S \supseteq \{i\}} \Sigma_{t_{N-i} \in T_{N-i}} \Sigma_{c_S \in C_S} \mu(S,c_S,t_S)\, p_{N-i}(t_{N-i})(u_i(c_S,t_N) - x_i(t_N)) \geq$

$\Sigma_{S \supseteq \{i\}} \Sigma_{t_{N-i} \in T_{N-i}} \Sigma_{c_S \in C_S} \mu(S,c_S,(t_{S-i},r_i))p_{N-i}(t_{N-i})(u_i(c_S,t_N) - x_i(t_N)), \quad \forall i \in N, \quad \forall t_i \in T_i, \quad \forall r_i \in T_i.$

We say that the status quo x is <u>strongly inhibitive</u> iff there do not exist any randomized blocking plans that are viable against μ. Extending the preceding result, we may naturally define the inner core to be the set of status-quo allocation rules that can be implemented by incentive-compatible mechanisms and are equal or Pareto-superior to limits of strongly inhibitive allocation rules. (Of course, such a core may be empty for many examples. However, Myerson [1988] showed a class of replicated games in which a related core concept is always nonempty.) Now we may state the generalization of Theorem 1 which shows us how to characterize such strongly inhibitive allocations in terms of fictitious transfers of <u>virtual utility</u>, as defined in Myerson [1984a, 1984b].

<u>Theorem 2.</u> An allocation rule $x:T_N \to \mathbb{R}^N$ is strongly inhibitive iff there exists nonnegative numbers $\lambda_i(t_i)$ and

$\alpha_i(r_i|t_i)$, for every player i in N and all possible types t_i and r_i in T_i, such that

$$\Sigma_{i\in S} \Sigma_{t_{N\backslash S}\in T_{N\backslash S}} p_{N\backslash S}(t_{N\backslash S}) \hat{x}_i(t_N,\lambda,\alpha)$$
$$> \Sigma_{i\in S} \Sigma_{t_{N\backslash S}\in T_{N\backslash S}} p_{N\backslash S}(t_{N\backslash S}) \hat{u}_i(c_S,t_N,\lambda,\alpha),$$
$$\forall S \subseteq N, \quad \forall c_S \in C_S, \quad \forall t_S \in T_S;$$

where

$$\hat{u}_i(c_S,t_N,\lambda,\alpha) = [(\lambda_i(t_i) + \Sigma_{r_i\in T_i} \alpha_i(r_i|t_i))u_i(c_S,t_N)$$
$$- \Sigma_{r_i\in T_i} \alpha_i(t_i|r_i)u_i(c_S,t_{N-i},r_i)]/p_i(t_i), \quad \text{and}$$
$$\hat{x}_i(t_N,\lambda,\alpha) = [(\lambda_i(t_i) + \Sigma_{r_i\in T_i} \alpha_i(r_i|t_i))x_i(t_N)$$
$$- \Sigma_{r_i\in T_i} \alpha_i(t_i|r_i)x_i(t_{N-i},r_i)]/p_i(t_i).$$

(Notice that, if player i has only one possible type t_i then $\hat{x}_i(t_N,\lambda,\alpha) = \lambda_i(t_i)x_i(t_N)$ and $\hat{u}_i(c_S,t_N,\lambda,\alpha) = \lambda_i(t_i)u_i(c_S,t_N)$, and so these virtual-utility formulas are a generalization of λ-weighted utility for games with incomplete information.)

Proof. Consider the linear programming problem of maximizing the sum in line (24) over all nonnegative vectors μ, subject to the constraints (25) and (26). The optimal value of this homogeneous linear program is either 0 or $+\infty$. The allocation rule x is strongly inhibitive iff the optimal value is zero. So, by the duality theory of linear programming, x is strongly inhibitive iff the dual problem has a feasible solution. For each i, t_i, and r_i, let $\lambda_i(t_i)$ and $\alpha_i(r_i|t_i)$ denote the dual variables for constraints (25) and (26) respectively. Then the constraints of the dual problem can be written

$$\Sigma_{i\in S} \Sigma_{t_{N\backslash S}\in T_{N\backslash S}} p_{N-i}(t_{N-i})[(u_i(c_S,t_N) - x_i(t_N))(\lambda_i(t_i)$$
$$+ \Sigma_{r_i\in T_i} \alpha_i(r_i|t_i)) - \Sigma_{r_i\in T_i} (u_i(c_S,t_{N-i},r_i)$$
$$- x_i(t_{N-i},r_i))\alpha_i(t_i|r_i)] \geq 1,$$

$$\forall S \subseteq N, \quad \forall t_S \in T_S, \quad \forall c_S \in C_S.$$

It is straightforward to show that these dual constraints can be satisfied iff the constraints in the theorem can be satisfied by some nonnegative vectors λ and α. Q.E.D.

References

Aumann, R. J. [1967]. "A Survey of Cooperative Games Without Side Payments." In M. Shubik, ed., Essays in Mathematical Economics, pp. 3-27. Princeton: Princeton University Press.

Aumann, R. J. [1985]. "An Axiomatization of the Non-Transferable Utility Value." Econometrica 53:599-612.

Aumann, R. J., and B. Peleg [1960]. "Von Neumann-Morgenstern Solutions to Cooperative Games Without Sidepayments." Bulletin of the American Mathematical Society 66:173-179.

Harsanyi, J. C. [1959]. "A Bargaining Model for the Cooperative n-Person Game." In R. D. Luce and A. W. Tucker, eds., Contributions to the Theory of Games IV, Princeton: Princeton University Press, pages 325-355.

Harsanyi, J. C. [1963]. "A Simplified Bargaining Model for the n-Person Cooperative Game." International Economic Review 4:194-220.

Harsanyi, J. C. [1967-8]. "Games with Incomplete Information Played by 'Bayesian' Players." Management Science 14:159-182, 320-334, 486-502.

Harsanyi, J. C., and R. Selten. [1972]. "A Generalized Nash Solution for Two-Person Bargaining Games with Incomplete Information." Management Science 18:80-106.

Hart, S. [1985]. "An Axiomatization of Harsanyi's Nontransferable Utility Solution." Econometrica 53:1295-1313.

Isbell, J. R. [1960]. "A Modification of Harsanyi's Bargaining Model." Bulletin of the American Mathematical Society 66:70-73.

Jentzsch, G. [1959]. "Some Thoughts on the Theory of Cooperative Games." In R. D. Luce and A. W. Tucker, eds., Contributions to the Theory of Games IV, Princeton: Princeton University Press, pages 407-442.

Kalai, E., and M. Smorodinsky [1975]. "Other Solutions to Nash's Bargaining Problem." Econometrica 45:513-518.

Luce, R. D., and H. Raiffa [1957]. Games and Decisions. New York: Wiley.

Myerson, R. B. [1980]. "Conference Structures and Fair Allocation Rules." International Journal of Game Theory 9:169-182.

Myerson, R. B. [1984a]. "Two-Person Bargaining Problems with Incomplete Information." Econometrica 52:461-487.

Myerson, R. B. [1984b]. "Cooperative Games with Incomplete Information." International Journal of Game Theory 13:69-96.

Myerson, R. B. [1988]. "Sustainable Matching Plans with Adverse Selection." Northwestern University discussion paper. To appear in Games and Economic Behavior.

Nash, J. F. [1950]. "The Bargaining Problem." Econometrica 18:155-162.

Nash, J. F. [1953]. "Two-Person Cooperative Games." Econometrica 21:128-140.

Neumann, J. von. 1928. "Zur Theorie der Gesellschaftsspiele." Mathematische Annalen 100:295-320. English translation in R. D. Luce and A. W. Tucker, eds., Contributions to the Theory of Games IV, Princeton: Princeton University Press, pages 13-42.

Neumann, J. von, and O. Morgenstern [1944]. Theory of Games and Economic Behavior. Princeton: Princeton University Press. 2nd ed., 1947.

Owen, G. [1972]. "A Value for Non-Transferable Utility Games." International Journal of Game Theory 1:95-109.

Qin, C.-Z. [1990]. "The Inner Core and the Strongly Inhibitive Set." CORE discussion paper, Louvain-la-Neuve, Belgium.

Roth, A. E. [1980]. "Values for Games Without Side-Payments: Some Difficulties with Current Concepts." Econometrica 48:457-465.

Shapley, L. S. [1953]. "A Value for n-Person Games." In H. Kuhn and A. W. Tucker, eds., Contributions to the Theory of Games 2, pp. 307-317. Princeton: Princeton University Press.

Shapley, L. S. [1964]. ""Values of Large Market Games: the Status of the Problem," RAND Corporation paper RM-3957PR.

Shapley, L. S. [1969]. "Utility Comparison and the Theory of Games." In La Decision, Editions du CNRS, Paris, pp 251-263. Reprinted in A. E. Roth, ed. The Shapley Value, pp. 307-319, Cambridge: Cambridge University Press.

Shapley, L. S., and M. Shubik [1953]. "Solutions of n-Person Games with Ordinal Utilities" (abstract). Econometrica 21:348.

Shubik, M. [1982]. Game Theory in the Social Sciences: Concepts and Solutions. Cambridge, MA: M.I.T. Press.

The Finagle Point for Characteristic Function Games

Guillermo Owen
Department of Mathematics, Naval Postgraduate School, Monterey, California 93943. This research was also supported in part by the Universities of Bergamo, Italy, and of Brescia, Italy.

Abstract

The Finagle point is that point which comes closest -- in the sense of distance -- to dominating any other point in a given space. In this article, the space is the set of imputations of a simple characteristic function game, and the distance is the l_1 metric in that set. It is shown that the finagle point can be computed by solving (in the minimax sense) a sequence of matrix games. Two examples are worked out in detail.

1. Rationale

We consider here the situation faced by a policy-maker (the incumbent) trying to please a set of voters in competition with a second policy-maker (the challenger). In general, there is no impregnable position, i.e., no matter what stand the incumbent takes, it is possible for the challenger to defeat him by a judicious choice of counter-stand. The incumbent must therefore hope to win by choice of a "flexible" strategy: his position must be flexible enough that (once the challenger has made himself known) it can be modified to defeat the challenger.

No incumbent is, of course, totally free of his past. While most politicians would no doubt agree that a foolish consistency is the hobgoblin of little minds, they also realize that large changes of position are at worst unbelievable and at best considered evidence of wishy-washy thinking. Thus the incumbent's true initial position should be such as to minimize the maximum necessary change. This is the mathematical problem involved.

In [Wuffle, Feld, Owen, and Grofman 1989] this same problem was considered in the context of spatial voting games. We treat it here in the context of von Neumann-Morgenstern [1944] TU games.

A slightly different approach to the problem is found in [Young 1978, 1979] and in [Shubik and Young 1978].

2. The Game-Theoretic Model

A simple game, here, is a finite set $N = \{1, 2, \ldots, n\}$ of voters, together with a collection $\mathbf{W}$ of subsets of N: the <u>winning coalitions</u>. It is assumed that $\mathbf{W}$ satisfies the conditions:

$$\phi \notin \mathbf{W} \tag{1}$$
$$N \in \mathbf{W} \tag{2}$$
$$\text{If } S \in \mathbf{W} \text{ and } S \subset T, \text{ then } T \in \mathbf{W} \tag{3}$$

The game is <u>proper</u> if $S, T \in \mathbf{W}$ means $S \cap T \neq \phi$. It is <u>decisive</u> if, for every S N, either $S \in \mathbf{W}$ or $N-S \in \mathbf{W}$. A voter $i \in N$ is a <u>dictator</u> if $\{i\} \in \mathbf{W}$; i is a <u>veto player</u> if $N-\{i\} \notin \mathbf{W}$.

An <u>imputation</u> is a vector $\mathbf{x} = (x_1, \ldots, x_n)$ satisfying

$$x_i \geq 0 \tag{4}$$
$$\sum_{i \in N} x_i = 1 \tag{5}$$

Given two imputations, $\mathbf{x}$ and $\mathbf{y}$, we define

$$P(\mathbf{x}, \mathbf{y}) = \{ i \mid x_i > y_i\} \tag{6}$$
$$P(\mathbf{y}, \mathbf{x}) = \{ i \mid y_i > x_i\} \tag{7}$$
$$I(\mathbf{x}, \mathbf{y}) = \{ i \mid x_i = y_i\}. \tag{8}$$

It is clear that the three sets $P(\mathbf{x},\mathbf{y})$, $P(\mathbf{y},\mathbf{x})$, and $I(\mathbf{x},\mathbf{y})$ determine a partition of N.

We say $\mathbf{x}$ <u>dominates</u> $\mathbf{y}$ ($\mathbf{x} \succ \mathbf{y}$) if $P(\mathbf{x},\mathbf{y}) \in \mathbf{W}$.

We use, moreover, the l_1 metric on the space of imputations:

$$d(\mathbf{x},\mathbf{y}) = \sum_{i \in N} |x_i - y_i| \tag{9}$$

Since $\mathbf{x}$ and $\mathbf{y}$ are both imputations, it is easy to see that

$$d(\mathbf{x},\mathbf{y}) = 2 \sum (x_i - y_i)^+ \tag{10}$$

where

$$t^+ = \max\{t,0\} = \begin{cases} t & \text{if } t \geq 0 \\ 0 & \text{if } t \leq 0 . \end{cases} \tag{11}$$

We now consider the following three-stage process:

(a) an <u>incumbent</u> chooses a position (imputation) $\mathbf{x}$.

(b) a <u>challenger</u> chooses a position $\mathbf{y}$ which dominates $\mathbf{x}$.

(c) the incumbent now chooses a new position, $\mathbf{z}$, <u>as close to $\mathbf{x}$ as</u>

possible, which dominates $\mathbf{y}$.

Since the incumbent is at liberty to choose both $\mathbf{x}$ and $\mathbf{z}$ from all imputations, and the challenger is free to choose $\mathbf{y}$ from all imputations, this gives rise to the mathematical problem

$$\min_{\mathbf{x}} \max_{\mathbf{y}} \inf_{\mathbf{z}} \{d(\mathbf{x},\mathbf{z}) \mid \mathbf{z} \succ \mathbf{y}\}$$

3. The Finagle Point and Radius

Let $\mathbf{x}$, $\mathbf{y}$ be imputations. We define the function

$$s(\mathbf{x},\mathbf{y}) = \tfrac{1}{2} \inf \{d(\mathbf{x},\mathbf{z}) \mid \mathbf{z} \succ \mathbf{y}\}. \tag{12}$$

If the game has a veto player, or a dictator, it may be that $\mathbf{y}$ belongs to the core of the game (the set of undominated imputations) and in this case $s(\mathbf{x},\mathbf{y}) = \infty$. Otherwise, $s(\mathbf{x},\mathbf{y})$ is a finite non-negative real number. (If $\mathbf{x} \succ \mathbf{y}$, then of course $s(\mathbf{x},\mathbf{y}) = 0$.)

Theorem 1. For a non-dictatorial, non-veto game, let $\mathbf{x}$ and $\mathbf{y}$ be imputations, and let $K = P(\mathbf{x},\mathbf{y}) \cup I(\mathbf{x},\mathbf{y})$. Then

$$s(\mathbf{x},\mathbf{y}) = \min_{Q} \sum_{i \in Q} (y_i - x_i) \tag{13}$$

where the minimum is taken over all Q such that $Q \cap K = \phi$ and $Q \cup K \in \mathbf{W}$.

Proof:

Suppose $\mathbf{z} \succ \mathbf{y}$. Then $P(\mathbf{z},\mathbf{y}) \in \mathbf{W}$. Let $Q = P(\mathbf{z},\mathbf{y}) - K$. Then, for $i \in Q$, we have $z_i > y_i > x_i$, and it follows that

$$d(\mathbf{x},\mathbf{z}) \geq 2 \sum_{i \in Q} (z_i - x_i) > 2 \sum_{i \in Q} (y_i - x_i).$$

Thus

$$d(\mathbf{x},\mathbf{z}) \geq 2 \min_{Q} \sum_{i \in Q} (y_i - x_i)$$

and so

$$s(\mathbf{x},\mathbf{y}) \geq \min_{Q} \sum_{i \in Q} (y_i - x_i)$$

Conversely, let R be the value of Q which minimizes the right side of (13). The game is not dictatorial; hence, R cannot equal N-K, for if R were equal to N-K, removing any player from R would still leave $K \cup R - \{j\} = N - \{j\} \in \mathbf{W}$, and would decrease the sum which R presumably minimizes. Thus, N-R-K is not empty.

Now, for small $\delta > 0$, choose $\mathbf{z}$ as follows:

$$z_i = y_i + \delta \qquad \text{if} \quad i \in R \cup I(\mathbf{x},\mathbf{y}) \tag{14}$$
$$z_i = x_i \qquad \text{if} \quad i \in P(\mathbf{y},\mathbf{x}) - R \tag{15}$$
$$y_i < z_i \leq x_i \qquad \text{if} \quad i \in P(\mathbf{x},\mathbf{y}) \tag{16}$$

To see that this last (16) can be done while preserving (14) and (15), w note that by (14) and (15), we will have

$$\sum_{N-P(\mathbf{x},\mathbf{y})} y_j > \sum_{N-P(\mathbf{x},\mathbf{y})} z_j > \sum_{N-P(\mathbf{x},\mathbf{y})} x_j$$

if only δ is small enough. It is clear, now, that

$$P(\mathbf{z},\mathbf{y}) = K \cup R$$

and so $\mathbf{z} \succ \mathbf{y}$. Now

$$d(\mathbf{x},\mathbf{z}) = 2 \sum_{R \cup I(\mathbf{x},\mathbf{y})} (z_i - x_i)$$

$$= 2 \sum_{R} (y_i - x_i) + m\delta$$

where m is the number of voters in $R \cup I(\mathbf{x},\mathbf{y})$. Since δ is arbitrarily small we find

$$\inf d(\mathbf{x},\mathbf{z}) \leq 2 \sum_{R} (y_i - x_i)$$

and so

$$s(\mathbf{x},\mathbf{y}) = \sum_{R} (y_i - x_i)$$

as desired.

Let $\mathbf{x}$ be the incumbent's position. We look, next, for the $\mathbf{y}$ whic maximizes $s(\mathbf{x},\mathbf{y})$. This represent's the challenger's optimal strategy, in tha it maximizes the distance which the incumbent must move.

For a fixed set K, we consider those $\mathbf{y}$ such that

$$P(\mathbf{x},\mathbf{y}) \cup I(\mathbf{x},\mathbf{y}) = K \tag{17}$$

and define

$$\sigma(\mathbf{x},K) = \max s(\mathbf{x},\mathbf{y}) \tag{18}$$

where the maximum is taken over all such $\mathbf{y}$.

By Theorem 1, it is clear that maximization of $s(\mathbf{x},\mathbf{y})$ requires that the $y_i - x_i$, for $i \in N-K$, be as large as possible. Now, since $\mathbf{x}$ and $\mathbf{y}$ are imputations,

$$\sum_{i \in K} (x_i - y_i) = \sum_{i \in N-K} (y_i - x_i) \tag{19}$$

and so y_i must be chosen as small as possible for $i \in K$. In other words, we must choose

$y_i = 0$ for $i \in K$.

In this case, we have

$$\sum_{i \in N-K} (y_i - x_i) = \sum_{i \in K} x_i \tag{20}$$

and it follows (setting $u_i = y_i - x_i$) that

$$\sigma(\mathbf{x},K) = \max \beta \tag{21}$$

subject to

$$\beta \leq \sum_{i \in Q} u_i \quad \text{for all } Q,\ \begin{matrix} Q \cap K = \phi \\ Q \cup K \in \mathbf{W} \end{matrix} \tag{22}$$

$$\sum_{i \in N-K} u_i = \sum_{j \in K} x_j \tag{23}$$

$u_i \geq 0$ for all $i \in N-K$

This is, of course, a linear program. The variables x_j ($j \in K$) appear only as parameters, and it is easy to see that we can rewrite this as

$$\sigma(\mathbf{x},K) = \left(\sum_{j \in K} x_j \right) \max \tau \tag{24}$$

subject to

$$\tau \leq \sum_{i \in Q} u_i \quad \text{for all } Q,\ \begin{matrix} Q \cap K = \phi \\ Q \cup K \in \mathbf{W} \end{matrix} \tag{25}$$

$$\sum_{i \in N-K} u_i = 1 \tag{26}$$

$$u_i \geq 0. \qquad (27)$$

This maximum value of τ can also be expressed as the minimax value of a matrix game, A. In fact, we can let the rows of A correspond to the voters i ϵ N-K; the columns correspond to the coalitions Q such that $Q \cap K = \phi$ and $Q \cup K$ ϵ **W**. The matrix is simply the incidence matrix, i.e.

$$a_{iQ} = \begin{cases} 1 & \text{if } i \epsilon Q \\ 0 & \text{if } i \notin Q \end{cases} \qquad (28)$$

and then

$$\sigma(\mathbf{x},K) = (\sum_{j \epsilon K} x_j)\, \tau(K) \qquad (29)$$

where $\tau(K)$ is the value of the matrix game A(K) given by (28).

The reader should note here that only minimal coalitions Q (subject to the conditions given) need be considered. In fact, the column corresponding to any Q' will be dominated by the column corresponding to any $Q \subset Q'$, and thus can be safely discarded in computing the value or optimal strategies of the matrix game.

For given $\mathbf{x}$, finally, we define

$$\pi(\mathbf{x}) = \max_K \sigma(\mathbf{x},K) \qquad (30)$$

By definition of σ, it is clear that

$$\pi(\mathbf{x}) = \max_{\mathbf{y}} s(\mathbf{x},\mathbf{y}) \qquad (31)$$

We will call $\pi(\mathbf{x})$ the <u>Finagle semi-radius</u> of $\mathbf{x}$: it is one half of the maximum distance that the incumbent will need to move, assuming that he starts at position $\mathbf{x}$, to defeat any challenger. (In fact, this is not quite true: he actually needs to move $2\pi(\mathbf{x})+\delta$, where δ is arbitrarily small but positive.) We look, finally, for the point which minimizes this.

Since, by (30), each $\sigma(\mathbf{x},K)$ is a linear function of $\mathbf{x}$, the function π is the maximum of several linear functions and is, therefore, a piecewise linear convex function. Its minimum can be obtained, then, as the solution of a linear program:

Minimize μ subject to (32)

$$\mu \geq \tau(K) \sum_{j \epsilon K} x_j \quad \text{for all } K \notin \mathbf{W} \qquad (33)$$

$$\sum_{j \in N} x_j = 1 \tag{34}$$

$$x_j \geq 0 \tag{35}$$

Once again, we see that this can be obtained as the minimax value of a matrix game. In fact, let B be a matrix with rows corresponding to coalitions $K \notin \mathbf{W}$ and columns corresponding to voters $j \in N$. Then $B = (b_{Kj})$, where

$$b_{Kj} = \begin{cases} \tau(K) & \text{if } j \in K \\ 0 & \text{if } j \notin K . \end{cases} \tag{36}$$

The minimizing vector $(x_1, \ldots, x_n)$ for this game is the <u>Finagle point</u>. The minimax value μ is, of course, the minimal Finagle semi-radius. We can also write

$$\mu = \min \ \max \ \inf \tfrac{1}{2} d(\mathbf{x}, \mathbf{z}) \tag{37}$$

where the min is taken over all imputations $\mathbf{x}$, the max is taken over all imputations $\mathbf{y}$, and the inf is taken over all imputations $\mathbf{z}$ which dominate $\mathbf{y}$.

4. Examples

1. Consider the 4-person apex game, i.e., the game with N = {1,2,3,4}, where **W** consists of coalitions {1,2}, {1,3}, {1,4}, and all coalitions with 3 or 4 voters.

For this game, the losing coalitions K are the one-voter sets {i}, as well as the coalitions $\{i,j\} \subset \{2,3,4\}$. We first compute $\tau(K)$ for each of these.

For K = {1}, the possible Q are the sets {2}, {3}, and {4}. (Only minimal Q need be considered.) The corresponding matrix is

$$A(\{1\}) = \begin{array}{c} \\ 2 \\ 3 \\ 4 \end{array} \begin{array}{c} \begin{array}{ccc} \{2\} & \{3\} & \{4\} \end{array} \\ \begin{pmatrix} 1 & 0 & 0 \\ 0 & 1 & 0 \\ 0 & 0 & 1 \end{pmatrix} \end{array}$$

with minimax value $\tau(\{1\}) = 1/3$.

For K = {2}, the possible (minimal) Q are {1} and {3,4}. Thus

$$A(\{2\}) = \begin{array}{c} \\ 1 \\ 3 \\ 4 \end{array} \begin{array}{c} \begin{array}{cc} \{1\} & \{3,4\} \end{array} \\ \begin{pmatrix} 1 & 0 \\ 0 & 1 \\ 0 & 1 \end{pmatrix} \end{array}$$

which has value 1/2. By symmetry, $\tau(\{2\}) = \tau(\{3\}) = \tau(\{4\}) = 1/2$.

Finally, let K = {2,3}. The possible Q are {1} and {4}. Then

$$A(\{2,3\}) = \begin{array}{c} \\ 1 \\ 4 \end{array} \begin{array}{c} \begin{array}{cc} \{1\} & \{4\} \end{array} \\ \begin{pmatrix} 1 & 0 \\ 0 & 1 \end{pmatrix} \end{array}$$

This also has value 1/2. Using symmetry, $\tau(\{2,3\}) = \tau(\{2,4\}) = \tau(\{3,4\}) = 1/2$.

The matrix B is then

$$\begin{array}{c} \\ \{1\} \\ \{2\} \\ \{3\} \\ \{4\} \\ \{2,3\} \\ \{2,4\} \\ \{3,4\} \end{array} \begin{array}{c} \begin{array}{cccc} 1 & 2 & 3 & 4 \end{array} \\ \begin{pmatrix} 1/3 & 0 & 0 & 0 \\ 0 & 1/2 & 0 & 0 \\ 0 & 0 & 1/2 & 0 \\ 0 & 0 & 0 & 1/2 \\ 0 & 1/2 & 1/2 & 0 \\ 0 & 1/2 & 0 & 1/2 \\ 0 & 0 & 1/2 & 1/2 \end{pmatrix} \end{array}$$

In this matrix, the rows labeled {2}, {3}, and {4} are dominated, reducing the matrix to

$$\begin{pmatrix} 1/3 & 0 & 0 & 0 \\ 0 & 1/2 & 1/2 & 0 \\ 0 & 1/2 & 0 & 1/2 \\ 0 & 0 & 1/2 & 1/2 \end{pmatrix}$$

and it is easy to see that this game has minimax value 1/6, while the optimal strategy for the minimizing player is

$$\mathbf{x} = (1/2,\ 1/6,\ 1/6,\ 1/6).$$

This is the Finagle point, with corresponding semi-radius 1/6.

We see that the incumbent should take position (1/2, 1/6, 1/6, 1/6). This guarantees that, whatever position challenger takes, the incumbent will need to move not more than $2(1/6) + \delta$ (where δ is arbitrarily small) to defeat him.

Suppose, however, that the incumbent makes a mistake, choosing instead position (.4, .2, .2, .2). This is not optimal in game matrix B, giving a payoff of .2 (which is of course larger than 1/6) against the rows labeled {2,3}, {2,4}, or {3,4}. This allows the challenger to choose his optimal position by using K equal to one of these three coalitions, say K = {2,3}. He should then choose $y_2 = y_3 = 0$. This gives him a "surplus" equal to $x_2 + x_3$ = .4, which he is now free to distribute to voters 1 and 4. For the optimal distribution of this surplus, we look at the game matrix A({2,3}), which has the optimal maximizing vector $u_1 = u_4 = 1/2$, i.e., challenger should give 1/2 of the surplus to each of voters 1 and 4. The challenger's optimal position is then

$$y_1 = x_1 + u_1(x_3+x_4) = .6$$
$$y_2 = y_3 = 0$$
$$y_4 = x_4 + u_4(x_3+x_4) = .4$$

or $\mathbf{y}$ = (.6, 0, 0, .4).

It can now be seen that, to defeat this $\mathbf{y}$, the incumbent, starting from (.4,.2,.2,.2), must move at least $2(0.2)+\delta$ units, i.e. he is definitely worse off than if he had chosen the Finagle point which required moves of only $2(1/6)+\delta$.

2. Consider now the 5-person apex game, with N = {1,2,3,4,5}, whose minimal winning coalitions are {1,2}, {1,3}, {1,4}, {1,5}, and {2,3,4,5}.

If the reader will consider the losing coalitions K, reader will easily calculate

$\tau(\{1\}) = 1/4$

$\tau(K) = 1/2$ for all other $K \notin W$.

The matrix B will have 15 rows and 5 columns. Of these rows, however, 10 are dominated, leaving the matrix

	1	2	3	4	5
{1}	1/4	0	0	0	0
{2,3,4}	0	1/2	1/2	1/2	0
{2,3,5}	0	1/2	1/2	0	1/2
{2,4,5}	0	1/2	0	1/2	1/2
{3,4,5}	0	0	1/2	1/2	1/2

Clearly, the minimizing strategy here is

$$\mathbf{x} = (.6, .1, .1, .1, .1)$$

which is the Finagle point. The Finagle semi-radius is the value of this matrix game, 0.15.

5. Postscript

The following points seem worth noting:

For the two games considered above, the Finagle point coincides with the Shapley value. On the other hand, for larger games, it seems to be closer to the nucleolus. Thus it would be interesting to study its general behavior from a purely mathematical point of view.

The technique used above for computation depends on the fact that v is a simple game. Techniques for computing the point for non-simple games would be of great interest.

The Finagle point need not be unique. It is, however, the optimal strategy in a matrix game; thus, if there is more than one Finagle point, then the set of all such points is a closed convex polyhedron.

The mathematics here follows closely that in Owen [1965] which however dealt with discriminatory stable sets. This suggests that some connection may exist between the Finagle point and stable sets. Obviously more study will be needed.

BIBLIOGRAPHY

Owen, G. "A Class of Discriminatory Solutions to Simple n-Person Games." Duke J. Math., 1965, 545-554.

Schmeidler, D. "The Nucleolus of a Characteristic Function Game." SIAM J. Applied Math., 1969, 1163-1170.

Shapley, L. S. "A Value for n-Person Games." Annals of Mathematics Studies, Study 28, ed. H. W. Kuhn and A. W. Tucker, Princeton University Press, 1953, 307-317.

Shubik, M., and P. Young. "The Nucleolus as a Noncooperative Game Solution." Game Theory and Politics, ed. P. Ordeshook, New York University Press, 1978, 511-527.

von Neumann, J., and O. Morgenstern. The Theory of Games and Economic Behavior. Princeton Univesity Press, 1944, 1947.

Wuffle, A, S. L. Feld, G. Owen, and B. Grofman. "Finagle's Law and the Finagle Point: A New Solution Concept for Spatial Voting Games." Amer. J. Poli. Sci., 1989, 348-375.

Young. P. "A Tactical Lobbying Game." Game Theory and Politics, ed. P. Ordeshook, New York University Press, 1978, 391-404.

Young, P. "Exploitable Surplus in n-Person Games." Applied Game Theory, ed. S. Brams, A. Schotter, and G. Schwödiauer, Physika- Verlag, 1979, 323-338.

Voting by Count and Account

Department of Mathematics,The Hebrew University

Jerusalem 91904, Israel

§1. Introduction

This paper is devoted to a game theoretic analysis of a voting system which was used by some Jewish communities in Europe in the seventeenth century. The system is described in M.M. Krochmal [1675, Responsum No. 1]. We now present an abridged translation of its description:

> When I came to Kremsir, to serve as rabbi, I found a controversy in the community about the election of a *Shamash* (Sexton). The community consisted of fifty taxpayers. Among them were a man, his two sons and his two sons-in-law, men of wealth who together paid three-fifths of the tax, the rest of the community paying two-fifths. Thus, for example, if fifty thaler were needed for the communal treasury, the five supplied thirty, while the other forty-five contributed twenty.
>
> The community had an ancient ordinance that for the election of a rabbi, cantor or shamash, there had to be agreement of " *the majority of the count* and the *majority of the account* ". (Here "majority of the count" means most of the members of the community and "majority of the account" means most of the money.) Now it happened that the wealthy man, his sons and sons-in-law, opposed the election of a *Shamash* which the other forty-five desired.
>
> The forty-five appealed to me, that the ordinance was a "decree which it is too difficult to comply with" (and could thus be abrogated in accordance with Jewish law) because the five always voted as one and, representing "the majority of wealth", they could effectively veto any community decision. The wealthy man countered that the ordinance was enacted by the greatest authorities of the generation and had been accepted and continues to be accepted as the "rule of the land" in all Jewish communities, and may not be altered in and by one community.
>
> I had to agree that the ordinance was a just one, for it followed the rule, "one abides by the decision of the majority". In this case, the number of taxpayers is a majority, but the wealthy are also a majority, because they supply the "majority of money" (which will support the rabbi, cantor or shamash). In any case, the force of the ordinance is that neither the poor who are *the majority of the count* can impose their will on the wealthy who are "*a majority of the account*", nor vice versa. A decision must be reached by agreement of both majorities: persons and wealth.

The formulation of the foregoing principle of "the majority of the count and the majority of the account" in game theoretic terms is very simple. Let $N = \{1, \ldots, n\}$ be the set of taxpayers in a community and let w^i be the tax paid by $i \in N$. The "majority of the count" rule corresponds to the symmetric simple game $u = (n, [\frac{n}{2}] + 1)$. We recall that u is defined by

$$u(S) = \begin{cases} 1, & \text{if } S \text{ contains at least } [\frac{n}{2}] + 1 \text{ members} \\ 0, & \text{otherwise.} \end{cases}$$

The "majority of the account" is represented by the weighted majority game $v = [w^1, \ldots, w^n]$, where

$$v(S) = \begin{cases} 1, & \text{if } \sum_{i \in S} w^i > \sum_{i \in N} w^i/2 \\ 0, & \text{otherwise.} \end{cases}$$

Finally, "the majority of the count and the majority of the account" method is given by the simple game v^*, where $v^*(S) = u(S)v(S)$ for all $S \subset N$.

W. Thomson, in a letter to R.J. Aumann, has raised the following problem: What is the relationship between the game-theoretic solutions of v and those of v^*? We prove in this paper that the Shapley value of v^* is more egalitarian than the Shapley value of v (see Corollary 3.4).

I am grateful to R. J. Aumann for communicating to me Thomson's problem and for several valuable discussions. Also, I am indebted to A. Neyman for a helpful conversation. Finally, I should thank G. Schwarz for suggesting the title of this paper.

§2. Preliminaries

Let $N = \{1, \ldots, n\}$ be a set of *players*. A *game on* N is a function v that associates a real number $v(S)$ with each subset S of N. We always assume that $v(\emptyset) = 0$. If u and v are games then the *product* of u and v, uv, is defined by $uv(S) = u(S)v(S)$ for all $S \subset N$. A game v is *monotonic* if $v(S) \leq v(T)$ when $S \subset T$. If $S \subset N$, then we denote by $|S|$ the number of members of S. A game v is *symmetric* if $S, T \subset N$ and $|S| = |T|$ imply that $v(S) = v(T)$. Thus, a symmetric game v may be considered as a function $v\colon \{0, 1 \ldots, n\} \to R$ that satisfies $v(0) = 0$, where R is the set of real numbers.

Let v be a game and let S and T be coalitions (i.e., nonempty subsets of N). S is *at least as desirable as* T *in* v, written $Sd(v)T$, if for all $B \subset N$ such that $B \cap (S \cup T) = \emptyset$, $v(S \cup B) \geq v(T \cup B)$. If $i, j \in N$ then we write $id(v)j$ if $\{i\}d(v)\{j\}$. $d(v)$ may be cyclic. However, the desirability relation between players is transitive (see Maschler and Peleg [1966]).

A game v is *simple* if $v(S) \in \{0, 1\}$ for all $S \subset N$. If v is simple then we denote $W = \{S | v(S) = 1\}$. W is the set of *winning* coalitions in v. A simple game is determined by its set of winning coalitions. Furthermore, a monotonic simple game $W \subset 2^N$ is determined by its set of minimal winning coalitions W^m. A simple game W is *proper* if $S \in W$ implies that $N - S \notin W$. W is *strong* if $S \notin W$ implies that $N - S \in W$. W is *weak* if $V = \cap\{S | S \in W\} \neq \emptyset$. Finally, a simple game W is a *weighted majority* game if there exist a quota $q > 0$ and weights $w^i \geq 0$, $i \in N$, such that $S \in W$ if and only if $\sum_{i \in S} w^i \geq q$. The $(n+1)$-tuple $[q; w^1, \ldots, w^n]$ is called a *representation* of W. A representation $[q; w^1 \ldots, w^n]$ of W is *homogeneous* if $\sum_{i \in S} w^i = q$ for all $S \in W^m$. W is a *homogeneous* weighted majority game if it has a homogeneous representation.

A *payoff vector* is a member of R^N, the n-dimensional Euclidean space. We define $\eta : R^N \to R^N$ as the function that arranges the components of each vector in a nondecreasing order. Thus, if $x \in R^N$, $x = (x^1, \ldots, x^n)$, then $\eta(x) = (x^{i_1}, \ldots, x^{i_n})$ where $x^{i_1} \le x^{i_2} \le \ldots \le x^{i_n}$. If $x, y \in R^N$, then x *Lorenz-dominates* y, written xLy, if $\sum_{i=1}^{j} \eta^i(x) \ge \sum_{i=1}^{j} \eta^i(y)$ for $j = 1, \ldots, n$. If $x, y \in R^N$, $\eta(x) \ne \eta(y)$, $\sum_{i=1}^{n} x^i = \sum_{i=1}^{n} y^i$, and xLy, then the payoff distribution x is more egalitarian than y (see, e.g., Weymark [1981]).

§3. On the Egalitarianism of the Shapley Value

In this section we prove a result which reflects the egalitarianism of the Shapley value. First, we recall that if v is a game then the *Shapley value* of v, $\varphi(v)$, is given by

$$\varphi^i(v) = \sum_{S \subset N} \frac{(|S|-1)!|N-S|!}{n!}(v(S) - v(S - \{i\})) \tag{3.1}$$

for all $i \in N$.

Theorem 3.1. *Let v, u, and w be games that satisfy the following conditions.*

(1) The desirability relation between players in v is complete.

(2) u and w are symmetric, $u(k) \ge w(k) \ge 0$ for $1 \le k \le n$ and $u(n) = w(n)$.

Then $\varphi(vw)L\varphi(vu)$.

The proof of Theorem 3.1 uses the following lemma.

Lemma 3.2. *Let v be a game on N, let $1 \le k \le n$, and let $i \in N$. We define*

$$f_v^i(k) = \sum\{v(S) | i \in S \text{ and } |S| = k\} / \binom{n-1}{k-1} \tag{3.2}$$

and

$$h_v(k) = \sum\{v(S) | \, |S| = k\} / \binom{n}{k} \tag{3.3}$$

Then

$$\varphi^i(v) = v(N)/n + \sum_{k=1}^{n-1} (f_v^i(k) - h_v(k))/(n-k) \tag{3.4}$$

The reader may prove by direct computation that (3.1) is equivalent to (3.4).

Proof of Theorem 3.1: Let $1 \le k < n$ and let $i \in N$. Then $f_{uv}^i(k) = u(k) f_v^i(k)$ and $h_{uv}(k) = u(k) h_v(k)$ (see (3.2) and (3.3)).

Thus

$$\varphi^i(uv) = u(n)v(N)/n + \sum_{k=1}^{n-1}(f_v^i(k) - h_v(k))u(k)/(n-k) \quad (3.5)$$

(see (3.4)). Similarly

$$\varphi^i(wv) = w(n)v(N)/n + \sum_{k=1}^{n-1}(f_v^i(k) - h_v(k))w(k)/(n-k) \quad (3.6)$$

By (3.5), (3.6), and the assumption that $u(n) = w(n)$

$$\varphi^i(wv) - \varphi^i(uv) = \sum_{k=1}^{n-1}(f_v^i(k) - h_v(k))(w(k) - u(k))/(n-k) \quad (3.7)$$

Because v has a complete desirability relation between players we may assume that $jd(v)j-1$ for $j = 2,\ldots,n$. Therefore, by the symmetry of u and w, $jd(uv)j-1$ and $jd(wv)j-1$ for $j = 2,\ldots,n$. By (3.1) the Shapley value preserves the desirability relation between players. Hence, $\varphi^j(uv) \geq \varphi^{j-1}(uv)$ and $\varphi^j(wv) \geq \varphi^{j-1}(wv)$ for $j = 2,\ldots,n$. Also, by (3.2), $f_v^j(k) \geq f_v^{j-1}(k)$ for $j = 2,\ldots,n$ and $k = 1,\ldots,n-1$. Finally, because $h_v(k) = \sum_{j=1}^n f_v^j(k)/n$ for $k = 1,\ldots,n-1$, it follows that

$$\sum_{j=1}^{t} f_v^j(k) - th_v(k) \leq 0 \text{ for } t = 1,\ldots,n \text{ and } k = 1,\ldots,n-1 \quad (3.8)$$

Now let $1 \leq j \leq n$. By (3.7), (3.8), and the assumption that $w(k) \leq u(k)$ for all k,

$$\sum_{i=1}^{j}(\varphi^i(wv) - \varphi^i(uv)) = \sum_{k=1}^{n-1}(\sum_{i=1}^{j} f_v^i(k) - jh_v(k))(w(k) - u(k))/(n-k) \geq 0$$

Thus, $\varphi(wv)L\varphi(uv)$. Q.E.D.

Theorem 3.1 has the following interpretation. Let $S \subset N$, $S \neq \emptyset$, and let $|S| = k$. The worth of S in uv is the product of $u(k)$, which depends only on the count of S, and $v(S)$, which is the "account" of S in the game v. A similar interpretation applies to wv. Because $0 \leq w(k) \leq u(k)$ the effect of the symmetric factor w is *stronger* than that of u. Therefore wv, as evaluated by its Shapley value, is more symmetric than uv. (Indeed, in the extreme case where $w(k) = 0$ for $k = 1,\ldots,n-1$, wv is a symmetric game.)

REMARK 3.3. The foregoing proof of Theorem 3.1 is due to A. Neyman. Our original proof was devised for simple games and used induction on the number of players.

Theorem 3.1 has an immediate application to voting by count and account. First we notice that a symmetric and monotonic simple game W is specified by a pair (n,k), where $1 \leq k \leq n$ is the size of a minimal winning coalition in W.

COROLLARY 3.4. *Let v be a weighted majority simple game and let $u \neq 0$ be a symmetric and monotonic simple game. Then $\varphi(uv)L\varphi(v)$.*

PROOF: Let $v = [q; w^1, \ldots, w^n]$. Then $w^i \geq w^j$ implies that $id(v)j$. Thus, the relation $d(v)$ between players is complete. Furthermore, if $u = (n, k)$, where $1 \leq k \leq n$, and $u^* = (n, 1)$, then $u(j) \leq u^*(j)$ for $j = 1, \ldots, n$, $u(n) = u^*(n) = 1$, and $u^*v = v$. Hence, by Theorem 3.1, $\varphi(uv)L\varphi(v)$.

REMARK 3.5. As in Section 1, v represents voting by account and u voting by count. By Corollary 3.4, voting by count ***and*** account is *more* egalitarian than voting by account.

Actually, the following result, which is stronger than Corollary 3.4, is true.

COROLLARY 3.6. *Let v be a simple game and let u be a symmetric simple game. If the desirability relation between players in v is complete and $u(n) = 1$, then $\varphi(uv)L\varphi(v)$.*

§4. Examples

We shall now show that Corollary 3.6 is a sharp result.

EXAMPLE 4.1. Let $N = \{1,2,3,4,5\}$, let $u = [8; 1,1,3,3,5]$, and let $v = [7; 1,1,2,2,5]$. Then $\varphi(u) = (3,3,13,13,28)/60$, $\varphi(v) = (2,2,7,7,42)/60$, and both u and v are homogeneous weighted majority games with the same desirability relation between players. Moreover, $\varphi(u)L\varphi(v)$. Nevertheless, $uv = [3; 0,0,1,1,2]$ and $\varphi(uv) = (0,0,1,1,4)/6$. Thus, $\varphi(uv)$ does not Lorenz-dominate $\varphi(v)$.

We conclude from Example 4.1 that in Corollary 3.6 the assumption that u is symmetric cannot be replaced by the following weaker condition: $\varphi(u)L\varphi(v)$ and $jd(v)i \Rightarrow jd(u)i$ for all $i, j \in N$.

EXAMPLE 4.2. Let $N = \{1,2,3,4\}$, let $v = [3; 2,1,1,1]$, and let $u = (N, W)$ be given by

$$W^m = \{\{1,2\},\{2,3\},\{3,4\},\{1,4\}\}.$$

Then $\varphi(v) = (3,1,1,1)/6$. Also, the cycle $(1,2,3,4)$ is a symmetry of u. Hence, the symmetry group of u is transitive. If $uv = (N, W_*)$ then

$$W_*^m = \{\{1,2\},\{1,4\},\{2,3,4\}\}.$$

As the reader may easily verify, $\varphi^3(uv) = 1/12$. Thus, $\varphi(uv)$ does not Lorenz-dominate $\varphi(v)$.

Example 4.2 has the following implication: In Corollary 3.6 it is impossible to replace the assumption that u is symmetric by the weaker assumption that u has a transitive symmetry group.

EXAMPLE 4.3. Let $N = \{1,2,3,4,5,6\}$ and let the monotonic and proper simple game $v = (N, W)$ be specified by

$$W^m = \{\{1,2,6\},\{3,4,5,6\},\{2,3,4,6\},\{2,3,5,6\},\{2,4,5,6\}\}.$$

Then $\varphi(v) = (4,13,5,5,5,28)/60$ and the desirability relation between players of v is incomplete. If $u = (6,4)$ and $v_* = uv$, then the set of minimal winning coalitions in v_* is

$$W_*^m = \{\{1,2,3,6\},\{1,2,4,6\},\{1,2,5,6\},\{3,4,5,6\},\{2,3,4,6\},\{2,3,5,6\},\{2,4,5,6\}\}.$$

As the reader can easily verify, $\varphi^1(v_*) = 1/20$. Thus, $\varphi(uv)$ does not Lorenz-dominate $\varphi(v)$.

Example 4.3 implies that the assumption that v has a complete desirability relation between players is essential for the validity of Corollary 3.6.

The following example shows that voting by count and account may not yield a weighted majority game.

EXAMPLE 4.4. Let $v = [25; 18,8,8,7,4,2,2]$ and $u = (7,4)$. Then uv is *not* a weighted majority game. Indeed, let W be the set of winning coalitions in uv. Then $\{1,5,6,7\} \in W$, $\{2,3,5,6,7\} \notin W$, $\{1,4,5\} \notin W$, and $\{2,3,4,5\} \in W$. Hence $\{1\}$ and $\{2,3\}$ are incomparable in the desirability relation $d(uv)$. Therefore uv is not a weighted majority game. Notice that the desirability relation *between players* in uv is complete. Also, both u and v are strong weighted majority games.

§5. Concluding Remarks

REMARK 5.1. Products of weighted majority games by coalitional games were investigated in Neyman [1985]. Our Corollary 3.4 might be related to Neyman's work. However, there is no direct relationship because Neyman's paper deals with games with a continuum of players.

REMARK 5.2. We proved in this paper that voting by count and account is more egalitarian than voting by account. We hope that our result will renew the interest in this old (double) majority rule.

REMARK 5.3. Let v be a weighted majority simple game and let $u \neq 0$ be a symmetric and monotonic simple game. We could not find any relationship between the solutions of v and those of uv beyond Corollary 3.4. In particular, we do not know whether the nucleolus of uv Lorenz-dominates the nucleolus of v.

References

Krochmal, M.M. [1675], Zemach Zedek, Responsa, Amsterdam.

Maschler, M. and B. Peleg [1966], "A Characterization, Existence Proof and Dimension Bounds for the Kernel of a Game," Pacific Journal of Mathematics, 18, 289–328.

Neyman, A. [1985], "Semivalues of Political Economic Games," Mathematics of Operations Research, 10, 390–402.

Weymark, J.A. [1981], "Generalized Gini Inequality Indices," Mathematical Social Sciences, 1, 409–430.

Fee Games: (N)TU-Games with Incomplete Information

Joachim Rosenmüller
Institut für Mathematische Wirtschaftsforschung (IMW)
Universität Bielefeld
Postfach 8640, D–4800 Bielefeld 1

ABSTRACT

A fee–game is a special version of an NTU–game with incomplete information exhibiting certain side payment properties. We analyze the structure of the set of incentive compatible and individually rational mechanisms and, based on this analysis, we propose a further axiom for a value for (N)TU–games with incomplete information.

PREFACE

The first to discuss cooperative games with incomplete information of the players concerning the other players "types" were HARSANYI and SELTEN [2]. However, the idea of the "Bayesian incentive compatible mechanism" does not appear in their work . This idea was introduced by MYERSON [6] [7] [8] into the framework of NTU–games with incomplete information. Both authors considered a version of a value for such games to be defined by a suitable set of axioms. While MYERSON argues against the HARSANYI–SELTEN value (on the grounds of his "probability invariance axiom") he provides his own version of a axiomatized solution concept. However, this value in turn may have discontinuity properties (cf. [2]).

MYERSON did not attempt to formulate axioms for the HARSANYI–SELTEN value in the presence of mechanisms. This was done by WEIDNER [10]. It turned out that there is a canonical axiomatization of HARSANYI's and SELTEN's value – however, it is limited to the case that the types of the players are distributed in a stochastically independent way.

These facts, it would seem, indicate that there is still room for discussion of a "value for NTU–games with incomplete information".

Our model – already introduced in [9], differs by an essential feature from the previous ones: we do assume that players can agree upon one of *continuously many* decisions. Hence we try to model the generalized version of a characteristic function of an NTU– game with incomplete information.

In particular, we focus on what is called a fee–game. Such a game exhibits some side– payment properties; these properties appear in the *ex ante* and *ex post* situations. However, the introduction of incentive compatible mechanisms somehow blurres the side– payment nature of the game. For fee–games, we give a detailed description of the structure of incentive compatible and *in mediis* individually rational mechanisms. In particular, for the case of incomplete information on one side, the structure of those mechanisms which are not constant is precisely described. Based on this analysis, within the final section, we introduce the "*expected contract axiom*", which may be considered as a starting point for a further discussion of an axiomatic version of values for NTU– games with incomplete information.

SECTION 1: Unanimous CII–Games

Within this section we introduce the general model as well as some basic notation.

Definition 1.1:

A (unanimous) cooperative game with incomplete information (*a CII–game*) is a 6–tupel

$$\Gamma = (I,T,p;\ \underline{X},\ \underline{x};U) \tag{1}$$

Here, $I = \{1,...,n\}$ is called the set of individuals, agents, or *players*. $T = \prod_{i \in T} T_i$ is the cartesian product of finite sets T_i, $i \in I$, and T_i is the set of possible *types* of player i.

Let $(\Omega,\underline{\underline{F}}.P)$ be an abstract probability space and let $\tau : \Omega \to T$ be a random variable selecting n–tupels of types; now p is a probability on T and our interpretation is that $p = P \circ \tau^{-1}$ is the distribution of τ. Of course, all relevant data will depend on p only (and not on the choice of $(\Omega,\underline{\underline{F}},P)$), hence p is called the *distribution of types*.

Let $e = (1,...,1) \in \mathbb{R}^n_+$ then $ex = \sum_{i \in I} x_i$ and

$$\underline{X} = \{x \in \mathbb{R}^n_+ \mid ex \leq 1\} \tag{2}$$

reflects the feasible *alternatives*, parameters, or "*contracts*" the players can agree upon. Within this paper we shall assume that contracts may only be reached unanimously, however, coalitions and their power can easily be introduced into the model. This, if players fail to agree upon some $x \in \underline{X}$, then they will have to bear the consequences of the status quo alternative $\underline{x}$. We assume $\underline{x} \in \underline{X}$.

Finally, the mapping

$$U : I \times T \times \underline{X} \to \mathbb{R} \tag{3}$$

reflects the utilities; if chance chooses $t \in T$ and the players agree upon $x \in \underline{X}$, then player i's utility is $U_i^t(x)$.

As yet, we have not specified the rules of the game, this will be possible once we have the notion of a "mechanism".

Definition 1.2:

Let Γ be a CII–game. A *mechanism is* a mapping

$$\mu : T \to \underline{X}.$$

A mechanism μ is called *incentive compatible* (IC) if, for all $i \in I$ and all $t_i, s_i \in T_i$, we have

$$E(U_i^\tau \circ \mu \circ \tau | \tau_i = t_i) \geq E(U_i^\tau \circ \mu(\tau_1,\ldots,s_i,\ldots\tau_n) | \tau_i = t_i). \tag{4}$$

(Later on we use $\mu^\tau = \mu \circ \tau$).

The concept of a ("Bayesian") IC mechanism goes back to HURWICZ [4], our version is close to the one adopted by MYERSON [7] in a different framework, see also [1]. The first authors to study CII–games were HARSANYI and SELTEN [2] – but in their context mechanisms are not present.

As particular version of a mechanism is the one we shall call *constant*, i.e., if $\mu^t = x$ for all $t \in T$ and some fixed $x \in \underline{X}$. E.g., the *status quo mechanisms* $\underline{\mu}$ is defined by

$$\underline{\mu}^t = \underline{x} \qquad (t \in T). \tag{5}$$

Now we are in the position to specify the rules according to which the game is to be played.

First of all, there is a *bargaining period*. Within this period players may agree upon a mechanism μ in which case they will register this mechanism with an agency – the referee or *court*. This agency is supposed to be very powerful in the sense that it is capable of eventually enforcing the mechanism agreed upon or – if no agreement has been reached – the status quo mechanism $\underline{\mu}$.

In the next step, *chance* chooses $t \in T$. Each player $i \in I$ observes his type $t_i \in T_i$.

Next, each player i *announces* his (alleged) type, say s_i *to the court*. This constitutes some $s \in T$, the court will choose $\mu(s) \in \underline{X}$ and player i receives utility $U_i^t(\mu(s))$.

According to the various stages of the development we consider basically three situations or "states of knowledge" with respect to the players and the court: "*ex ante*" – before the chance move, "*in mediis*" ("mediis in rebus" indicates the not so purely temporal aspect) – after the chance move and prior to the announcements, and "*ex post*" – after the announcements.

Now, whenever players agree upon – and register – some mechanism μ, then a game (in extensive/-strategic form) Γ^{μ} arises in which players behave strategically by announcing types in view of the types they observed. Thus, in this game, strategies are mappings $\sigma_i : T_i \to T_i$. If "telling the truth" (i.e., the identity mapping) is a Nash equilibrium in Γ^{μ} the μ is IC and it is not hard to see that this definition is equivalent to the formal one presented in Definition 1.2 (i.e., to (4)). For the details, see [5] or [9].

Existence of IC–mechanisms does not constitute a problem (e.g. constant mechanisms are IC).

As a part of our story, we assume that the court will register only IC mechanisms. That is, registering mechanisms which in a sense induce players to lie about their true situation are considered to be against societies basic rules ("*contra bonos mores*"). From this it follows that the court *ex post* will be fully informed about the players true type. And, since "law enforcement", i.e. assigning $\mu(t)$ and the resulting utilities, takes place *ex post*, this provides implicitly a further argument for the restriction to IC mechanisms: the court wants to make sure that, when it enforces people to accept the results induced by μ, this is based on knowledge of the types and on the knowledge of truthfull announcement of these types.

There are further restrictions which the court wants to observe. If a player *in mediis*, i.e., after observing his own type, realizes that his conditional expectation of utility is worse than if no agreement had been reached at all, he will cry foul and try to get out of the agreement. Indeed, the court may not be able to enforce such mechanisms and hence it will restrict itself to the registration of *in mediis individually rational* (IR) IC mechanisms.

The formal definition is, of course, as follows.

Definition 1.3:

Let Γ be a CII–game, the IC mechanism μ is *individually rational in mediis*, (for short: IR) if, for all $i \in I$ and all $t_i \in T_i$

$$E(U_i^{\tau} \circ \mu^{\tau} | \tau_i = t_i) \geq E(U_i^{\tau}(\underline{x}) | \tau_i = t_i) = E(U_i^{\tau} \circ \underline{\mu}^{\tau} | \tau_i = t_i) \tag{6}$$

Let $\mathfrak{M}$ denote the set of all mechanisms obeying this definition, i.e.

$$\mathfrak{M} = \mathfrak{M}(\Gamma) = \{\mu : T \to \underline{X} \mid \mu \text{ is IC and IR}\}. \tag{7}$$

Note that we have implicitly adopted the viewpoint that recontracting in mediis is not possible. Nevertheless the players, when bargaining ex ante about mechanisms, will have the situation *in mediis* in mind. This may cause them in advance to reach mechanisms that are elements of $\mathfrak{M}$. So, by assuming that mechanisms outside of $\mathfrak{M}$ are "immoral" (*contra bonos mores*) we restrict ourself to situations in which the players in principle would not decide differently if bargaining took place after the chance move. Some authors seem to indicate that it makes no difference as to whether bargaining takes place ex ante or in mediis. In our present setup we avoid discussing this question; nevertheless it might be argued that the model does not change when bargaining takes place *in mediis*.

A similar discussion could center around the question as to whether a mechanism should be ex post individually rational. (Formally this would mean that for all $i \in I$ and for all $t \in T$ $U_i^t(\mu^t) \geq U_i^t(\underline{x})$). Indeed, if IC mechanisms are employed then not only the court might be aware of the true types of the players as selected by chance. It could very well happen that the court enforces a payoff upon a player which *ex post* is worse than if he had not joined an agreement at all and, as the power of the court is much higher than compared to an ordinary bargaining situation with complete information, enforcing such kind of outcome again could be considered immoral. Much can be said in favor of this point. On the other hand it is clearly seen that a lot of situations can occur in which players value their *in mediis* expected payoff much higher than the expected payoff of the status quo alternative and nevertheless end up regretting the agreement with a positive probability. In such cases agreements would not be reached *ex ante* or *in mediis* even so players would like to contract. We have adopted the viewpoint that, presently, we do not want to insist on *ex post* individually rational mechanisms. Maybe, this can be supported by the assumption that the court keeps the complete information of the true types of the players as a secret. Of course, this is a rather weak argument: after all, some agreements are not in accordance with good customs just because some agency keeps the violation of certain principles underlying the idea of "*boni mores*" as a secret. On the other hand, disclosing the players' type might also violate certain basic principles of society.

SECTION 2: FEE–GAMES

A fee–game is a special type of a CII–game. Some of its properties reflect the fact that "side payments" be permitted: *ex post* it is seen that a game with transferable utility is at hand once the types are revealed. *Ex ante*, the expected utilities of parameters reveal the T.U.–character. However, once IC mechanisms are agreed upon, the side payment character is blurred.

Recall that $e = (1,\dots,1) \in \mathbb{R}_+^n$ and $\underline{X} = \{x \in \mathbb{R}_+^n \mid ex \leq 1\}$. We write

$$\partial\underline{X} = \{x \in \underline{X} \mid ex = 1\}. \tag{1}$$

Definition 2.1:

1. $b^0 \in \mathbb{R}^n_+$ is called a *fee* if $eb^0 < 1$ holds true.

2. If b^0 is a fee, then the *Standard T.U.−game* (*STU−game*) or bargaining situation (à la NASH) *generated by* b^0 is the pair (O, V_{b^0}) with

$$V_{b^0} := \{x - (ex)\, b^0 \mid x \in \underline{X}\}. \tag{2}$$

The interpretation is obvious: a unit of money can be distributed among the players by agreement. However, registration bears some cost which is proportional to the total amount of money. And each player i is allotted a specific share of this cost, more precisely, for a unit of money which he obtains by agreement he has to pay a "fee" b^0_i.

Note that (O, V_{b^0}) formally is a (unanimous) NTU–game in the ordinary sense (full information) but players register a contract $x \in \underline{X}$ (and not a utility n–tupel).

In this setup, fees depend on characteristics of the players. It may be that player i pays an expert or lawyer to work out the details of the contract and that this payment depends on the size of the contract. It could also be a feasible interpretation to assume that the total payment $ex = \sum_{i \in I} x_i$ goes to the court (as to cover the expenses of registration) and that the court charges players according to circumstances which are based on certain properties/characteristics of players. In German courts at least it is possible to obtain a special tariff at court by pointing out (and proving!) that personal circumstances are justifying such remedy ("Armenrecht").

Now to some trivial technical proceedings.

Remark 2.2:

Given a fee b^0, the mapping

$$U^0 : \underline{X} \rightarrow \mathbb{R}^n \quad , \quad U^0(x) = x - (ex)\, b^0$$

yields $V^{b^0} = \{U^0(x) \mid x \in \underline{X}\}$; thus $U^0_i(x)$ reflects player i's utility if $x \in \underline{X}$ is agreed upon. In particular, $U^0(\underline{x}) = U^0(0) = 0$ is the utility vector of the *status quo point*. The individually rational part of the Pareto frontier, i.e., the set $\{U^0(x) \mid x \in \partial\, \underline{X},\ U^0(x) \geq 0\}$ is a simplex with extreme points

$$(1 - eb^0)\, e^i \quad (i \in I), \tag{4}$$

where $e^i = (0,\dots,1,\dots,0)$ denotes the i'th unit vector. The inverse images of these extremals (under $(U^0)^{-1}$, that is) are the vectors

$$a^{0i} := (1 - eb^0)\, e^i + b^0, \tag{5}$$

the convex hull of which, say

$$C^0 := \mathrm{cvH}(\{a^{0i} \mid i \in I\}) \tag{6}$$

is as well a simplex. All faces of C^0 are parallel to the corresponding faces of $\partial \underline{X}$. The vectors a^{0i} convey all necessary information, thus they may equally be used to describe the STU–game generated by b^0. Indeed, we have

Lemma 2.3:
Let $(\overset{\circ}{a}{}^{i})_{i \in I}$ be n linearly independent vectors, $\overset{\circ}{a}{}^{i} \in \partial\, \underline{X}$ ($i \in I$), such that the faces of the simplex generated are parallel to the corresponding faces of $\partial \underline{X}$. Then there is a unique fee $b^0 \in \mathbb{R}^n_+$ such that

$$\overset{\circ}{a}{}^{i} = (1 - eb^0)\, e^i + b^0 \qquad (i \in I)$$

holds true.

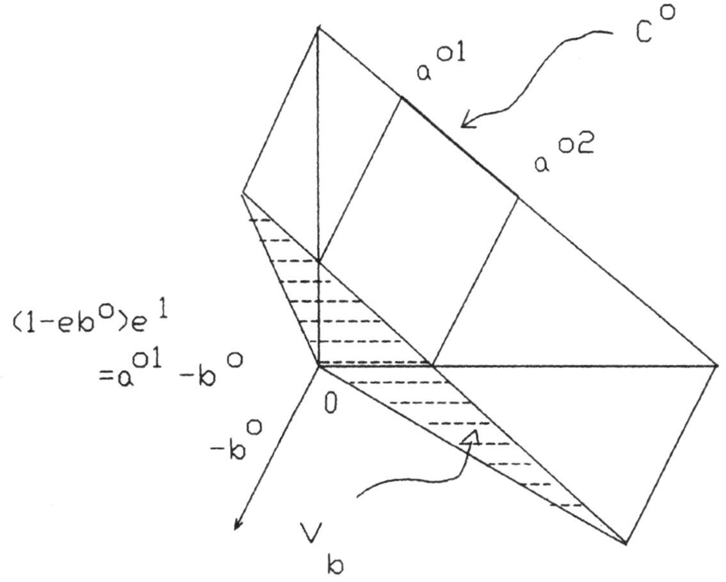

Fig.1
The STU–game generated by a fee

Proof:

Trivial; b^0 is obtained via the unique affine mapping that throws the a^i into multiples of unit vectors, i.e., we have

$$b^0 = \frac{1}{n} \sum_{i \in I} \mathring{a}^i - \lambda \frac{e}{n} \tag{7}$$

with

$$\lambda = \mathring{a}^1_1 - \mathring{a}^2_1 = \ldots = \mathring{a}^i_i - \mathring{a}^k_i \quad (i \in I, k \in I, k \neq i) \tag{8}$$

Without making this a formal definition we shall call a simplex $C^0 \subseteq \underline{X}$ which is spanned by its vertices $(\mathring{a}^i)_{i \in I}$, satisfying the conditions of 2.3, via (6) an *admissible subsimplex* of $\underline{X}$.

We are now in the position to formally describe "fee games" – CII–games that are generated by type depending fees.

Definition 2.4:

1. A *fee schedule* is a mapping

   $b : I \times T \to \mathbb{R}_+$

   (i.e., a matrix) such that, for every $t \in T$, b^t is a fee.

2. A CII–game $\Gamma = (I,T,p;\underline{X},\underline{x};U)$ is a *fee–game* if

   1. $\underline{x} = 0$ (8)

   2. There is a fee schedule b such that for every $t \in T$

   $$U^t(x) = x - (ex)b^t. \tag{9}$$

First let us pause for some interpretation.

In this model chance chooses $t \in T$ and, eventually, player $i \in I$ will pay $(ex)b^t_i$ towards the "expenses of the contract" if the agreement $x \in \underline{X}$ is reached, thus his utility is $x_i - (ex)b^t_i$. Again we may assume that these are his costs for his expert or lawyer or else that the court is charging a fee for in order to register – and enforce – a contract (or mechanism).

Nevertheless it might be necessary to justify the selection of the fee vector by a chance move in a more detailed manner: e.g., the "expenses of the contract" might result from a more elaborate economical activity which makes the money to be distributed available, thus expenses or costs may as well result from economical activities and depend on random influences.

Note that each player observes only his own b^t_i; in particular the court does not know the actual fee that has been selected by chance. However, if the court registers a mechanism $\mu \in \mathfrak{M}$, then obviously the advantages are (at least) two–fold: *in mediis*, when players announce their observed type to the court, none of them will start arguing on the grounds that their basic rights are violated – since μ is individually rational. And *ex post*, when the court knows all the types announced, it is clear that these types are the true ones – thus the court can indeed act out the agreement specified by μ, μ^t that is, based on the knowledge of the true t. We may, therefore, attempt to further complete our story of how the game is played by assuming that, as a part of the registered agreement, the total amount of money $ex = \sum_{i \in I} x_i$ is first transferred into the courts custody, where after the amount $x_i - (ex)b_i^t$ is allotted to player i. (Of course the term "*in mediis*" does not apply from the viewpoint of the court as it does not observe any results of the chance move.)

If we feel that the court cannot be assumed to have enough power in order force player to *pay* towards the final distribution of money ($x_i - (ex)b_i^t$ might be negative), then it might choose to *collect* $\max_t b_i^t$ from player i together with the registration – and repay whatever necessary eventually.

Now, in order to add some geometrical interpretation, observe that in the *ex post* situation players have been involved in the STU–game $(0, V^t)$ with $V^t = V_{b^t} = \{U^t(x) \mid x \in \underline{X}\}$ (see Definition 2.1).

In view of Lemma 2.3 and the subsequent discussion it follows that a fee–game is equivalently described by a system $(C^t)_{t \in T}$ of admissible subsimplices of $\underline{X}$. E.g., the paradigm of a two person CII game is a system of intervals via $\underline{X}$, (see Figure 2) each of which gives rise to an STU–game (as indicated in Figure 1).

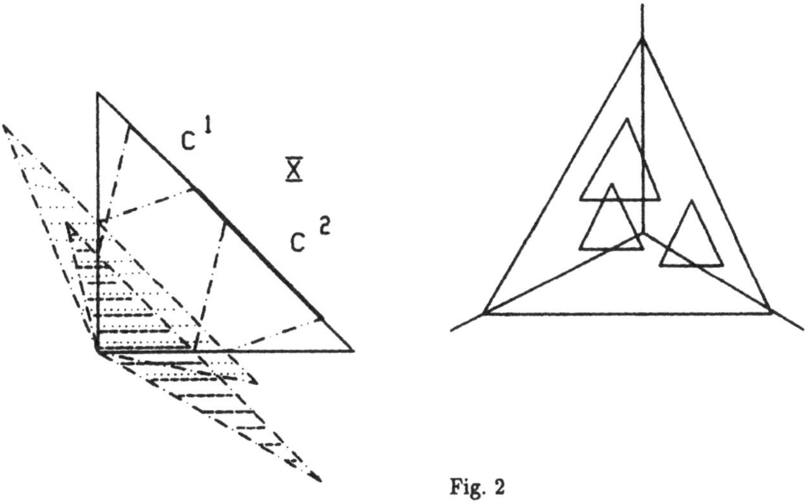

Fig. 2
Paradigm of the fee–game

Next, it is obvious that *ex ante* and *in mediis* players visualize an STU–game as well. For, if we define

$$V^{\underline{X}} := \{E\,U^{\tau}(x) \mid x \in \underline{X}\} \tag{10}$$

and

$$V^{\underline{X}|t_i} = \{E(U^{\tau}(x) \mid \tau_i = t_i) \mid x \in \underline{X}\}, \tag{11}$$

then clearly, we have

Lemma 2.5:

$(0,V^{\underline{X}})$ and $(0,V^{\underline{X}|t_i})$ are STU–games generated by Eb^{τ} and $E(b^{\tau} \mid \tau_i = t_i)$ respectively. Thus, $V^{\underline{X}} = V_{Eb^{\tau}}$ and $V^{\underline{X}|t_i} = V_{E(b^{\tau}|\tau_i = t_i)}$.

Proof:

Trivial E.g., $E(b^{\tau}(\tau_i = t_i)$ is at once recognized to be a fee and we have, for $x \in \underline{X}$

$$E(U^{\tau}(x) \mid \tau_i = t_i) = E(x - (ex)b^{\tau} \mid \tau_i = t_i) = x - (ex)\,E(b^{\tau} \mid \tau_i = t_i).$$

Finally, it turns out that in our present model the introduction of IC–mechanisms in general amounts to a restriction of the available utilities. To this end define

$$V^{\mathfrak{M}} := \{E\,U^{\tau} \circ \mu^{\tau} \mid \mu \in \mathfrak{M}\} \tag{12}$$

and

$$V^{\mathfrak{M}|t_i} := \{E(U^{\tau} \circ \mu^{\tau} \mid \tau_i = t_i) \mid \mu \in \mathfrak{M}\}. \tag{13}$$

Now, if $W_+ = W \cap \mathbb{R}^n_+$ denotes the nonnegative part of a subset W of $\mathbb{R}^n$, then we have

Lemma 2.6:

$V^{\mathfrak{M}}$ and $V^{\mathfrak{M}|t_i}$ are nonempty, compact, convex polyhedra satisfying

$$V^{\mathfrak{M}} \subseteq V^{\underline{X}}_{+},\ V^{\mathfrak{M}|t_i}_{+} \subseteq V^{\underline{X}|t_i}_{+}.$$

Proof:

Both sets are nonempty containing the utility vector $0 \in \mathbb{R}^n$ which is obtained via the status quo mechanism $\underline{\mu}$.

Next, the set of mechanisms has the canonical linear structure of $\underline{X}^T \subseteq (\mathbb{R}^n)^T$ and the inequalities defining IC–mechanisms ((4) in SEC.1, that is) can be rewritten as

$$\sum U_i^{(t_i,t_{-i})}(\mu^{(t_i,t_{-i})})\, p(t|t_i) \geq \sum_{t_{-i}\in T_{-i}} (U_i^{(t_i,t_{-i})}(\mu^{(s_i,t_{-i})})\, p(t|t_i). \tag{14}$$

Here $t = (t_i,t_{-i}) \in T_i \times \prod_{j\neq i} T_j =: T_i \times T_{-i}$ and $(p(\cdot\,|t_i)$ is the conditional distribution of τ given $\tau_i = t_i$, i.e. $p(\cdot\,|t_i) = P(\tau = \cdot\,|\tau_i = t_i)$. Since U^t is affine, (14) constitutes a system of linear inequalities ("in μ"). Similarly the conditions defining mechanisms in $\mathfrak{M}$ (i.e. (4) in Definition 1.2 and (6) in Definition 1.3) constitute a system of linear inequalities. Hence $\mathfrak{M}$ is a compact convex polyhedron. The same is now true for $V^{\mathfrak{M}}$ and $V^{\mathfrak{M}|t_i}$ since e.g. the latter one is the image of $\mathfrak{M}$ under the composition of the affine mappings

$$\underline{X}^T \to (\mathbb{R}^n)^T;\ (\mu^t)_{t\in T} \to U_1^t(\mu^t),\dots,U_n^t(\mu^t))_{t\in T}$$

and

$$(\mathbb{R}^n)^T \to \mathbb{R}^n;\ (u_1^t,\dots,u_n^t) \to (\sum_{T^{-i}} u_1^t\, p(t|t_i),\dots, \sum_{T^{-i}} u_n^t\, p(t|t_i)).$$

Next, let us prove that $V^{\mathfrak{M}} \subseteq V_{+}^{\underline{X}}$. The elements of $V^{\mathfrak{M}}$ are positive, since by definition for $\mu \in \mathfrak{M}$

$$E(U_i^\tau \circ \mu^\tau) = \sum_{t_i\in T_i} E(U_i^\tau \circ \mu^T \mid \tau_i = t_i)\, P(\tau_i = t_i) \geq 0 \cdot \Sigma P(\tau_i = t_i)$$

(cf. Definition 1.3). Since $V_{+}^{\underline{X}}$ is a simplex spanned by 0 and $(1 - e\, E\, b^\tau)\, e^i$ $(i \in I)$ (cf. Lemma 2.5 and Fig.1), it is sufficient to show that, for $\mu \in \mathfrak{M}$, $E\, U^\tau \circ \mu^\tau$ is dominated by some element on the Pareto surface $\partial V^{\underline{X}}$ of $V^{\underline{X}}$. This, however, follows at once since

$$E\, U^\tau \circ \mu^\tau = E(\mu^\tau - (e\,\mu^\tau)\, b^\tau) = E(e\,\mu^\tau)\left[\frac{\mu^\tau}{e\mu^\tau} - b^\tau\right] \leq E\left[\frac{\mu^\tau}{e\mu^\tau} - b^\tau\right] = E\,\frac{\mu^\tau}{e\mu^\tau} - E\, b^\tau \in \partial V^{\underline{X}},$$

with an obvious modification for the case that $P(\mu^\tau = 0) > 0$. This proves the first relation of Lemma 2.6. The second can be done analogously. Note, however, that $V^{\mathfrak{M}|t_i}$ may contain elements with negative coordinates $j \neq i$.

Definition 2.7:

1. μ is Pareto efficient (P.E.)

   a) *ex ante*, if $E\, U^\tau \circ \mu^\tau$ is P.E. in $V^{\mathfrak{M}}$

b) *in mediis*, if there is no $\tilde{\mu} \in \mathfrak{M}$ such that, for all $i \in I$ and $t_i \in T_i$

$$E(U^\tau \circ \mu^\tau \mid \tau_i = t_i) \leq E(U^\tau \circ \tilde{\mu}^\tau \mid \tau_i = t_i)$$

holds true with *strict* inequality for at least some i and t_i

c) *ex post*, if there is no $\tilde{\mu} \in V^{\mathfrak{M}}$ such that, for all $i \in I$ and $t_i \in T_i$

$$U_i^t(\mu^t) \leq U_i^t(\tilde{\mu}^t)$$

holds true with *strict* inequality for at least one i and t_i.

2. μ^t is Pareto efficient in V^t if $U^t(\mu^t)$ is located on the Pareto surface of V^t, i.e., $\mu^t \in \partial V^t$ or, equivalently if $e\mu^t = 1$.

Note that (as in other models) P.E. *ex ante* implies P.E. *in mediis* implies P.E. *ex post*. For, e.g., if μ is not P.E. ex post, then $\tilde{\mu}$ as in 3. yields $\tilde{\mu}$ as in 2. by integrating with $p(\cdot \mid t_i)$.

Note also, that is much more to ask for "$\mu^t \in \partial V^t$ for all $t \in T$" as compared to 3. Indeed, this is a very strong requirement and, as it turns out, if it is satisfied than, generically, μ is a constant mechanism.

Theorem 2.8:

Assume n = 2. Let I,T; $\underline{X}, \underline{x}$ and U be fixed. There is an open and dense set of probability distributions with the following properties: if $\mu \in \mathfrak{M}$ is such that μ^t is P.E. in V^t for every $t \in T$, then there is $\hat{x} \in \underline{X}$ with $\mu^t = \hat{x}$ $(t \in T)$.

Proof:

First, fix $i \in I$, $t_i, s_i \in T_i$. By IC we know

$$E(U_i^\tau \circ \mu^\tau \mid \tau_i = t_i) \geq E(U_i^\tau \circ \mu^{\tau_{-i}, s_i} \mid \tau_i = t_i) \tag{15}$$

and as $\mu^t \in V^t$ for all t we have $e\mu^t = 1$ $(t \in T)$, hence (15) reads

$$E(\mu_i^\tau - b_i^\tau \mid \tau_i = t_i) \geq (\mu^{\tau_{-i}, s_i} - b_i^\tau \mid \tau_i = t_i)$$

i.e.,

$$E(\mu_i^\tau \mid \tau_i = t_i) \geq (\mu^{\tau_{-i}, s_i} \mid \tau_i = t_i) \tag{16}$$

which implies

$$E\,\mu_i^\tau \geq E(\mu_i^{\tau_{-i}, s_i})$$

for all $i \in I$ and all $s_i \in T_i$. Since $\mu^t \in V^t$ $(t \in T)$, summation over i yields

$$1 \geq E\,\mu_1^{s_1,\tau_2} + E_{\mu_2}^{\tau_1,s_2} = \sum_{t_2 \in T_2} \mu_1^{s_1,t_2}\, p^{t_2} + \sum_{t_1 \in T_1} \mu_2^{t_1,s_2} \qquad (s_1 \in T_1,\ s_2 \in T_2)$$

Multiplying each of these inequalities with $p^{s_1} = P(\tau_1 = s_1)$ and then summing over $s_1 \in T_1$ yields

$$1 \geq \sum_{s_1 \in T_1} \sum_{t_2 \in T_2} \mu_1^{s_1,t_2}\, p^{s_1}\, p^{t_2} + \sum_{t_1 \in T_1} \mu_2^{t_1,s_2}, \qquad (s_2 \in T_2)$$

so that the same procedure performed with p^{t_2} finally results in

$$1 \geq \sum_{s_1 \in T_1} \sum_{t_2 \in T_2} \mu_1^{s_1,t_2}\, p^{s_1}\, p^{t_2} + \sum_{t_1 \in T_1} \sum_{s_2 \in T_2} \mu_2^{t_1,s_2},\, p^{t_1}\, p^{s_2}$$

$$= \sum_{t \in T} (\mu_1^t + \mu_2^t)\, p^{t_1,t_2} = 1. \tag{17}$$

Now, in view of (17) it follows that non of the inequalities employed can be a strict one, that is, in view of (16) we may conclude that

$$E(\mu_i^\tau \mid \tau_i = t_i) = E(\mu^{\tau_{-i},s_i} \mid \tau_i = t_i) \quad (i \in I,\ s_i \in T_i,\ t_i \in T_i)$$

hence

$$\sum_{t_2 \in T_2} \mu_1^{t_1,t_2}\, p^{t_1,t_2} = \sum_{t_2 \in T_2} \mu_1^{s_1,t_2}\, p^{t_1,t_2} \tag{18}$$

and

$$\sum_{t_1 \in T_1} \mu_2^{t_1,t_2}\, p^{t_1,t_2} = \sum_{t_1 \in T_1} \mu_1^{t_1,s_2}\, p^{t_1,t_2}. \tag{19}$$

Let us focus on (18). For any s_1 and t_1 the $|T_2|$-vectors $\mu_1^{s_1\cdot}$ and $p^{t_1\cdot}$ satisfy

$$(\mu_1^{s_1\cdot} - \mu_1^{s_1'\cdot})\, p^{t_1\cdot} = 0. \tag{20}$$

These are for fixed s_1 and s_1' exactly $|T_1|$ equations. If $|T_1| \geq |T_2|$ then 0 is the only solution of (20) (i.e. of the linear system $x\,p^{t_1\cdot} = 0$ ($t_1 \in T_1$)) if the matrix $p^{\cdot\cdot}$ has full rank $|T_2|$ – which is the case for an open and dense set of matrices (distributions). If $|T_2| \geq |T_1|$, we draw the same conclusion from (19). Thus, (20) yields $\mu_1^{s_1\cdot} = \mu_1^{s_1'\cdot}$ and, as $\mu_1^{\cdot\cdot} + \mu_2^{\cdot\cdot} = 1$, we see that all $\mu_i^{t_1 t_2}$ are equal, q.e.d.

Remark 2.9:

Suppose that in addition to $n = 2$ we require player 2 to have just one type, i.e.

$$T = T_1 \times \{*\} . \tag{21}$$

Then it can be verified that the statement of Theorem 2.8 holds true for *all* probability distributions (instead of almost all).

SECTION 3: Incomplete information on one side

In this section we consider the special case where only two players are involved in the game, such that player 2's type is deterministic. Thus, player 1 *in mediis* has full information since he observes his own type and knows his opponents type.

Of course this is a severe restriction. In particular, various results of other authors apply to this restricted model – e.g. WEIDNER [10] (who is dealing with the case that types are distributed independently) should be taken into account.

Nevertheless, we feel that open questions can first be attached by restricting ourselves to this simple (?) case.

To be more precise, we assume throughout this section that T is of the special shape $T = T_1 \times \{*\}$.

Then, it makes sense to write μ^{t_1} instead of $\mu^{(t_1,*)}$ $(t_1 \in T_1)$. Similarly, since player 1 is fully informed *in mediis* it is clear that $V^{\underline{X}|t_1} = V^{(t_1,*)}$ $(t_1 \in T_1)$, again we write V^{t_1} for this quantity. Similarly we use p^{t_1} instead of $p^{(t_1,*)}$ $(= p(t_1|*) = P(\tau_1 = t_1) = \cdots)$ etc.

Next, note that our special case allows for player 1 to induce an "ordering" of two types: If $\alpha,\beta \in T_1$ and $b_1^\alpha < b_1^\beta$ (i.e. he pays a smaller fee in case his type is α) then, for all $x \in \underline{X}$, $x \neq 0$

$$U_1^\alpha(x) = x_1 - (ex)\, b_1^\alpha < x_1 - (ex)\, b_1^\beta = U_1^\beta(x) .$$

Lemma 3.1:

Let $T = \{\alpha,\beta\} \times \{*\}$ and assume that $b_1^\alpha < b_1^\beta$. Suppose that $\mu \in \mathfrak{M}$ is *ex ante* P.E. and satisfies $\mu^\alpha \neq \mu^\beta$. Then the following statements hold true

1. μ^{α} is V^{α}–Pareto efficient 2. $U_1^{\alpha}(\mu^{\beta}) = U_1^{\alpha}(\mu^{\alpha})$ -

3. $U_1^{\beta}(\mu^{\alpha}) < U_1^{\beta}(\mu^{\beta})$

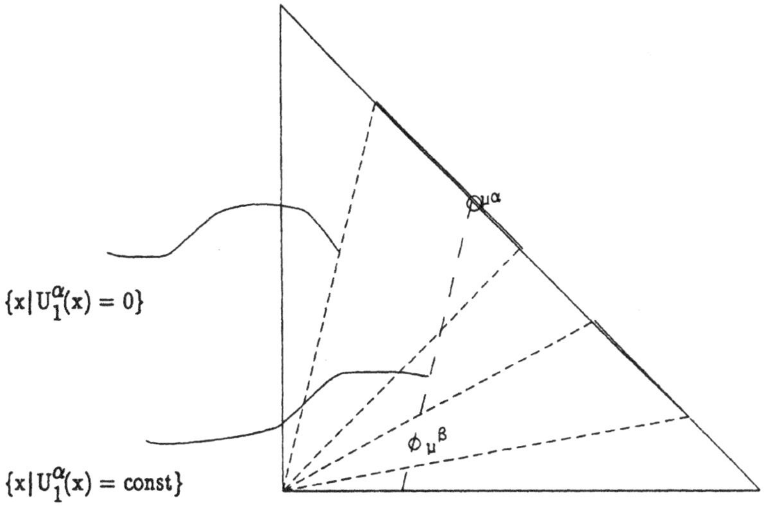

Fig.3
General position of non–constant $\mu \in \mathfrak{M}$

Proof:

Observe that U_{α}^{1} is linear with gradient $h_1^{\alpha} = (1 - b_1^{\alpha}, - b_1^{\alpha})$, thus the vector $\epsilon_1^{\alpha} := (b_1^{\alpha}, 1 - b_1^{\alpha}) > 0$ satisfies $U_1^{\alpha}(\epsilon_1^{\alpha}) = h_1^{\alpha}\epsilon_1^{\alpha} = 0$ (cf. Fig.4).

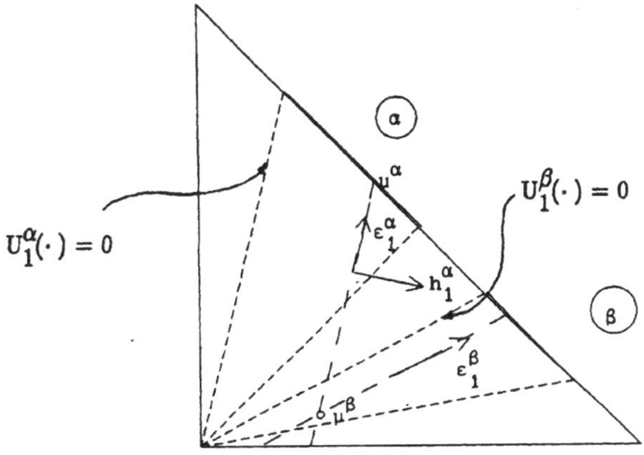

Fig.4
Gradients of utilities

Moreover,

$$U_1^\beta(\epsilon_1^\alpha) = h_1^\beta \, \epsilon_1^\alpha$$

$$= (1-b_1^\beta, -b_1^\beta)\,(b_1^\alpha, 1-b_1^\alpha) \tag{1}$$

$$= b_1^\alpha - b_1^\beta < 0 \,.$$

In order to prove the first statement, assume μ^α is not P.E. in V^α, then, for $\epsilon > 0$ sufficiently small let

$$\mu^{\alpha,\epsilon} = \mu^\alpha + \epsilon \, \epsilon_1^\alpha \qquad \in \underline{X}$$

As μ is IC we observe that

$$U_1^\alpha(\mu^{\alpha,\epsilon}) = U_1^\alpha(\mu^\alpha) \geq U_1^\alpha(\mu^\beta) \tag{2}$$

and in view of (1)

$$U_1^\beta(\mu^{\alpha,\epsilon}) = U_1^\beta(\mu^\alpha) + \epsilon \; U_1^\beta(\epsilon_1^\alpha) \tag{3}$$

$$< U_1^\beta(\mu^\alpha) \leq U_1^\beta(\mu^\beta)$$

(again using that μ is IC.) Now, (2) and (3) show that $(\mu^{\alpha,\epsilon}, \mu^\beta) =: \mu^\epsilon$ is IC. Moreover, as

$$U_2^\alpha(\epsilon_1^\alpha) = (-b_2^\alpha, 1-b_2^\alpha)\,(b_1^\alpha, 1-b_1^\alpha) = 1 - b_1^\alpha - b_2^\alpha = 1 - eb^\alpha > 0 \tag{4}$$

it is seen that $\mu^\epsilon \in \mathfrak{M}$ and (again viewing (2) and (4)) that (no matter what P) μ^ϵ is an *ex ante* Pareto improvement compared to μ.

In order to prove the second statement, assume

$$U_1^\alpha(\mu^\beta) < U_1^\alpha(\mu^\alpha) \,.$$

Now, for small $\epsilon > 0$, consider

$$\mu^{\beta,\epsilon} := \mu^\beta + \epsilon \, \epsilon_1^\beta$$

Then

$$U^\alpha{}_1(\mu^{\beta,\epsilon}) < U_1^\alpha(\mu^\alpha) \tag{5}$$

is still true for small $\epsilon > 0$. Moreover

$$U_1^\beta(\mu^\alpha) \leq U_1^\beta(\mu^\beta) = U_1^\beta(\mu^\beta + \epsilon\epsilon_1^\beta) \tag{6}$$

(as $U_1^\beta(\epsilon_1^\beta) = 0$). Again, from (5) and (6) it is derived at once that $\mu^\epsilon := (\mu^\alpha, \mu^{\beta,\epsilon}) \in \mathfrak{M}$.

Finally, in order to verify the third statement, assume that

$$U_1^\beta(\mu^\alpha) = U_1^\beta(\mu^\beta).$$

Then in view of (1), we have

$$h_1^\beta(\mu^\alpha - \mu^\beta) = 0$$

$$h_1^\alpha(\mu^\alpha - \mu^\beta) = 0.$$

But h_1^β and h_1^α are linear independent in view of

$$\begin{vmatrix} h_1^\beta \\ h_1^\alpha \end{vmatrix} = \begin{vmatrix} 1 - b_1^\beta, & -b_1^\beta \\ 1 - b_1^\alpha, & -b_1^\alpha \end{vmatrix} = b_1^\beta - b_1^\alpha < 0.$$

Lemma 3.2:

Let $T = T_1 \times \{*\}$ and let $\mu \in \mathfrak{M}$ be *ex ante* P.E. If, for $\alpha, \beta \in T_1$, $b_1^\alpha = b_1^\beta$ then $\mu^\alpha = \mu^\beta$.

Proof:

In this case, $U_1^\alpha(\cdot) = U_1^\beta(\cdot)$. Hence, using IC, yields

$$U_1^\alpha(\mu^\alpha) \geq U_1^\alpha(\mu^\beta) = U_1^\beta(\mu^\beta) \geq U_1^\alpha(\mu^\beta) = U_1^\beta(\mu^\beta) \geq U_1^\beta(\mu^\alpha) = U_1^\alpha(\mu^\alpha) \tag{7}$$

where all inequalities have to be equations.

It follows that μ^α and μ^β both are multiples of $\epsilon_1^\alpha = \epsilon_1^\beta$; assume (w.l.g.) that

$$\mu^\alpha = \lambda_1 \epsilon_1^\alpha > \lambda_2 \epsilon_2^\alpha = \mu^\beta.$$

Then

$$U_2^\alpha(\epsilon_1^\alpha) = (-b_2^\alpha, 1-b_2^\alpha)(b_1^\alpha, 1-b_1^\alpha) = 1 - b_2^\alpha - b_1^\alpha > 0 \tag{8}$$

and

$$U_2^\beta(\epsilon_1^\alpha) = U_2^\beta(\epsilon_1^\beta) > 0, \tag{9}$$

thus

$$U_2^{t_1}(\mu^\alpha) > U_2^{t_1}(\mu^\beta) \tag{10}$$

for $t_1 = \alpha, \beta$. It follows that $\bar{\mu} = (\mu^\alpha, \mu^\alpha)$ is IC (by (7)) and an *ex ante* improvement for player 2 (by

(10)), thus $\bar{\mu} \in \mathfrak{M}$ follows at once as well as the fact that $\bar{\mu}$ Pareto dominates μ *ex ante*. The assumption $L_1 > \lambda_2$ must, therefore be wrong, q.e.d.

Theorem 3.3:

Let $T = T_1 \times \{*\}$ and consider $\mu \in \mathfrak{M}$, μ *ex ante* P.E.. Let

$$T_1^l \quad (l = 1,\dots,L)$$

denote the disjoint subsets of T_1 given by the requirements

$$b_1^\alpha = b_1^\beta \qquad (\alpha,\beta \in T_1^l) \qquad (l = 1,\dots,L)$$

$$b_1^\alpha < b_1^\beta \qquad (\alpha \in T_1^l, \beta \in T_1^{l+1}) \qquad (l = 1,\dots,L-1)$$

such that

$$T_1 = \sum_{l=1}^{L} T_1^l$$

(using Σ instead of $\cup$ for disjoint unions). Then the following holds true:

1. μ^α is V^α Pareto efficient $(\alpha \in T_1^l)$

2. $U_1^\alpha(\mu^\beta) = U_1^\beta(\mu^\beta) \quad (\alpha \in T_1^l, \beta \in T_1^{l+1}, l = 1,\dots,L-1)$

3. $U_1^\beta(\mu^\alpha) < U_1^\beta(\mu^\beta) \quad (\alpha \in T_1^l, \beta \in T_1^{l+1}, l = 1,\dots,L-1)$

Proof:

Use the techniques supplied by 3.1 and 3.2 (cf. Fig.5).

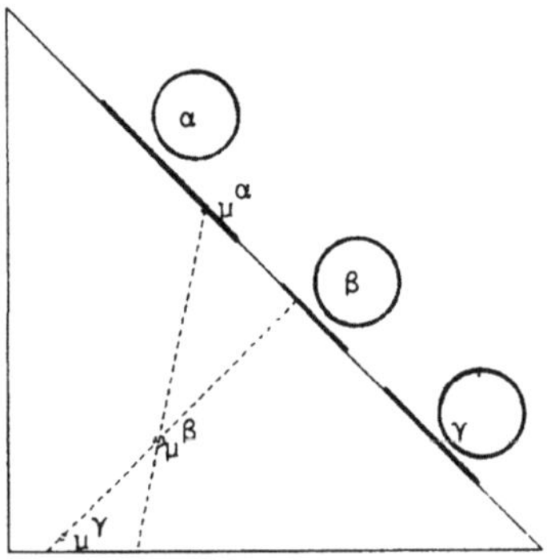

Fig.5
Three types in Figure 3

Theorem 3.4:

Let $T = \{\alpha,\beta\} \times \{*\}$ and $b_1^\alpha < b_1^\beta$. Suppose $\mu \in \mathfrak{M}$ is *ex ante* P.E. and satisfies $\mu^\alpha \neq \mu^\beta$. Then the following holds true: If $E\, U_2^\tau \circ \mu^\tau > 0$, then $U_1^\beta(\mu^\beta) = 0$.

Proof:

For small $\eta > 0$ define

$$\mu^{\alpha,\eta} := \mu^\alpha + p^\beta (\eta, -\eta) \tag{11}$$

$$\mu^{\beta,\eta} := \mu^\beta + p^\alpha \frac{\eta}{-U_1^\beta(\epsilon_1^\alpha)} \epsilon_1^\alpha + p^\beta \frac{\eta}{-U_1^\alpha(\epsilon_1^\beta)} \epsilon_1^\beta \tag{12}$$

Assuming $U_2^\alpha(\mu^\alpha) > 0$, $U_1^\beta(\mu^\beta) > 0$ it is clear that $\mu^{\cdot,\eta} = (\mu^{\alpha,\eta}, \mu^{\beta,\eta})$ is individually rational *in mediis* for player 1 ($U_1^\alpha(\mu^{\alpha,\eta}) > U_1^\alpha(\mu^\alpha)!$) (for small $\eta > 0$).

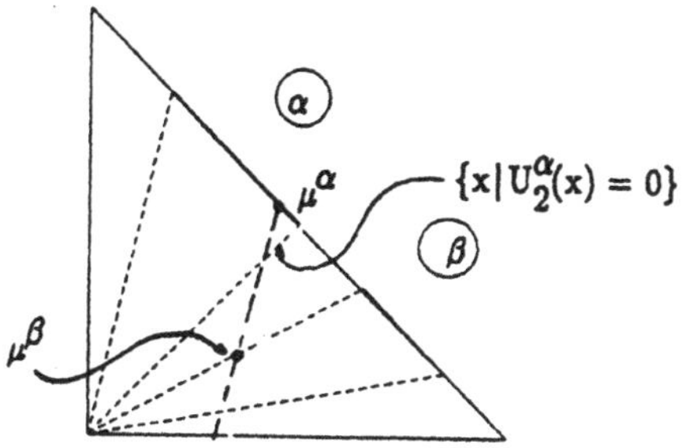

Fig.6
General position of non-constant $\nu \in \mathfrak{M}$
– the final version –

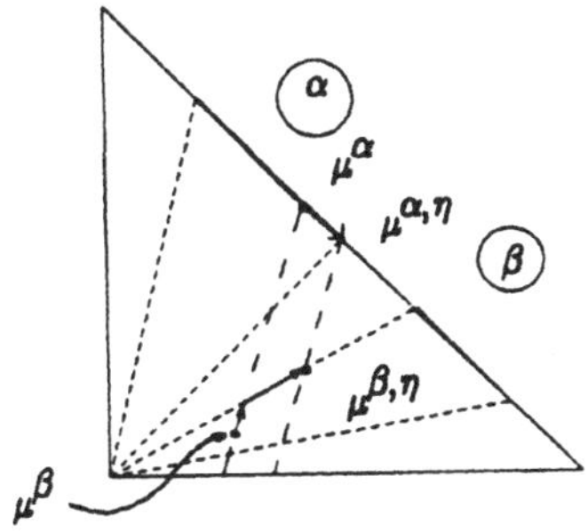

Fig.7
The proof for Fig.6

Moreover, $\mu^{\cdot,\eta}$ is IC. For, in view of 3.1, we know that

$$U_1^\beta(\mu^\alpha) < U_1^\beta(\mu^\beta)$$

and this relation is preserved with respect to $\mu^{\cdot,\eta}$ for small η. In addition (again using 3.1)

$$U_1^\alpha(\mu^{\beta,\eta}) = U_1^\alpha(\mu^\beta) + p^\beta\eta = \ldots$$

(since $U_1^\alpha(\epsilon_1^\alpha) = 0$)

$$\ldots = \quad U_1^\alpha(\mu^\alpha) + p^\beta\,\eta = \ldots$$

(by 3.1)

$$\ldots = \quad U_1^\alpha(\mu^{\alpha,\eta}) \,.$$

Thus, $\mu^{\cdot,\eta}$ is IC and hence $\mu^{\cdot,\eta} \in \mathfrak{M}$. We are now going to show that $\mu^{\cdot,\eta}$ is an *ex ante* improvement (strictly for player 2) for both players – this proves also $\mu^{\cdot,\eta} \in \mathfrak{M}$ (since for player 2 *ex ante* = *in mediis*) and hence establishes the Theorem.

Now

$$E\,U_1^\tau \circ \mu^{\tau,\eta} = E\,U_1^\tau \circ \mu + p^\alpha p^\beta\,\eta + p^\beta p^\alpha \frac{\eta}{-U_\beta^1(\epsilon_1^\alpha)} \cdot U_\beta^1(\epsilon_1^\alpha) = E\,U_1^\tau \circ \mu \tag{13}$$

and

$$E\,U_2^\tau \circ \mu^{\tau,\eta} = E\,U_2^\tau \circ \mu - p^\alpha p^\beta\,\eta + p^\beta p^\alpha \frac{\eta U_2^\beta(\epsilon_1^\alpha)}{-U_1^\beta(\epsilon_1^\alpha)} + (p^\beta)^2\,\frac{\eta U_2^\beta(\epsilon_1^\beta)}{-U_1^\alpha(\epsilon_1^\beta)} \tag{14}$$

Next, compute

$$U_1^\alpha(\epsilon_1^\beta) = b_1^\beta - b_1^\alpha > 0 \tag{15}$$

and

$$U_2^\beta(\epsilon_1^\beta) = 1 - b_2^\beta - b_1^\beta = 1 - eb^\beta > 0\,, \tag{16}$$

hence, it suffices to show

$$\frac{U_2^\beta(\epsilon_1^\alpha)}{-U_1^\beta(\epsilon_1^\alpha)} \geq 1\,. \tag{17}$$

However,

$$U_2^\beta(\epsilon_1^\alpha) = (-\,b_2^\beta, 1 - b_2^\beta)\,(b_1^\alpha, 1 - b_1^\alpha) = 1 - b_2^\beta - b_1^\alpha > b_1^\beta - b_1^\alpha = -\,U_1^\beta(\epsilon_1^\alpha)$$ q.e.d.

There are a few conclusions we can easily draw in view of this result.

Corollary 3.5:

Let $T = \{\alpha,\beta\} \times \{*\}$ and $b_1^\alpha < b_1^\beta$. Suppose $\mu^\alpha \in \underline{X}$ is V^α efficient (i.e. $e\mu^\alpha = 1$) and $U_1^\alpha(\mu^\alpha) \geq 0$. If $\mu^\beta \in \underline{X}$ exists such that $(\mu^\alpha, \mu^\beta) \in \mathfrak{M}$ is *ex ante* P.E., then it is uniquely defined.

Proof:

As the geometrical situation is cleared by Theorem 3.4 (and represented by Fig.6), we shall restrain ourselves to some verbal–geometrical argument. Clearly, if $\mu^\beta \neq \mu^\alpha$, then μ^β is uniquely defined as the intersections of the two lines $\{x \mid U_1^\alpha(x) = U_1^\alpha(\mu^\alpha)\}$ and $\{x \mid U_1^\beta(x) = 0\}$ as depicted in Fig.6. If these lines do not intersect in $\underline{X}$ then $\mu^\beta = \mu^\alpha$ constitutes $\mu \in \mathfrak{M}$ if and only if $E\ U_2^T(\mu^\alpha) \geq 0$. Thus, it is not hard to give precise conditions so that μ^β exists and to actually provide a formula for its computation.

Corollary 3.6: ("Unique implementation")

Let the conditions of 3.4 and 3.5 be satisfied. Let $u \in V^{\mathfrak{M}}$ be $V^{\mathfrak{M}}$–Pareto efficient (i.e., located on the Pareto surface of $V^{\mathfrak{M}}$). Then there exists *uniquely* $\mu \in \mathfrak{M}$ such $E\ U^T \circ \mu^T = u$.

Proof:

Again we shall argue verbally and claim that the details are obtained by inspecting 3.4 and Fig.6. We distinguish three cases.

Case 1:

Assume that u is in $V^{\underline{X}}$ Pareto efficient. Then, clearly, μ^t must be V^t efficient for $t = \alpha,\beta$. By Theorem 2.8 – or rather Remark 2.9 – μ is constant, i.e., $\mu^t = \hat{x}$ for some $\hat{x} \in \underline{X}$. Since

$$u = E\ U^T(\hat{x}) = \hat{x} - (e\hat{x})\ Eb^T = \hat{x} - Eb^T, \tag{18}$$

we obtain $\hat{x} = u - Eb^T$.

Case 2:

Now, if u is not efficient in $V^{\underline{X}}$, then $\mu^\alpha \neq \mu^\beta$. We shall distinguish Case 2a and Case 2b according to Theorem 3.4.

Case 2a:

$U_2^\alpha(\mu^\alpha) > 0$. Then $U_1^\beta(\mu^\beta) = 0$.

It is not hard to see that the four linear equations

$$u = U^\alpha(\mu^\alpha) + U^\beta(\mu^\beta) = \mu^\alpha - (e\mu^\alpha)\ b^\alpha + \mu^\beta - (e\mu^\beta)\ b^\beta \tag{19}$$

$$e\,\mu^\alpha = 1 \qquad \text{(by 3.3)} \tag{20}$$

$$U_1^\alpha(\mu^\beta) = U_1^\alpha(\mu^\alpha) \quad \text{(by 3.3)} \tag{21}$$

$$U_1^\beta(\mu^\beta) = 0 \quad \text{(by 3.4)} \tag{22}$$

determine μ^α and μ^β uniquely.

Case 2b:

If $U_2^\alpha(\mu^\alpha) = 0$, then we argue analogously to 2a. q.e.d.

Corollary 3.7:

Let the conditions of 3.4 and 3.5 be satisfied. Let $\hat{x} \in \underline{X}$ be such that $\hat{\mu}$, as defined by $\hat{\mu}^t = \hat{x}$ ($t \in T$) satisfies $\hat{\mu} \in \mathfrak{M}$. Then

$$U_1^\alpha(\mu^\alpha) < U_1^\alpha(\hat{x}),\ U_1^\beta(\mu^\beta) = 0 \text{ and } E\ U_1^\tau \circ \mu^\tau < E_1^\tau\ U_1^\tau \circ \hat{\mu}^\tau.$$

Thus, in the present situation, the expectation for player 1 with respect to a nonconstant mechanism in $\mathfrak{M}$ is worse than compared to any constant mechanism in $\mathfrak{M}$.

Again, the proof rests on a geometrical argument: inspection of Fig.6 shows that, for player 1, $U_1^\alpha(\mu^\alpha) < U_1^\alpha(\hat{x})$ (for $\hat{x}$ must be "in mediis IR" in situation β) while $U_1^\beta(\mu^\beta) = 0 < U_1^\beta(\hat{x})$. q.e.d.

It is now easy to visualize the shape of $V^{\mathfrak{M}} \subsetneq V_+^{\underline{X}}$ (cf. Fig.8).

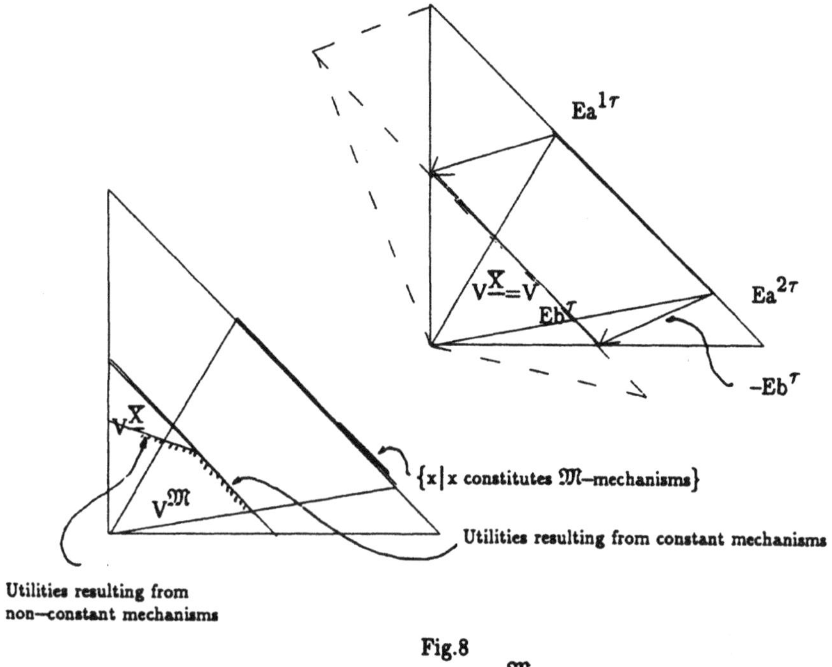

Fig.8
The shape of $V^{\mathfrak{M}}$

Theorem 3.8:

Let $T = T_1 \times \{*\}$ and let T_1^l (l=1,...,L) denote the number of the decomposition of T_1 indicated in 3.3, i.e. $b_1^\alpha < b_1^\beta$ ($\alpha \in T_1^l$, $\beta \in T_1^{l+1}$ l=1,...,L−1). Let $\mu \in \mathfrak{M}$ be *ex ante* P.E. and suppose $\mu^\alpha \neq \mu^\beta$ ($\alpha \in T_1^l$, $\beta \in T_1^{l+1}$ l=1,...,L−1). Now, if $E\, U_2^\tau \circ \mu^\tau > 0$ then $U_1^\beta(\mu^\beta) = 0$ for $\beta \in T_1^L$.

Proof:

For notational convenience, we shall restrict ourselves to the case that $T_1 = \{\alpha,\beta,\gamma\}$ and $b_1^\alpha < b_1^\beta < b_1^\gamma$.

For sufficiently small $\eta > 0$ take $\mu^{\cdot,\eta}$ to be defined by the following

$$\mu^{\alpha,\eta} := \mu^\alpha + p^\beta p^\gamma(\eta, -\eta)\ ,\quad \mu^{\beta,\eta} := \mu^\beta + p^\beta p^\gamma \eta \frac{\epsilon_1^\beta}{U_1^\alpha(\epsilon_1^\beta)}\ ,\quad \mu^{\gamma,\eta} := \mu^\gamma + p^\alpha p^\beta \eta \frac{\epsilon_1^\beta}{U_1^\gamma(\epsilon_1^\beta)}\ .$$

It is not hard to see that this constitutes an *in mediis* i.r. mechanism.

In order to show that the IC–property is preserved during the transition from μ to $\mu^{\cdot,\eta}$, we have to distinguish 3 cases.

a) ("α,β") Player 1 does not pretend to be β when α is his type; because

$$U_1^\alpha(\mu^{\alpha,\eta}) = U_1^\alpha(\mu^\alpha) + p^\beta p^\gamma \eta \quad\text{and}\quad U_1^\alpha(\mu^{\beta,\eta}) = U_1^\alpha(\mu^\beta) + p^\beta p^\gamma \eta$$

b) ("α,γ") Player 1 does not pretend to be γ when α is his type. This follows for small $\eta > 0$ since $U_1^\alpha(\mu^\alpha) > U_1^\alpha(\mu^\gamma)$ holds true – which is a consequence of Theorem 3.3.

c) ("β,γ") Player 1 does not pretend to be γ when β is his type, because

$$U_1^\beta(\mu^{\beta,\eta}) = U_1^\beta(\mu^\beta) \quad\text{and}\quad U_1^\beta(\mu^{\gamma,\eta}) = U_1^\beta(\mu^\gamma)\ .$$

Thus, we have established $\mu^{\cdot,\eta} \in \mathfrak{M}$. It remains to show that this mechanism constitutes a Pareto improvement versus μ.

Now, writing $\pi := p^\alpha p^\beta p^\gamma$ we observe that player 1 has expectation

$$E\, U_1^\tau \circ \mu^{\tau,\eta} = E\, U_1^\tau \circ \mu^\tau + \pi\eta + 0 + \pi\eta \cdot (-1) = E\, U_1^\tau \circ \mu^\tau\ ,$$

while player 2 has expectation

$$\mathrm{E}\ \mathrm{U}_2^\tau \circ \mu^{\tau,\eta} = \mathrm{E}\ \mathrm{U}_2^\tau \circ \mu^\tau - \pi\,\eta + (\mathrm{p}^\beta)^2\, \mathrm{p}^\gamma\, \eta \frac{\mathrm{U}_2^\beta(\epsilon_1^\beta)}{\mathrm{U}_1^\alpha(\epsilon_1^\beta)} + \pi\,\eta \frac{\mathrm{U}_2^\gamma(\epsilon_1^\beta)}{-\mathrm{U}_1^\gamma(\epsilon_1^\beta)}.$$

However, the quotient $\frac{\mathrm{U}_2^\beta(\epsilon_1^\beta)}{\mathrm{U}_1^\alpha(\epsilon_1^\beta)}$ has already been established to be positive (see (15) and (16), this term appears in the proof of 3.4 as well). Thus it remains to show that

$$\frac{\mathrm{U}_2^\gamma(\epsilon_1^\beta)}{-\mathrm{U}_1^\gamma(\epsilon_1^\beta)} \geq 1 \tag{23}$$

– which is the exact analogue to (17), q.e.d.

Note that, structurally, the case $|\mathrm{T}_1| = 3$ is slightly different compared to $|\mathrm{T}_1| = 2$. While the proof seems a bit easier and more straight forward, the result is of course slightly weaker: If player 2 in his "worst type" has positive utility, than player 1 in *his* worst type has zero utility – nothing is said about the intermediate types.

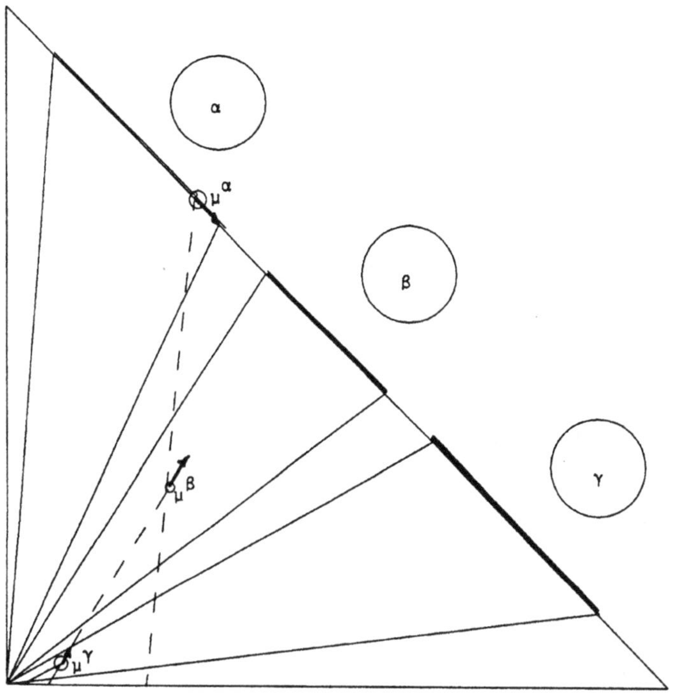

Fig.9
Three types in Figure 6

SECTION 4: The expected contract

Consider the Standard T.U–game $(0,V^t)$, i.e., the *ex post* version when $t \in T$ has been realized by the chance move. A "value" (e.g., the NASH–bargaining solution) in the "classical" context of complete information is a mapping, say

$$\chi : \{(u,W) \mid \mathbb{R}^n \supseteq W \text{ compact, convex; } u \in W\} \ .$$

Given some elementary axioms (symmetry etc.) such a value (e.g. the NASH–solution) will satisfy

$$\chi(0,V^t) = \frac{e}{n}(1-eb^t) =: \bar{u}^t \tag{1}$$

(where $V^t = V_{b^t}$ (see 2.1) and $e = (1,...,1)$).

$\bar{u}^t$ can be considered a (rather undisputable) solution for the *ex post* situations . And the *ex ante* expectation of players if in each *ex post* situation $\bar{u}^t$ could be chosen (i.e. in the case of full information) is given by

$$E\,\chi(0,V^\tau) = E\,\bar{u}^t = \frac{e}{n}(1-e\,Eb^\tau) + \chi(0,V_{Eb^t}) = :\bar{u}\,. \tag{2}$$

This formula (which is of course resting heavily on linearity) shows that the above mentioned *ex ante* expectation is as well the value of the "expected" STU game $(0,V_{Eb^\tau})$. (Of course $V_{Eb^\tau} = V^{\underline{X}}$ (cf. 2.5)).

Clearly, given complete information, players would register the mechanism implied by

$$\begin{aligned} \bar{x}^t :&= \frac{e}{n}(1-b^t) + b^t \\ &= \bar{u}^t + b^t = (U^t)^{-1}(\bar{u}^t) \\ &= \frac{1}{n}\sum_{k=1}^{n} a^{tk} \end{aligned} \tag{3}$$

(cf. 2.3 for the meaning of the a^{tk}). This motivates the following.

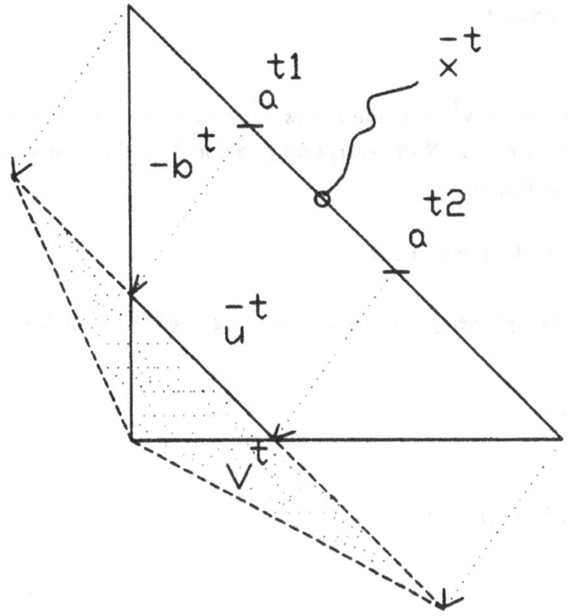

Fig.10
The expected contract

Definition 4.1:

The *expected contract* is given by

$$\bar{x} = E\, x^{\tau} = \frac{e}{n}(1-e\, Eb^{\tau}) + Eb^{\tau}\,.$$

Note that the expected utility of $\bar{x}$ equals $\bar{u}$, i.e.

$$E\, U^{\tau}(\bar{x}) = E(\bar{x} - (e\,\bar{x})b^{\tau}) = \bar{x} - Eb^{\tau} = \frac{e}{n}(1-e\, Eb^{\tau}) = \bar{u} \tag{4}$$

that is, the expected contract implements the ex ante expectation of a situation with complete information.

Thus, if players had no information, they could register $\bar{x}$ and have the same expected utility (ex ante) as in a world of truth speaking individuals in which, after each choice of t, everybody would reveal his type and the fair $\chi(0,V^t)$ would result by an obvious agreement.

Now, suppose players, being aware that the world is not the ideal place they (we all) would like it to be, agree upon the implementation of BIC mechanisms.

When bargaining *ex ante*, they might want to have a mechanism which is *ex ante* Pareto efficient. In that case, there should be no reason for deviating from the result that – in *ex ante* expectation – was presented by the fairness considerations "with full information" – provided of course, such BIC mechanism exists at all.

Now we have

Lemma 4.2:

Assume $n = 2$. Let I,T; $\underline{X}$, $\underline{x}$ and U be fixed. There is an open and dense set of probabilities (distribution of τ) such that the following holds true:

Let μ be BIC such that

$$E\, U^{\tau}(\mu^{\tau}) = \bar{u}.$$

Then $\mu^t = \bar{x}$ $(t \in T)$.

That is, the only way to implement a fair *ex ante* utility is by means of the (constant) mechanism that chooses the expected contract.

Proof:

It is easily seen that μ is V^t–Pareto efficient for all $t \in T$. Therefore, we may choose the dense and open set according to the result of Theorem 2.8. Then, as μ is BIC there is $\hat{x} \in \underline{X}$ such that

$$\mu^t = \hat{x} \quad (t \in T)$$

holds true. Now, as

$$\bar{x} - Eb^{\tau} = \bar{u} = E\, U^{\tau} \circ \mu^{\tau} = E\, U^{\tau}(\hat{x}) = \hat{x} - (e\hat{x})\, Eb^{\tau} = \hat{x} - Eb^{\tau},$$

we conclude that $\hat{x} = \bar{x}$, q.e.d.

The simple result of Lemma 4.2 bears some consequences in view of a discussion regarding the concept of a "bargaining solution" or "value for CII–games". For our present purpose it suffices to fix I,T,$\underline{X}$ and $\underline{x} = 0$, hence a value would be a mapping

$$\psi: \{(p, U) \mid \Gamma = (I, T; p; \underline{X}, \underline{x}; U) \text{ is CII}\} \rightarrow \underline{X}^T$$

satisfying certain axioms.

We do not want to discuss the axioms of symmetries. Also, in the present ("side payment") context, an axiom of covariance with respect to linear rescaling of utility cannot be formulated – note, however, that rescaling the elements of $\underline{X}$ and U is also covariant, thus there is no problem in discussing "Fee–games with varying currency".

However, we want a solution to yield, at any instant, an ex ante Pareto efficient mechanism $\mu \in \mathfrak{M}$. More precisely (if we write $\psi(\Gamma)$ instead of $\psi(p,U)$ by obvious reasons and use $\mathfrak{M}(\Gamma)$ as an obvious extension of notation):

Axiom $\mathfrak{M}$:

"For any Γ, $\psi(\Gamma) \in \mathfrak{M}(\Gamma)$ and $\psi(\Gamma)$ is *ex ante* Pareto efficient"

Furthermore, we believe that the expected contract $\bar{x}$ should be chosen if it induces a mechanism in $\mathfrak{M}$. That is

Axiom $\bar{x}$: (The expected contract axiom)

"Let $\bar{\mu} = \bar{\mu}(\Gamma)$ be the (constant) mechanism induced by $\bar{x}$ $(= \bar{x}(\Gamma))$, i.e.,

$$\bar{\mu}^t = \bar{x} \quad (t \in T) .$$

If $\bar{\mu} \in \mathfrak{M}$, then $\psi(\Gamma) = \bar{\mu}$.

There are two motivations we may offer for this axiom, and we shall discuss them in the light of the case $T = \{\alpha, \beta\} \times \{*\}$.

In this case, player 2 (since *ex ante* and *in mediis* is the same for him) initially viewed $V^{\underline{X}}$. However, player 1 convinced him, that he would not accept anything but IR *in mediis* and this led them to discuss mechanisms – IC and IR mechanisms, that is. So now, player 2 views $V^{\mathfrak{M}}$. Suppose it turns out that $\bar{u} \in V^{\mathfrak{M}}$. Player 2, a firm supporter of the IIA–Axiom, argues that $V^{\mathfrak{M}} \subseteq V_{+}^{\underline{X}}$ (Lemma 2.6) and it turns out that $\bar{u}$ is uniquely "implemented" by $\bar{\mu}$ (Lemma 4.2).

Player 1, if he wants to argue against this, may propose different *constant* contracts, to start with. However, any constant μ is generated by some $x \in \underline{X}$. But then the discussion may as well be conducted in the ex ante situation. The only difference (with respect to the "no information" case) is that player 1 knows he will not regret his decision to contract with player 2 *in mediis* – but the IC property is not needed at all. Thus, all arguments directed against $\bar{x}$ can be discussed from the *ex ante* viewpoint (this is an IIA argument) – and it is hard to see any such arguments as $\bar{x}$ satisfies all symmetry, efficiency (and covariance) requirements.

Next, player 1 could argue for nonconstant mechanisms.

Lemma 4.3:

Let $T = \{\alpha,\beta\} \times \{*\}$ and $b_1^{\alpha} < b_1^{\beta}$. Assume $\mu \in \mathfrak{M}$ to be *ex ante* Pareto efficient and $\mu^{\alpha} \neq \mu^{\beta}$. Also assume that $\bar{\mu} \in \mathfrak{M}$. Then

$$U_1^{\alpha}(\mu^{\alpha}) < U_1^{\alpha}(\bar{x}),\ U_1^{\beta}(\mu^{\beta}) = 0 \quad \text{and} \quad E\, U_1^{\tau} \circ \mu^{\tau} < E\, U_1^{\tau} \circ \bar{\mu}^{\tau} = \bar{u}_1 .$$

Proof:

This is a consequence of Corollary 3.7.

Clearly, this settles all attempts of player 1 to argue towards nonconstant mechanisms: he will be worse of in any situation. no matter whether he considers *ex ante or in mediis* situations.

Are there values satisfying **Axiom** $\overline{\mathbf{x}}$? As is easily seen, any mechanism "implementing" the NASH–value of $V^{\mathfrak{M}}$ will do this (and in our "standard case", $T = \{\alpha,\beta\} \times \{*\}$ this mechanism is uniquely defined). However, it needs some more axiomatic field work to characterize a value for CII games. We will deal with this question in a further discussion.

LITERATURE

[1] D'Aspremont, C. and Gérard–Varet, L.–A.: Incentives and incomplete information. Journal of Public Economics 11 (1979), pp. 25 – 45

[2] Harsanyi, J.C. and Selten, R.: A generalized Nash solution for two–person bargaining games with incomplete information. Management Science 18 (1972), pp. 80 – 106

[3] Holmström, B. and Myerson, R.B.: Efficient and durable decision rules withincomplete information. Econometrica 51 (1983), pp. 1799 – 1819

[4] Hurwicz, L.: On informationally decentralized systems. Decision and Organisation. (R. Radner and B. McGuire eds.) North Holland (1972), pp. 297 – 336

[5] Myerson R.B.: Incentive compatibility and the bargaining problem. Econometrica 47 (1979), pp. 61 – 73

[6] Myerson, R.B.: Two–person bargaining problems with incomplete information. Econometrica 52 (1984), pp. 461 – 487

[7] Myerson, R.B.: Cooperative games with incomplete information. International Journal of Game Theory 13 (1984), pp. 69 – 96

[8] Myerson, R.B. and Satterthwaite, M.A.: Efficient mechanisms for bilateral trading. Journal of Economic Theory 29 (1983), pp. 265 – 281

[9] Rosenmüller, J.: Remarks on cooperative games with incomplete information. Working paper 166, Inst. of Math. Econ. (IMW), University of Bielefeld (1988), 44 p.

[10] Weidner, F.: On the Harsanyi–Selten Value. Working paper 180, Inst. of Math. Econ. (IMW), University of Bielefeld (1989), 44 p.

A Non-Cooperative Interpretation of Value and Potential[1]

Sergiu Hart
Tel-Aviv University and Harvard University

and

Andreu Mas-Colell
Harvard University

1. Introduction

One of the most successful cooperative solution concepts in game theory is that of *value*. Originally defined for n-person games with transferable utility ("TU") by Shapley [1953b], it has been extended to general games with nontransferable utility ("NTU") by Harsanyi [1959, 1963] and Shapley [1969] and applied to numerous models. These applications, in turn, have always yielded important insights (to mention just one: the "value equivalence principle" in purely competitive economies).

In Hart and Mas-Colell [1989], we have introduced the concept of the *potential* of a game, which is shown to lead precisely to the Shapley value in the TU case. We refer to that paper for the details. The definition of the potential (which is just a combination of the basic principles of "efficiency" and "marginal contributions") has proved to be a (very useful) technical tool. However, a direct game theoretical interpretation of the potential is still lacking. To have one would be highly desirable.

We propose here one such interpretation. More precisely, we will present an auxiliary two-person zero-sum game, whose minimax = maximin value is precisely the potential. What is perhaps more important is that the optimal strategy of player I turns out to be precisely to choose the Shapley value! This yields a (noncooperative) interpretation for both the Shapley value and the potential. Moreover, the potential emerges as a kind of sequential price of the original game. A permanent theme in Harsanyi's work is the connection between non-cooperative and cooperative concepts. This paper provides, therefore, one more such link.

---

[1] Dedicated with great admiration to John C. Harsanyi. Research partially supported by grants from the National Science Foundation and the U.S.-Israel Binational Science Foundation.

There are a few other noncooperative approaches to the Shapley TU-value. Specifically, Gul [1989], O. Hart and Moore [1990] and Mas-Colell [1988]. One of the advantages of our model is that it can be directly extended to general nontransferable utility games. (See Hart and Mas-Colell [1992] for another non-cooperative approach to n-person NTU games.)

We start Section 2 with a simple example that introduces our model; the case of finite TU-games is treated there. The general NTU case is then analyzed in Section 3. In this context, the potential leads to the *egalitarian solution*, which is of interest on its own but which also constitutes the first step in the construction of the Harsanyi [1963] NTU-value (the second step is the intrinsic determination of the relative "weights" of the players' payoffs, according to which the egalitarian solution is determined). Section 4 is devoted to the study of "large" games; more precisely, (NTU) games with a continuum of players falling into finitely many types.

2. Finite Games with Transferable Utility

To explain the idea behind our approach, let us start with a simple example. There are 2 players $i = 1, 2$; player 1 by himself can get 10, player 2 by himself 30, and together they can get 100. There is also a "raider" R, an agent who wants to buy the game. Thus if 1 and 2 have agreed on, say, a (50, 50) split, then it appears that R will have to give 50 to each one. But suppose R wants to minimize the total amount he pays. He may then do the following (whenever the agreement (50, 50) emerges): offer[2] 51 to player 1 if he will break the agreement (recall that we are in a noncooperative setup where agreements are not binding). Then 1 will take 51 and "leave the game," after which R will have to pay only 30 to player 2, for a total of 81 (which is of course less than 100). Is this the best R can do? No, since he can give 51 to player 2 to leave the game, and then 10 to player 1, in total 61. Now is (50, 50) the worst agreement from R's point of view? A little thought will show that the most R will have to pay is 71, which happens whenever the two players choose (40, 60). Note that (40, 60) is precisely the outcome obtained by the principle of "split the surplus equally" (1 gets 10, 2 gets 30, and the surplus 60 is evenly divided among them). It is therefore the *Shapley value* of the game (actually, any single-valued solution that is symmetric and covariant with linear transformations will yield the same).

This example generalizes. Let (N, v) be an n-person TU-game, where $N = \{1, 2, ..., n\}$ is the set of players and $v : 2^N \to \mathbf{R}$ is the coalitional (or characteristic) function, with $v(\emptyset) = 0$.

[2] Assume for simplicity that 1 is the smallest unit.

We will consider an *auxiliary two-player zero-sum game* $\Gamma \equiv \Gamma(N,v)$. Player I chooses a payoff configuration $\mathbf{x} = (x_S)_{S\subset N}$, where $x_S = (x_S^i)_{i\in S} \in \mathbb{R}^S$ is a feasible payoff vector for the coalition S for each $S \subset N$; i.e., $\sum_{i\in S} x_S^i \leq v(S)$. Player II chooses an order π over N(i.e., $\{\pi(1), \pi(2), \ldots, \pi(n)\} = N$). The payoff from II to I is then

$$\sum_{t=1}^{n} x_{S(t)}^{\pi(t)}, \tag{1}$$

where $S(t) := \{\pi(t), \ldots, \pi(n)\}$. To understand this, assume for simplicity that π is the natural order $1, 2, \ldots, n$. The payoff of II to I is the sum of the following amounts: First, the payoff of player 1 in the grand coalition $N = S(1)$, i.e., x_N^1; second, the payoff $x_{N\backslash\{1\}}^2$ of player 2 in the remaining coalition $N\backslash\{1\} = S(2)$; and so on, up to the last player's payoff $x_{\{n\}}^n$.

The interpretation is as follows: Think of I as the "owner" of (the players of) the game (N,v); then II is a "raider" that "buys off" the players (in N) from I one by one. For each coalition S, I chooses a feasible payoff vector x_S (the final imputation if S "forms" — whenever S is the set of players that are still in the game), while II chooses the order in which he buys off the players. We assume that II wants to minimize the total payoff (by choosing the right order), and I wants to maximize it (by choosing the right payoff vectors).

The result is:

Proposition A: *In the two-person zero-sum game* $\Gamma(N,v)$*:*

(i) *The unique optimal strategy of player I is to choose x_S equal to the Shapley value of the game (S,v), for all $S \subset N$.*

(ii) *The unique optimal strategy of player II is to choose all $n!$ orders of N with equal probability.*

(iii) *The minimax (= maximin) value equals $P(N)$, the potential of the game*[3] (N,v).

Proof. Note that the payoff function of Γ is linear in player I's strategy, and his strategy set—the set of feasible payoff configurations—is convex. Therefore player I has no need for mixed strategies (as we will see below, he has a unique pure optimal strategy, therefore there are no mixed optimal strategies).

For every $S \subset N$, let $Q(S)$ be the minimax value of $\Gamma(S,v)$; clearly[4] $Q(\phi) = 0$ and $Q(\{i\}) = v(i)$ for all $i \in N$. Assume inductively that $Q(T) = P(T)$ for all $T \subsetneq S$.

---

[3] Since we are dealing with only one coalitional function v, we will write $P(S)$ for $P(S,v)$; see the "Moreover" statement in Hart and Mas-Colell [1989, Theorem A].

[4] $\Gamma(\phi, v)$ is an "empty" game (the strategy sets are trivial); as for $\Gamma(\{i\}, v)$, the result is immediate.

One obtains the following recursive formulas:

$$Q(S) = \operatorname*{Max}_{x_S} \operatorname*{Min}_{p_S} \sum_{i \in S} p_S^i \left[x_S^i + Q(S \backslash \{i\})\right] \tag{2}$$

$$Q(S) = \operatorname*{Min}_{p_S} \operatorname*{Max}_{x_S} \sum_{i \in S} p_S^i \left[x_S^i + Q(S \backslash \{i\})\right] \tag{3}$$

where x_S ranges over feasible S-payoff vectors (i.e., $x_S = (x_S^i)_{i \in S} \in \mathbf{R}^S$ and $\sum_{i \in S} x_S^i \leq v(S)$) and p_S is a probability distribution over S (i.e., $p_S = (p_S^i)_{i \in S} \in \mathbf{R}_+^S$ and $\sum_{i \in S} p_S^i = 1$; if $S = S(t) = \{\pi(t), \ldots, \pi(n)\}$, then p_S^i is the conditional probability that $\pi(t) = i$ given $\pi(1), \ldots, \pi(t-1)$).

Now $Q(S \backslash \{i\}) = P(S \backslash \{i\})$ by assumption; taking $p_S = \left(\frac{1}{|S|}, \ldots, \frac{1}{|S|}\right)$ in (2) yields

$$Q(S) \leq \operatorname*{Max}_{x_S} \frac{1}{|S|} \left[\sum_{i \in S} x_S^i + \sum_{i \in S} P(S \backslash \{i\})\right] \leq \frac{1}{|S|} \left[v(S) + \sum P(S \backslash \{i\})\right] = P(S)$$

(recall (2.2) in Hart and Mas-Colell [1989]). Taking[5] $x_S = Sh(S, v)$ in (3) gives

$$x_S^i + Q(S \backslash \{i\}) = Sh^i(S, v) + P(S \backslash \{i\}) = P(S)$$

for all $i \in S$, thus

$$Q(S) \geq \sum_{i \in S} p_S^i P(S) = P(S) \quad .$$

If $x_S \neq Sh(S, v)$, then there is $i \in S$ such that $x_S^i < Sh^i(S, v)$ (since $\sum x_S^i \leq v(S) = \sum Sh^i(S, v)$), and then $p_S^i = 1$, $p_S^j = 0$ for all $j \neq i$ makes the payoff $x_S^i + P(S \backslash \{i\}) < P(S)$. If $p_S \neq \left(\frac{1}{|S|}, \ldots, \frac{1}{|S|}\right)$, say $p_S^1 > p_S^2$, then choosing $x_S = Sh(S, v) + (\epsilon, -\epsilon, 0 \ldots, 0)$ for $\epsilon > 0$ yields a payoff of $P(S) + \epsilon(p_S^1 - p_S^2) > P(S)$. This shows the uniqueness of the optimal strategies. □

Thus one may view the potential $P(N)$ as a kind of *sequential price* of the game (N, v): how much player II has to pay to "buy" all the players in N in sequence. It is interesting to note that in a *market game* (i.e., when (N, v) is totally balanced), one has $P(N) \leq v(N)$ (see Hart and Mas-Colell [1988, Corollary 2]), thus II indeed prefers buying off the players in N

[5] We write "Sh" for the Shapley value payoff vector.

one by one (and paying $P(N)$) rather than all at once (and paying $v(N)$); see also the example at the beginning of this section.

It is possible to give an interpretation of the previous construction from a slightly different point of view. Suppose player I (the owner, or social planner) wishes to assign payoffs according to the egalitarian principle. How to do this? If the game was one of pure bargaining (for games normalized to $v(\{i\}) = 0$ for all $i \in N$, this means that $v(S) = 0$ for any $S \subsetneq N$) it is clear how to proceed: simply give every player $\frac{v(N)}{N}$. In more strategic terms we can think of the equal split as the outcome of an attempt by the planner to "keep the game together" by maximizing the utility of the worst-off player (the cheapest one to raid). This, in turn, determines the "sequential price of the game", as paid by an hypothetical player II, to be $P(N) = \underset{x_N}{\operatorname{Max}}\left\{\underset{i}{\operatorname{Min}}\{x_N^1, \ldots, x_N^n\} : \sum_i x_N^i = v(N)\right\} = \frac{v(N)}{N}$. The strategic reinterpretation gives then a hint on how to apply the egalitarian principle in the general case: let player I maximize the amount that a raider will have to pay sequentially in order to buy all the players (call this maximal amount the sequential price). It is worthwhile noting that in doing so player I can proceed recursively. If the sequential price $P(N\backslash\{i\})$ has already been determined for all i then $P(N)$ will obviously be given by

$$P(N) = \underset{x_N}{\operatorname{Max}}\left\{\underset{i}{\operatorname{Min}}\{x_N^1 + P(N\backslash\{1\}), \ldots, x_N^n + P(N\backslash\{n\})\} : \sum_i x_N^i \leq v(N)\right\},$$

which has as unique solution the $\bar{x}_N$ satisfying $\sum \bar{x}_N^i = v(N)$ and $P(N) = \bar{x}_i + P(N\backslash\{i\})$ for all i. Hence $P(N)$ is indeed the potential of the game (N, v). Because of the maxmin = minmax theorem we do not actually need to think of the Raider as moving after the Owner. They can move simultaneously. By choosing the next player to be bought at random (with equal probability over the remaining i) the Raider can guarantee that $P(N)$ is his maximal payment.

3. Finite Games with Nontransferable Utility

Let us move now to the general NTU case. Let (N, V) be an NTU-game, with $V(S) \subset \mathbf{R}^S$ being the set of feasible payoff vectors for the coalition S. We will make the standard assumptions that each one of the sets $V(S)$ is convex, closed, comprehensive, nonempty and not the whole space $\mathbf{R}^S$. Moreover, we assume that for all S the set $V(S)$ is nonlevel (i.e., x, $y \in bdV(S)$ = the boundary of $V(S)$ and $x \geq y$ imply $x = y$) and has a smooth boundary (i.e., at every point on its boundary there is a unique supporting hyperplane). Define the two-player zero-sum game $\Gamma(N, V)$ exactly as in the TU-case: I chooses a feasible payoff configuration $\mathbf{x}$

(i.e., $\mathbf{x} = (x_S)_{S \subset N}$ with $x_S \in V(S)$ for all S) and II chooses an order π on N; the payoff is then given by (1).

Proposition B. *In the two-person zero-sum game* $\Gamma(N, V)$:

(i) *The unique optimal strategy of player I is to choose* x_S *equal to the egalitarian solution of the game* (S, v), *for all* $S \subset N$.

(ii) *The unique optimal strategy of player II is to choose the order* π *with probability* $\prod_{t=1}^{n} \lambda_{S(t)}^{\pi(t)}$, *where* $\lambda_S = (\lambda_S^i)_{i \in S} \in \mathbb{R}_{++}^S$ *is the normal to* $bdV(S)$ *at the egalitarian solution of* (S, V), *normalized by* $\sum_{i \in S} \lambda_S^i = 1$.

(iii) *The minimax (= maximin) value equals* $P(N)$, *the potential of*[6] (N, V).

The *egalitarian value* is the first step in the construction of the *Harsanyi* [1963] NTU-value. It may be viewed as the outcome generated by the extension of the principle of "split the surplus equally" to more than 2 players. For TU-games, it coincides with the Shapley value. For general NTU-games, it has been studied and axiomatized by Kalai and Samet [1985]. In Hart and Mas-Colell [1989, Section 6], it is obtained via the potential approach, as follows: $P(\phi) = 0$ and $(P(S) - P(S \backslash \{i\}))_{i \in S} \in bdV(S)$ for all S, uniquely determine the potentials $P(S)$; then $P(S) - P(S \backslash \{i\})$ is $Eg^i(S, V)$, the egalitarian solution payoff of i in (S, V).

Note that our smooth boundary assumption implies that λ_S in (ii) is uniquely determined (however, see the Remark following the Proof of Proposition B). The probability distribution over the $n!$ orders of N may be described as follows: Assume that $\pi(1), \ldots, \pi(t-1)$ have been already determined and "bought off," in the interpretation of Section 2, and let $S := N \backslash \{\pi(1), \ldots, \pi(t-1)\}$ be the set of remaining players. Let $x_S = Eg(S, v) \in bdV(S)$ and $\lambda_S \in \mathbb{R}_+^S$ a supporting normal there (i.e., $\lambda_S \cdot x_S \geq \lambda_S \cdot y$ for all $y \in V(S)$). Normalize λ_S to be a probability vector (i.e., $\sum_{i \in S} \lambda_S^i = 1$). Then the next player $\pi(t)$ is chosen according to the distribution λ_S, i.e., $\text{Prob}(\pi(t) = i | \pi(1), \ldots, \pi(t-1)) = \lambda_S^i$ for all $i \in S$.

Proof of Proposition B. As in the proof of Proposition A, let $Q(S)$ be the minimax value of $\Gamma(S, V)$; we will show that $Q(S) = P(S)$ by induction.

The recursive formulas (2) and (3) hold here too (with $x_S \in V(S)$). Choosing $x_S = Eg(S, V)$ in (3) implies $Q(S) \geq P(S)$. Next note that $x_S^i + Q(S \backslash \{i\}) = x_S^i - (P(S) - P(S \backslash \{i\})) + P(S) = x_S^i - Eg^i(S, v) + P(S)$; choosing $p_S = \lambda_S$ in (2) therefore yields $Q(S) \leq P(S)$. Moreover, if $p_S \neq \lambda_S$, then p_S is *not* a supporting normal at $Eg(S, V)$, thus there is $x_S \in V(S)$ with $p_S \cdot x_S > p_S \cdot Eg(S, V)$, and the payoff is $> P(S)$. Finally, if $x_S \neq Eg(S, V)$ then there is

[6] We again write $P(N)$ for $P(N, V)$.

$i \in S$ such that $x_S^i < Eg^i(S, V)$ (here we use the nonlevelness assumption); player *II* choosing $p_S^i = 1$ (and $p_S^j = 0$ for $j \neq i$) yields a lower payoff. □

Remark. From the proof it follows that if we drop the assumption of the smoothness of the boundary of $V(S)$, then the optimal strategy of player *II* is no longer unique; any supporting λ_S at $Eg(S, V)$ will work. Similarly, if we drop the nonlevelness assumption, then $Eg(S, V)$ is no longer the unique optimal choice of player *I* for S; any $x_S \geq Eg(S, V)$ on the boundary of $V(S)$ will do. However, note that the minimax value is not affected, i.e., $Q(S) = P(S)$ for all S holds always (i.e., *without* nonlevelness and smooth boundaries).

If the game is a TU-game, then we always have $\lambda_S = \left(\frac{1}{|S|}, \ldots, \frac{1}{|S|}\right)$ which yields the uniform distribution over orders. For an NTU-game however, the distribution will be in general *not uniform*. An interesting (and surprising) implication is that *the principle of "all orders equally likely" does not follow from considerations of players' symmetry ("anonymity"), but rather from having transferable utility!* Indeed, the egalitarian solution is, of course, "anonymous" — permuting the names of the players does not matter; however, the probability distribution over orders is in general not uniform. This may shed some further light on the differences between egalitarian solutions and other NTU concepts based on the principle of all orders being equally likely (e.g., the Shapley NTU-value).

An interesting question is whether one can obtain the *Harsanyi NTU-value* rather than the egalitarian solutions. Of course, for every given vector of weights $w = (w^i)_{i \in N} \in \mathbf{R}^N_{++}$, if we define the payoff of the two-person zero-sum game by

$$\sum_{t=1}^{n} \frac{1}{w^{\pi(t)}} x_{S(t)}^{\pi(t)}$$

instead of (1), then in Proposition B we will get the w-egalitarian solution Eg_w (and, in Proposition A for the TU-case, the w-Shapley value of Shapley [1953a]). This follows immediately by the potential approach — see Section 6 in Hart and Mas-Colell [1989]. Note that λ_S will now satisfy $\sum_{i \in S} \lambda_S^i \cdot \frac{1}{w^i} y^i \leq \sum_{i \in S} \lambda_S^i \cdot \frac{1}{w^i} Eg_w^i(S, V)$ for all $S \subset N$ and $y \in V(S)$. It follows that a Harsanyi NTU-value is obtained if and only if $\lambda_N = (\frac{1}{n}, \ldots, \frac{1}{n})$ (see Theorems 6.2 and D in Hart and Mas-Colell [1989]). It would be worthwhile[7] to try and incorporate the choice of the weights w into the game (without, for instance, adding one more player).

[7] This is a suggestion of Ehud Kalai.

4. Infinite Games

The construction of the previous two sections extends in a natural way to games with an infinite number of players. We will consider here the setup of Hart and Mas-Colell [1991]: There is a continuum of players of finitely many (n) types. A coalition is represented by a profile $x = (x^1, \ldots, x^n) \in \mathbb{R}^n_+$, with x^i being the total mass of players of type i in the coalition. The grand coalition is $\bar{x} >> 0$. The coalitional function V associates to every profile x a set $V(x) \subset \mathbb{R}^n_+$ of feasible per-capita payoff vectors (i.e., $a \in V(x)$ whenever the payoff vector a, where each player of type i receives a^i, is attainable by a coalition of profile x; note that we only consider type-symmetric payoffs). The TU-case corresponds, of course, to $V(x) = \{a \in \mathbb{R}^n : \sum_{i=1}^n x^i a^i \leq v(x)\}$.

We will make the basic assumptions of Hart and Mas-Colell [1991], namely A.1–A.6 there.

Given such a game $(\bar{x}, V)$, consider the following two-person zero-sum game $\Gamma(\bar{x}, V)$: player I chooses, for each $x \leq \bar{x}$, a feasible payoff vector $\alpha(x) \in V(x)$ (we assume α to be measurable). Player II chooses a path $\mathbf{x} : [0,1] \to \mathbb{R}^n_{++} \cup \{0\}$, where $\mathbf{x}(0) = 0$, $\mathbf{x}(1) = \bar{x}$ and $\mathbf{x}$ is absolutely continuous; let $\Pi(\bar{x})$ denote the set of all such paths (this is denoted $\Gamma_0(\bar{x})$ in Hart and Mas-Colell [1991]). The payoff from II to I is

$$\int_0^1 \alpha(\mathbf{x}(t)) \cdot \dot{\mathbf{x}}(t)\, dt\,.$$

The interpretation is as follows: think of t (= "time") as going from 1 to 0, then $\mathbf{x}(t)$ is (the profile of) the set of remaining players at time t. As t moves to $t - dt$, $\mathbf{x}(t)$ changes to $\mathbf{x}(t) - \dot{\mathbf{x}}(t)\, dt$: player II buys off some additional players, namely $\dot{\mathbf{x}}^i(t)\, dt$ players of type i, for each $i = 1, \ldots, n$. He has to pay according to the current payoff vector $\alpha(\mathbf{x}(t))$ (chosen by I), in total $\sum_{i=1}^n \alpha^i(\mathbf{x}(t))\dot{\mathbf{x}}^i(t)\, dt = \alpha(\mathbf{x}(t)) \cdot \dot{\mathbf{x}}(t)\, dt$.

In the current setting, it is shown in Hart and Mas-Colell [1991] that the *potential* is given by a variational problem, namely

$$P(x) := \inf \left\{ \int_0^1 v(\mathbf{x}(t), \dot{\mathbf{x}}(t))\, dt : \mathbf{x} \in \Pi(x) \right\}$$

for every $x \in \mathbb{R}^n_{++}$. where $v(x, \cdot)$ is the supporting function of the convex set $V(x)$, i.e., $v(x,p) := \sup\{p \cdot a : a \in V(x)\}$ for all $p \in \mathbb{R}^n_+$. A path $\mathbf{x} \in \Pi(x)$ is called *minimal* if it is a minimizer above, that is $P(x) = \int v(\mathbf{x}, \dot{\mathbf{x}})$. The basic result (Theorem A) of Hart and Mas-Colell [1991a] is that minimal paths exist and that P as defined above is Lipschitz, thus differentiable almost

The construction of the previous two sections extends in a natural way to games with an infinite number of players. We will consider here the setup of Hart and Mas-Colell [1991]: There is a continuum of players of finitely many (n) types. A coalition is represented by a profile $x = (x^1, \ldots, x^n) \in \mathbb{R}^n_+$, with x^i being the total mass of players of type i in the coalition. The grand coalition is $\bar{x} >> 0$. The coalitional function V associates to every profile x a set $V(x) \subset \mathbb{R}^n_+$ of feasible per-capita payoff vectors (i.e., $a \in V(x)$ whenever the payoff vector a, where each player of type i receives a^i, is attainable by a coalition of profile x; note that we only consider type-symmetric payoffs). The TU-case corresponds, of course, to $V(x) = \left\{a \in \mathbb{R}^n : \sum_{i=1}^n x^i a^i \leq v(x)\right\}$.

We will make the basic assumptions of Hart and Mas-Colell [1991], namely A.1–A.6 there.

Given such a game $(\bar{x}, V)$, consider the following two-person zero-sum game $\Gamma(\bar{x}, V)$: player I chooses, for each $x \leq \bar{x}$, a feasible payoff vector $\alpha(x) \in V(x)$ (we assume α to be measurable). Player II chooses a path $\mathbf{x} : [0,1] \to \mathbb{R}^n_{++} \cup \{0\}$, where $\mathbf{x}(0) = 0$, $\mathbf{x}(1) = \bar{x}$ and $\mathbf{x}$ is absolutely continuous; let $\Pi(\bar{x})$ denote the set of all such paths (this is denoted $\Gamma_0(\bar{x})$ in Hart and Mas-Colell [1991]). The payoff from II to I is

$$\int_0^1 \alpha(\mathbf{x}(t)) \cdot \dot{\mathbf{x}}(t)\, dt\,.$$

The interpretation is as follows: think of t (= "time") as going from 1 to 0, then $\mathbf{x}(t)$ is (the profile of) the set of remaining players at time t. As t moves to $t - dt$, $\mathbf{x}(t)$ changes to $\mathbf{x}(t) - \dot{\mathbf{x}}(t)\, dt$: player II buys off some additional players, namely $\dot{\mathbf{x}}^i(t)\, dt$ players of type i, for each $i = 1, \ldots, n$. He has to pay according to the current payoff vector $\alpha(\mathbf{x}(t))$ (chosen by I), in total $\sum_{i=1}^n \alpha^i(\mathbf{x}(t))\dot{\mathbf{x}}^i(t)\, dt = \alpha(\mathbf{x}(t)) \cdot \dot{\mathbf{x}}(t)\, dt$.

In the current setting, it is shown in Hart and Mas-Colell [1991] that the *potential* is given by a variational problem, namely

$$P(x) := \inf \left\{ \int_0^1 v(\mathbf{x}(t), \dot{\mathbf{x}}(t))\, dt : \mathbf{x} \in \Pi(x) \right\}$$

for every $x \in \mathbb{R}^n_{++}$, where $v(x, \cdot)$ is the supporting function of the convex set $V(x)$, i.e., $v(x,p) := \sup\{p \cdot a : a \in V(x)\}$ for all $p \in \mathbb{R}^n_+$. A path $\mathbf{x} \in \Pi(x)$ is called *minimal* if it is a minimizer above, that is $P(x) = \int v(\mathbf{x}, \dot{\mathbf{x}})$. The basic result (Theorem A) of Hart and Mas-Colell [1991a] is that minimal paths exist and that P as defined above is Lipschitz, thus differentiable almost everywhere, and its gradient is efficient, i.e., $\nabla P(x) \in bdV(x)$, whenever P is differentiable at x.

everywhere, and its gradient is efficient, i.e., $\nabla P(x) \in bdV(x)$, whenever P is differentiable at x.

We now describe the optimal strategies of the two players in $\Gamma(\bar{x}, V)$. For player II, it is just a minimal path $\mathbf{x} \in \Pi(\bar{x})$. For player I it is more complicated. First, there exist optimal strategies $\alpha(\cdot)$ with $\alpha(x) = \nabla P(x)$ for almost every x. And second, every optimal strategy must satisfy $\alpha(\mathbf{x}(t)) = \nabla P(\mathbf{x}(t))$ for almost every $t \in (0,1]$ whenever $\mathbf{x} \in \Pi(\bar{x})$ is a minimal path.

Proposition C. *In the two-person zero-sum game $\Gamma(\bar{x}, V)$:*

(i) *Optimal strategies for player I are described above.*

(ii) *The optimal strategies for player II are to choose minimal paths, i.e., $\mathbf{x} \in \Pi(\bar{x})$ that minimize $\int_0^1 v(\mathbf{x}(t), \dot{\mathbf{x}}(t))\, dt$ over $\Pi(\bar{x})$.*

(iii) *The minimax (= maximin) value equals $P(\bar{x})$, the potential at $\bar{x}$.*

Proof. If player II chooses an optimal path $\mathbf{x}$, then $\int_0^1 \alpha(\mathbf{x}(t)) \cdot \dot{\mathbf{x}}(t)\, dt \leq \int_0^1 v(\mathbf{x}(t), \dot{\mathbf{x}}(t))\, dt = P(\bar{x})$ since $\alpha(x) \in V(x)$ for all x. If $\mathbf{x}$ is not an optimal path, then $\int v(\mathbf{x}, \dot{\mathbf{x}}) > P(\bar{x})$; let α be such that for all $t \in (0,1)$ we have $\alpha(\mathbf{x}(t)) \in bdV(\mathbf{x}(t))$ and $\alpha(\mathbf{x}(t)) \cdot \dot{\mathbf{x}}(t) = v(\mathbf{x}(t), \dot{\mathbf{x}}(t))$ (and arbitrary for other x); then the payoff is $\int v(\mathbf{x}, \dot{\mathbf{x}}) > P(\bar{x})$. This proves (ii), and also that the minimax value is $\leq P(\bar{x})$.

Consider now player I. Assume first that the potential function happens to be C^1. Letting $\alpha(x) := \nabla P(x)$ for all x yields, for any path $\mathbf{x} \in \Pi(\bar{x})$, a payoff of

$$\begin{aligned}
\int_0^1 \alpha(\mathbf{x}(t)) \cdot \dot{\mathbf{x}}(t)\, dt &= \int_0^1 \nabla P(\mathbf{x}(t)) \cdot \dot{\mathbf{x}}(t)\, dt \\
&= \int_0^1 \frac{d}{dt} P(\mathbf{x}(t))\, dt \\
&= P(\mathbf{x}(1)) - P(\mathbf{x}(0)) = P(\bar{x})\,.
\end{aligned}$$

Thus $\alpha = \nabla P$ guarantees $P(\bar{x})$ to player I.

In general, P is not necessarily differentiable — only Lipschitz. But the essence of the above argument still goes through. We skip the details. (By approximating P by a smooth P_ε in an appropriate way we get an $\alpha_\varepsilon(\cdot)$ which guarantees $P(\bar{x}) - \varepsilon$. An accumulation point of $\{\alpha_\varepsilon : \varepsilon > 0\}$, which can be shown to exist, then yields an $\alpha(\cdot)$ guaranteeing $P(\bar{x})$.) □

References.

Gul, F. [1989], "Bargaining Foundations of Shapley Value," *Econometrica* **57**, 81–95.

Harsanyi, J. C. [1959], "A Bargaining Model for the Cooperative n-Person Game," in *Contributions to the Theory of Games IV*, A. W. Tucker and R. D. Luce (eds.), Princeton University Press, 325–355.

Harsanyi, J. C. [1963], "A Simplified Bargaining Model for the n-Person Cooperative Game," *International Economic Review* **4**, 194–220.

Hart, O. and J. Moore [1990], "Property Rights and the Nature of the Firm," *Journal of Political Economy* **98**, 1119–1158.

Hart, S. and A. Mas-Colell [1988], "The Potential of the Shapley Value," in *The Shapley Value: Essays in Honor of Lloyd S. Shapley*, A. E. Roth (ed.), Cambridge University Press, 127–137.

Hart, S. and A. Mas-Colell [1989], "Potential, Value and Consistency," *Econometrica* **57**, 589–614.

Hart, S. and A. Mas-Colell [1991], "Egalitarian Solutions of Large Games: I. A Continuum of Players," (mimeo).

Hart, S. and A. Mas-Colell [1992], "n-Person Non-Cooperative Bargaining," (mimeo).

Kalai, E. and D. Samet [1985], "Monotonic Solutions to General Cooperative Games," *Econometrica* **53**, 307–327.

Mas-Colell, A. [1988], "Algunos Comentarios sobre la Teoria Cooperativa de los Juegos," *Cuadernos Economicos* **40**, 143–161.

Shapley, L. S. [1953a], "Additive and Non-Additive Set Functions," Ph.D. thesis, Princeton University.

Shapley, L. S. [1953b], "A Value for n-Person Games," *Annals of Mathematics Studies* **28** (*Contributions to the Theory of Games II*), H. W. Kuhn and A. W. Tucker (eds.), Princeton University Press, 307–317.

Shapley, L. S. [1969], "Utility Comparison and the Theory of Games," in *La Decision*, Editions du CNRS, Paris, 251–263.

II. Mechanism Design

Types Decomposition and Perfect Implementations*

Ehud Kalai
Kellogg Graduate School of Management
Northwestern University
Evanston, Illinois 60208

and

Dov Samet
Recanati Graduate School of Business Administration
Tel Aviv University
Tel Aviv, Israel

Abstract

A revelation (or delegation) game is Nash incentives compatible if and only if it is perfectly incentives compatible whenever the types of the players are personal (independently drawn with private valuations of actions). An outcome in a personal-types Bayesian environment is perfectly implementable, if and only if it is Nash implementable. These observations are shown to be immediate consequences of a general types decomposition theorem regarding the perfect equilibria of Bayesian games.

1. Introduction

Social scientists use game theoretic solutions to analyze the performance of social implementation mechanisms involving information and incentives. Gibbard (1973) and Satterthwaite (1975) attempted to use dominant strategy incentive compatible revelation games in order to deal with information and incentives issues arising in Arrow's (1951) social choice problem. As they and many other authors discovered, the dominant strategy approach was too demanding and failed to bring about meaningful results. Given the failure of this approach, Hurwicz (1972), Peleg (1977), Groves and Ledyard (1977), Hurwicz and Schmeidler (1978), Kalai and Rosenthal (1978), and many others (see Maskin (1985), Postlewaite (1985), and Roberts (1986) for general studies and references) have replaced the solution concept of dominant strategy by the weaker notion of Nash equilibrium and applied it to general classes of strategic economic games. Their approach paralleled the game theoretic development of the notion of correlated equilibrium due to Aumann (1974) and has proven to be more successful.

More recently, researchers wishing to model asymmetries of information in a more explicit manner recognized that many economic situations can be viewed as the games of incomplete information discussed by Harsanyi (1967-8). Consequently, we have seen many economic and managerial problems being studied and analyzed as Bayesian games of incomplete information. A sample of studies of this type is Wilson (1967), d'Aspremont

---

*This research was partly sponsored by a grant from the U.S. National Science Foundation, NSF No. SES-9011790. An earlier version of the paper, entitled "Are Bayesian-Nash Incentives and Implementations Perfect?" appeared as Discussion Paper No. 680 in the Center for Mathematical Studies in Economics and Management Science, Northwestern University, 1986.

and Gerard-Varet (1979), Milgrom and Weber (1982), Myerson (1981), Matthews (1979), Milgrom and Roberts (1982), Harris and Raviv (1981), Maskin and Riley (1984), Riley and Samuelson (1981), Crawford and Sobel (1982), Myerson and Satterthwaite (1983), Postlewaite and Schmeidler (1986), and Palfrey and Srivastava (1986). For a comprehensive survey and study of Bayesian games, implementations, and incentives, we refer the reader to Myerson (1985).

An active direction of research in game theory deals with the weaknesses of the Nas equilibrium concept. Selten (1975) has argued that this concept may predict outcomes which are not likely to occur and proposed a strengthening of it which he called perfect (Nash) equilibrium. Following Selten (1975), Harsanyi and Selten (1980), Myerson (1978) Kreps and Wilson (1982), Kalai and Samet (1984), Mertens and Kohlberg (1982), Bernheim (1984), Pearce (1984), McLennan (1985), and others have studied a variety of these types of modifications. We refer the reader to van Damme (1983) for a comprehensive study and survey of this literature. Also, Forges (1984) and Myerson (1984) developed these concepts further in order to adapt them better for incentive problems involving communications and other considerations of this type. However, a major portion of the implementation literature uses the unmodified notion of Nash (i.e., Bayesian) equilibrium, and thus suffers from the weaknesses discussed by Selten. To illustrate an example of such a weakness and for further discussion, we introduce here the following hypothetical game.

<u>The Confession Game</u>. Two players might have jointly committed an illegal act. No one but the two of them knows whether they did it. Society would like to implement the following outcomes. If they did commit the crime then they should each be penalized (say, impose an \$M fine on each of them), but otherwise no penalty should be imposed. I is suggested that this conditional outcome be implemented by the following mechanism. A representative of the society will ask each one of them, separately, whether he has participated in the illegal act. If any one of them admits to it then the penalty will be imposed on both of them. However, if they both deny the accusation then they will be let go without penalty.

It is argued that the revelation game suggested above is "Nash incentives compatible" and that the players will therefore reveal the truth. The underlying logic is that every player will reason to himself that if his opponent tells the truth then he cannot gain by lying, and thus telling the truth does not contradict his incentives.

Obviously it would be naive to assume that the scheme proposed is satisfactory. While it is possible that a criminal player would not lose by telling the truth, it is very likely that he would. If there is any possibility that his criminal partner would lie then he would also prefer to lie (we assume here that there is no inherent satisfaction from telling the truth). Lying is a dominant strategy for a criminal playe in this game. When we apply Selten's perfect equilibrium notion to the analysis of this game it selects what we consider to be the more reasonable (undominated strategies) prediction, that they would deny having committed the crime under such a scheme.

A main purpose of this paper is to report general circumstances under which incentives and implementations are perfect in the sense of Selten. In the process of doing that we introduce a general decomposition property of Bayesian games and exhibit a structural theorem relating perfect equilibrium to this decomposition. The prefect equilibrium notion that we use is sometimes referred to as "trembling hand" perfection as opposed to "subgame perfection." In other words, we use Selten's original terminology. Perfection of an equilibrium is a stronger requirement than subgame perfection. For example, in the confession game discussed earlier, the "bad" equilibrium is still subgame perfect. The same relationship holds with Kreps and Wilson's notion of sequentiality.

For a strategy to be a Nash equilibrium it is required that simultaneously every player's equilibrium action be an optimal reaction to his opponent's equilibrium actions. For a strategy to be a perfect equilibrium it is further required that every player's equilibrium action be an optimal reaction to some interior strategies of his opponent's which are arbitrary close to their equilibrium actions. As usual, a player's strategy is called interior if there is a positive (however small) probability that he would take every available action at every state of his information. This can be viewed as a minimal continuity or robustness requirement on the proposed equilibrium concept. An important implication of this condition is that optimality is tested not only at the equilibrium point, but also at some interior points in every neighborhood of it. This is important because it rules out knife-edge strategic stability which is based upon players assigning certainty (probability one) to their conjectures on the actions of their opponents. Under such idealized certainty the players can get locked in unbelievable behavior which is really not credible if any small doubt enters their considerations. Consider, for instance, the confession game discussed earlier. True revelation is optimal to a criminal player if he is certain that his opponent reveals truthfully. However, if there is any small positive probability that his criminal opponent does not reveal truthfully then his only optimal reaction is to lie. Thus the true revelation could not be a perfect equilibrium behavior of the criminal player.

In the next section we formally introduce the notions of Bayesian game, Nash equilibrium, and perfect (Nash) equilibrium. We basically follow the notations and conventions as in Myerson (1985) for the Bayesian games and their Nash equilibria. For the perfect equilibrium concept we use the definition described in Kalai and Samet (1985). It is easy to verify that this definition is equivalent to the original one given in Selten (1975).

The Bayesian games studied here may be thought of as n person two-stage extensive form games. In the first stage nature draws a type for every one of the n players using a commonly known probability distribution on the finite set consisting of all type combinations (n-tuples of types). After the draw every player is informed about the realization of his type and only his type. In the second stage, simultaneously, every player chooses an action from a finite set of actions available to him. The set of actions available to a player is the same regardless of the particular type realization. At this point the game is over and every player is rewarded a von Neumann-Morgenstern utility level which is determined by the realized types and action combinations.

The third section of the paper is devoted to the study of special families of

Bayesian games and the relationship between their Nash and perfect equilibria. The family of revelation games consists of the ones in which the set of actions available to every player consists of his set of types. Thus in such a game we may think of an action of a player as being a declaration (possibly a false one) of the type that he is. For example, the confession game discussed previously is such a game. A revelation game is called incentive compatible if declaring one's true type by every one of the players turns out to be a Nash equilibrium. The revelation games have special importance in the incentives literature because of the "revelation principle." This useful principle states that to every Bayesian game with a choice of a Nash equilibrium outcome, we can associate a derived incentive-compatible revelation game in which the truthful play yields the same outcome. Thus, in classifying the set of outcomes that may result from the Nash equilibria of a given family of games, it suffices to restrict our attention to the set of outcomes that result from truthful play in the derived incentive-compatible revelation games.

A somewhat broader family of games consists of what we call delegation games. In a delegation game every player is prescribed an action for every one of his type realizations. We refer to these prescriptions as guidelines. We say that a delegation game (and its guidelines) are incentive compatible if the total obedience strategy by all the players turns out to be a Nash equilibrium. Mathematically these games turn out to be very similar to the revelation games. Indeed, formally, every revelation game can be viewed as a delegation game (with the guidelines being for every player to reveal his true type). Because of this fact the "revelation principle" induces immediately a "delegation principle." This latter principle says that to every Bayesian game with a choice of Nash equilibrium outcome, we can associate a derived incentive-compatible delegation game in which obedient playing yields the same outcome.

Because of the importance of revelation and delegation games we are interested in knowing when such incentive-compatible games are perfectly incentive compatible. By perfect incentive compatibility we mean that the truthful or the obedient strategies are perfect, and not just Nash, equilibria.

As we report in Section 3, for a significant class of revelation and delegation games, being incentive compatible is equivalent to being perfectly incentive compatible. This class is characterized by the property that the types of each player are what we call personal. For a given Bayesian game we say that it has personal types if two conditions hold. The first condition is that the types of the various players are drawn independently (in the probabilistic sense). The second condition is that the individual player payoffs are private. This means that a player's payoff depends only on his type and the action combination of the group of all players. However, given his own fixed type and given a fixed group action combination his payoff will not be affected if we vary the types of his opponents (provided that their actions are not changed). The confession game discussed earlier violates the probabilistic independence assumption (the types--criminal or not--are completely correlated) but satisfies the independent payoff condition. Thus, the types are not personal.

As it turns out, many Bayesian games satisfy the two conditions just discussed. For example, most of the Bayesian game papers cited earlier make use of these conditions. The two main results of Section 3 state that if an incentive-compatible revelation or delegation game has only personal types then it must be perfectly incentive compatible.

In Section 4 we discuss the relationships between Nash implementability and perfect implementability of an outcome function. We define the notion of an n-person Bayesian environment to consist of four components. The first component describes again the finite sets of possible type realizations for every one of the n players. The second component is a commonly known probability distribution on the set consisting of all possible type combinations. The third component is a finite set describing all the possible outcomes in the environment. The fourth component of the Bayesian environment consists of a von Neumann-Morgenstern utility function for every one of the players. We assume that a player's utility depends on the choice of an outcome in the environment and on the type combination of the n players. In such an environment it is assumed that after the initial drawing of a type combination, every player is informed of his own particular type. Based on this information, and his updated probabilistic belief regarding the type combination of his opponents, a player has an ex post preference over the possible outcomes in the environment. As in the case of Bayesian games we say that the environment consists of personal types if the types are independently drawn and every player's utility depends on the outcome and only on his own type.

An implementor in a Bayesian environment wishes to bring about an outcome for every combination of types. The implementor may be a public agent with an interest in maximizing some social welfare or he may be a private agent pursuing his own private interests. For example, a public agent may want to bring about a Pareto optimal outcome for every combination of type realization. A private implementor may be a seller selling an object to the n players, and his interest may be to bring about the outcome that will maximize the total revenues that he can receive from every configuration of buyer types. Generally, what the implementor wishes to implement is an outcome function, which assigns a probability distribution over the set of outcomes for every type combination. However, the implementor may face two difficulties. First, because of their own strategic considerations, the players may not reveal their true types to the implementor. The second difficulty may come up in situations where in order to bring about a certain outcome the players themselves have to cooperate by taking some necessary actions, and the implementor does not have the authority to control their actions. In order to have a unified model that covers all the different types of implementors, we take the following approach.

We define a Bayesian game in a given Bayesian environment as a pair consisting of two components. The first component has a set of actions available to every one of the n players. The second component is a result function which assigns a probability distribution over the set of the environmental outcomes to every action combination of the players. Now the utility functions that the players have over the environment's outcomes induces a utility function over the players' actions. Thus every Bayesian game in the given Bayesian environment induces a regular Bayesian game in one natural way. In this way the notions of Nash and perfect equilibrium get extended to games in the environment.

We assume that an implementor in a given Bayesian environment is described by a family of environmental games from which he has the authority to select one game. This family describes the amount of control that he has. For example, if the family of games available to him is a singleton, then he has no real control and the players will play this one given game. At the other extreme, a social implementor (for example, a legislator) can have much control in the sense that his family of games may be large and he can choose the "rules of the social game" from many available ones. Interesting implementors are intermediary ones with partial control. For example, a seller wishing to sell an object can consider many different auction methods. Each choice of an auction method would mean one choice of a game from his available set. It would be reasonable to assume that if he is a private seller then his family contains all the legal auction methods provided that they include the action "do not participate" in the feasible action set of every one of the players (buyers).

We say that an outcome function is Nash implementable in a given Bayesian environment by a given implementor if his family of implementing games includes one with a Nash equilibrium that yields this outcome function. Similarly, an outcome function is perfectly implementable if the implementor has a game with a perfect equilibrium yielding the outcome function.

The main result of Section 4 is in illustrating that in a Bayesian environment with personal type, an implementor can Nash implement a given outcome if and only if he can perfectly implement it. This equivalence is illustrated under the assumptions that the family of implementing games is complete (as defined there), and is a mathematical corollary of the results in the previous section.

In Section 5 we discuss briefly correlated equilibria that may be viewed as equilibria in Bayesian games. Our results for general Bayesian games show that those correlated equilibria that correspond to totally mixed strategy equilibria of the game in normal form are perfect. This result is not just a corollary of the trivial fact that totally mixed strategy equilibria of games in normal form are perfect.

In Section 6 we deal with Bayesian games having a general type structure (not necessarily personal types). We show that the type combinations in such a game can be decomposed into cells of personal types. Such a decomposition separates the types of every player by two characteristics. One characteristic is public and its realization may affect the other players through the probability of their own type selection and through their payoffs. The other characteristic is personal and its realization does not affect the other players in either of these two ways. We show that every Bayesian game has such a unique coarsest personal types decomposition. It turns out that the perfect equilibrium notion on Bayesian game factors nicely through the type decomposition. For a strategy combination to be a perfect equilibrium it suffices that it be a best reply to an arbitrarily close strategy combination which is interior but only relative to the personal decomposition. By being interior relative to the decomposition we mean that in every cell of personal types, the cell types use all the available strategies among them (rather than every type uses all of the strategies). It is then obvious to see that the results in Sections 3 and 4 follow immediately from this general result when the decomposition involved is the trivial coarse decomposition, i.e., consists of only one such cell.

In Section 7 we give two examples of Bayesian games which demonstrate that neither independence of types nor private payoffs is enough to guarantee perfection even when the probability of each type combination is positive. One of the examples shows in particular that totally mixed correlated equilibria are not necessarily perfect.

2. Definitions and Notations

We define an n-person Bayesian game G by a set of players $N = \{1,2,\ldots,n\}$ and by a four tuple

$$G = (T = \times_{i\in N} T_i,\ A = \times_{i\in N} A_i,\ p,\ u = (u_1,u_2,\ldots,u_n)).$$

The interpretations of these symbols are as follows.

a. T_i is a finite set describing the possible types of player i.

b. A_i is a finite set describing the set of actions that player i may take. He can then take any of these actions, regardless of his type.

c. p is a probability distribution on T. Thus, for $t = (t_1,t_2,\ldots,t_n)$, p(t) is the prior probability that player one would be of type t_1, player two of type t_2, etc. We assume throughout this paper that for every player i and every $t_i \in T_i$ the marginal probability of type t_i being chosen is positive ($P(t_i) > 0$).

d. u_i describes the utility of player i as a function of the collective action that is taken and of the types of all the players. Thus $u_i\colon A \times T \to \mathbb{R}$.

We may think of the game G as an extensive form game being played in two stages. First, nature selects a type for each player resulting in a vector $t \in T$ according to the probability distribution p. Now, simultaneously, every player discovers his own type and proceeds to choose any one action from his set of actions A_i. In his choice of an action he can use randomizations over actions. This process results in a pair (a,t) which yields to the players a vector of utilities $u(a,t) = (u_1(a,t),\ldots,u_n(a,t))$.

Since the actions of this game involve randomizations, we extend the utility functions by the usual expected utility rules. If r is a probability distribution on $A \times T$ then

$$u_i(r) = \sum_{(a,t)\in A\times T} u_i(a,t)r(a,t).$$

For a vector $v = (v_1,v_2,\ldots,v_i,\ldots,v_m)$ and a scalar w_i we use the conventions

$(v_{-i}) = (v_1,v_2,\ldots,v_{i-1},v_{i+1},\ldots,v_m)$, and

$(v_{-i}\colon w_i) = (v_1,v_2,\ldots,v_{i-1},w_i,v_{i+1},\ldots,v_m)$.

Also, for any finite set M we use the symbol $\Delta(M)$ to denote the set of probability distributions on M. Thus, $g \in \Delta(M)$ means that $g\colon M \to \mathbb{R}$, $g(m) \geq 0$ for every $m \in M$, and $\sum_{m\in M} g(m) = 1$. The int $\Delta(M)$ denotes the distributions that put positive probabilities on every element of M ($g(m) > 0$ for every $m \in M$).

A strategy of player i is a function

$s_i: T_i \to \Delta(A_i)$.

We denote the set of all strategies of player i by S_i and we let $S = \times_{i\in N} S_i$ denote the set of strategy tuples of the n players.

A strategy s_i is called completely mixed for type t_i if $s_i(t_i) \in \text{Int}\,\Delta(A_i)$. s_i is called completely mixed if it is completely mixed for every $t_i \in T_i$. A strategy tuple s is completely mixed if all its s_i's are completely mixed.

For a given strategy tuple $s \in S$ and a types vector $t \in T$ we use $\overline{s(t)}$ to denote the probability distribution induced on A when each player plays his randomized strategy independently of the others. Thus, for every $a \in A$

$$\overline{s(t)}\,(a) = \Pi_{i\in N}\,(s_i(t_i))(a_i).$$

Now we can extend the utility functions u_i to be defined on $S \times T$, in the natural way as follows

$$u_i(s,t) = \Sigma_{a\in A}\,\overline{s(t)}\,(a)u_i(a,t)$$
$$u_i(s) = \Sigma_{t\in T}\,u_i(s,t)p(t).$$

We also define $p(t|t_i)$ to be the conditional distribution on t given that player i is of type t_i, $p(t_i)$ is the marginal distribution of p on T_i and

$$u_i(s|t_i) = \Sigma_{t\in T}\,u_i(s,t)p(t|t_i).$$

his expected utility conditioned on his type. We will also use other standard notations for conditional and marginal probabilities as necessary. Observe that

$$u_i(s) = \Sigma_{t_i\in T_i}\,u_i(s|t_i)p(t_i) \qquad (*)$$

In general, if $\{L_1, L_2, \ldots, L_m\}$ is a partition of T then

$$u_i(s|L_k) = \Sigma_{t\in L_k}\,u_i(s,t)p(t|L_k)$$

and

$$u_i(s) = \Sigma_{k=1}^{m}\,u_i(s|L_k)p(L_k).$$

We define the best reply (b.r.) strategies as follows. For a strategy tuple s and player i-th's strategy $r_i \in S_i$ we say that $r_i \in b.r_i(s)$ (really of s_{-i}) if $u_i(s_{-i}: r_i) = \max\{u_i(s_{-i}: q_i): q_i \in S_i\}$. For $r,s \in S$ we say that $r \in b.r(s)$ if $r_i \in b.r_i(s)$ for every $i \in N$. We say that a strategy tuple f* is a Nash equilibrium if $f^* \in b.r.(f^*)$.

We say that f* is a perfect (Nash) equilibrium if there is a sequence of completely mixed strategies $f^r \to f^*$ and $f^* \in b.r.(f^r)$ for $r = 1,2,\ldots$. In other words, f* is a best reply to a sequence of completely mixed strategies approaching it. It follows immediately that a perfect equilibrium is a Nash equilibrium.

3. Games, Revelation Games and Delegation Games with Personal Types

We say that the types in a Bayesian game are personal if two conditions hold:

1. Statistical independence, i.e., $p(t) = \Pi_{i \in N} p(t_i)$ for every $t \in T$; and
2. Private payoffs, i.e., if $t, \bar{t} \in T$ with $t_i = \bar{t}_i$ then $u_i(a,t) = u_i(a,\bar{t})$ for every $a \in A$.

Observe that many games are of this type. For example, all games involving exchange of commodities for which players have private valuations and having types that are drawn independently fall into this category.

A strategy $s \in S_i$ is completely mixed relative to T_i if for every $a_i \in A_i$ there is a type $t_i \in T_i$ with $(s_i(t_i))(a_i) > 0$. Thus, being completely mixed relative to T_i is much weaker than being completely mixed because it is not required that every type of player i completely mixes his actions but only that the group of types as a whole completely mix their actions.

Theorem 3.1: Let G be a game with personal types. A strategy tuple s* is a perfect equilibrium if and only if there is a sequence of strategy tuples $s^r \to s^*$ with $s^* \in$ b.r.(s^r) and each s_i^r being completely mixed relative to T_i.

Proof: This theorem follows immediately from its generalization, which is proved in the following sections.

Corollary 3.1: In a game with personal types, if a Nash equilibrium is completely mixed relative to the T_i's, then it is perfect.

Proof: Choose $s^r = s^*$ for each r.

A game G is called a revelation game if $A_i = T_i$ for every player i. In other words, the strategies of the players are to declare their type (with the option to lie).

A revelation game is called Nash incentive compatible if playing honestly is a Nash equilibrium. In other words, the strategy defined by $(s_i(t_i))(t_i) = 1$ for every $i \in N$ and every $t_i \in T_i$ is a Nash equilibrium.

A revelation game is called perfectly incentive compatible if playing honestly is a perfect Nash equilibrium.

Corollary 3.2: A revelation game with personal types is Nash incentive compatible if and only if it is perfectly incentive compatible.

Proof: It is obvious that the honest strategy is completely mixed relative to T_i for every player i. Thus, if honesty is a Nash equilibrium then it must be perfect equilibrium.

By a delegation game we mean a Bayesian game with a vector of functions $g = (g_1, \ldots, g_n)$ such that each g_i is a function from T_i onto A_i. We refer to the functions in g_i as the individual guidelines. Since each g_i is onto it follows that in a delegation game $|A_i| \leq |T_i|$ for every player i.

A delegation game is (or its guidelines are) Nash incentive compatible if obedience to the guidelines is a Nash equilibrium. In other words, the strategy tuple s defined by $(s_i(t_i))(g_i(t_i)) = 1$ is a Nash equilibrium. Similarly, it is perfectly incentive compatible if s is a perfect equilibrium.

Observe that every revelation game is a delegation game in one natural way when we define the guidelines to be honest revelation. Also with these guidelines every Nash incentive compatible (respectively, perfectly incentive compatible) revelation game is a Nash incentive compatible (and, respectively, perfectly incentive compatible) delegation game. However, not every delegation game is a revelation or even strategically equivalent to one. If every guideline function of the delegation is one-to-one then it is clear that the delegation game is essentially a revelation game (the only difference is that the names of the actions may not coincide with the names of the types). But if the guideline functions are not one-to-one then we have a delegation game with $|A_i| < |T_i|$ for some players and thus it cannot be a revelation game.

Observe, however, that the well-known "revelation principle" induces immediately a "delegation principle" which states that every Nash equilibrium of a Bayesian game induces a Nash incentive compatible delegation game with the same payoff distribution. This follows immediately from the revelation principle and the fact that every revelation game is a delegation game.

Corollary 3.3: A delegation game with personal types is Nash incentive compatible if and only if it is perfectly incentive compatible.

Proof: This is also an immediate consequence of Corollary 3.1.

4. Nash Implementability and Perfect Implementability

In this section it would be convenient to introduce a universal set of outcomes over which the types of the players have preferences. As the players play games, outcomes in this set will result, yielding utility to the players through their preferences over the outcomes.

We first describe an n-person Bayesian environment by

$$E = (T = \times_{i \in N} T_i,\ P,\ C,\ u = (u_1, u_2, \ldots, u_n)).$$

Each T_i is a finite set describing the types of player i, p is a probability distribution on T with $p(t_i) > 0$ for every player i and every $t_i \in T_i$, C is a finite set of outcomes, and every u_i is a von Neumann-Morgenstern utility function representing the preferences of player i, u_i: $C \times T \to \mathbb{R}$.

To illustrate such an environment consider the confession game described in the Introduction and the following Bayesian environment. $N = \{1,2\}$ are the two players, $T_1 = T_2 = \{H,C\}$ denote whether a player is the honest or criminal type, and $p(H,H) = 1 - \varepsilon$, $p(C,C) = \varepsilon$, $p(H,C) = p(C,H) = 0$ describes the probabilities of the type combinations.

The relevant set of outcomes (penalties) here may be modeled as $C = R_-^2$ $(= -R_+^2)$ and the utility of player 1 (and similarly, player 2) may be viewed as

$$u_1((-x,-y),t) = -x$$

for every one of the possible four type combinations.

Now we define a Bayesian game in the environment E by a set of actions $A = \times_{i \in N} A_i$ and a resulting outcome assignment $R: A \to \Delta(C)$. While we have now defined the notion of a Bayesian game twice we observe that a Bayesian game in the environment E described by a pair (A,R) induces a Bayesian game (T,A,P,u) in one natural way. The T, A and p components of the Bayesian game are the same while the utility functions u_i are extended from the environment to the game by

$$u_i(a,t) = u_i(R(a),t)$$

Returning to the confession game environment described above we could describe the scheme of the introduction as the following game G:

$A_1 = A_2 = \{A,D\}$, standing for Admit and Deny,

$R(D,D) = (0,0)$ with probability one and,

$R(D,A) = R(A,D) = R(A,A) = (-M,-M)$

with probability one.

The induced utilities for $i = 1,2$, and for every type combination t are:

$u_i((D,D),t) = 0$, and

$u_i((D,A),t) = u_i((A,D),t) = u_i((A,A),t) = -M.$

We are interested in the question of what outcome functions can be implemented by an implementor. An outcome function is a function $O: T \to \Delta(C)$. The implementor may be a public or private agent and may have a variety of incentives in implementing various outcome functions. There are two problems that the implementor may face. First, he may not fully know the types of the players. Second, he may not be able to fully control their actions. We assume that he has some control over the rules or the choice of the game that will be played. The implementor hopes that by choosing the game cleverly and assuming a type of solution regarding the players' behavior he may be able to bring about the outcome function that he chooses.

We let Γ be a set of Bayesian games in the Bayesian environment E and refer to them as the implementation games. The interpretation is that these are the games from which the implementor can choose.

We say that an outcome function O is Nash implementable by Γ if there exists a game in Γ with Nash equilibrium strategy tuple that induces the outcome function O. Formally, there should be a game (A,R) in Γ with a Nash equilibrium strategy tuple s^* such that for every $t \in T$

$$\Sigma_{a \in A} R(a)\ \overline{s^*(t)}\ (a) = O(t).$$

Namely, for every type combination t at the Nash equilibrium s^* the players should take actions that induce the distribution $O(t)$.

Similarly, we say that an outcome function O is perfectly implementable by Γ if there is a perfect Nash equilibria strategy s^* which implements O.

What is implementable obviously depends on the solution concept. It is clear that every perfectly implementable outcome is Nash implementable but not the converse. Also, what is implementable depends on the set of games Γ available to the implementor. The larger Γ is the more outcomes it may implement. For example, returning to the confession games, the outcome function that we want to implement there is

$$O(C,C) = O(H,C) = O(C,H) = (-M,-M) \text{ with probability } 1,$$
$$\text{and } O(H,H) = (0,0) \text{ with probability } 1.$$

If we let $\Gamma = \{G\}$ consists of the game described above (i.e., the scheme from the Introduction), then since true revelation is a Nash equilibrium of G we would say that Γ Nash implements O (Γ also Nash implements the total denial strategy). However, Γ does not perfectly implement O. It is interesting to see that by enlarging Γ society can perfectly implement this desired O. The trade-off in doing this is that the implementor may punish the innocent players harder than before (only off the equilibrium path). Consider the game G_2 defined as follows:

$$A_1 = A_2 = \{A,D\} \text{ as before but}$$
$$R(A,D) = (-M,-1.5M) \text{ with probability } 1,$$
$$R(D,A) = (-1.5M,-M) \text{ with probability } 1,$$
$$R(A,A) = (-M,-M) \text{ with probability } 1,$$
$$R(D,D) = (0,0) \text{ with probability } 1.$$

In other words, G_2 is designed to punish a single denier harder than a confessor. If we let $\Gamma = \{G,G_2\}$ then Γ perfectly implements O by using G_2 in which true revelation is a perfect equilibrium.

One assumption that is necessary for our implementability equivalence theorem is that Γ is rich enough. It will state that the implementor, starting from every feasible game in Γ can reduce it by crossing out actions that are not used at a Nash equilibrium of this feasible game. More formally, we say that Γ is <u>complete</u> if for every $G = (A,R) \in \Gamma$ and every Nash equilibrium strategy s^* of G the game

$$G^-(s^*) = (\times_{i\in N} A_i^-,R) \in \Gamma$$

where

$$A_i^- = \{a_i \in A_i\colon \text{ for some } t_i \in T_i(s_i^*(t_i))(a_i) > 0\}.$$

(See Remark 4.1 for further discussion).

We say that a Bayesian environment E <u>consists of only personal types</u> if as before the types of the players are statistically independent, and their preferences are private, i.e., for every $i \in N$ and $c \in C$, $u_i(c,t) = u_i(c,\bar{t})$ whenever $t_i = \bar{t}_i$.

<u>Theorem 4.1: The Equivalence of Nash and Perfect Implementability</u>: In a Bayesian environment consisting only of personal types with a complete set of implementation games the sets of Nash implementable outcome functions and perfectly implementable outcome functions coincide.

<u>Proof</u>: We have only to argue that every Nash implementable outcome is perfectly implementable. But if O is Nash implementable by a game G and a Nash equilibrium s^* then it is Nash implementable by the game $G^-(s^*)$ where the strategy tuple s^* (with a little

abuse of notations) is also a Nash equilibrium inducing O. But by Corollary 3.1, s^* is a perfect equilibrium of $G^-(s^*)$. Thus O is perfectly implementable by Γ.

Remark 4.1: The condition that Γ is complete means that the implementor can impose a game where actions that are not used at a given equilibrium are omitted. This assumption is consistent with the philosophy of the implementation literature that we follow in this paper. In this literature it is implicitly assumed that in the case of multiple Nash equilibrium the implementor would have a way of choosing the final one. Our completeness requirement is a weak version of this assumption because it only assumes that the implementor chooses the family of actions that may be used at this equilibrium strategy (rather than the exact strategy). A stronger assumption would have been to let the implementor choose any subset of the players' actions. This would have been too strong because one of its implications would have been that the implementor (remembering that he may be a private agent) can force the players to actions which would yield less than the individually rational payoffs. For example, a seller of an item with such a power can force his buyers to pay for his item much more than their valuations for the item. This, of course, could not happen under our completeness assumption.

We also note that a similar theorem could have been attained by other notions of completeness. For example, starting with a game $G = (A,R)$ and a Nash equilibrium s^*, we could have required that Γ include the games $\hat{G} = (\hat{A},R)$ where the actions available to player i are the mixed strategies induced by his types, i.e.,

$$\hat{A}_i = \{(s^*_i(t_i): t_i \in T_i\}.$$

Notice that the game G is equivalent (up to names of actions) to what is known as the revelation game induced by G and s^*. We then in effect could apply the revelation principle and Corollary 3.2 to obtain the same result.

5. Correlated Equilibria

The implementation problem, described in Section 4, is that of finding sets of actions A_i for the players and an outcome assignment R defined for these actions given the types T_i and the probability distribution p over types. The designed actions and outcome assignment with the given structure of types generate a Bayesian game as described above. The mediation problem treated in this section is reversed. The actions of the players A_i and the utilities resulting from these actions are given to the mediator (i.e., the mediator faces a game in normal form). What has to be designed is a message space T_i for each player i and a probability distribution p on $T = \times_{i=1}^n T_i$. The Bayesian game is then defined as follows. The mediator chooses t in T according to the probability distribution p, each player i is informed of his type t_i and chooses accordingly an action $a_i \in A_i$. One can easily verify that an "instruction principle" analogous to the "revelation principle" holds for the mediation problem. That is, any distribution over outcomes that can be achieved by such games can be achieved also if we confine ourselves to message spaces $T_i = A_i$ and to equilibria of the Bayesian game in which each player chooses the action given to him as a message.

In terms of the definition of Section 3 such Bayesian games are Nash incentive compatible. (Of course, the interpretation of these games is different here; "types" should be replaced by "instructions" and "revelation" by "obedience.")

Now let $g = ((A_i)_{i \in N}, (U_i)_{i \in N})$ be a game in normal form. Let p be a probability distribution over $A = \times_{i \in N} A_i$ and let G(p) be the Bayesian game described above. We say that p is a <u>correlated equilibrium</u> in g if G(p) is Nash incentive compatible, i.e., if "obeying" is a Nash equilibrium in G(p). Similarly, p is a <u>perfectly correlated equilibrium</u> of G(p) if it is perfectly incentive compatible.

Applying Theorem 3.1 to correlated equilibrium yields the following.

<u>Corollary 5.1</u>: Let p be a correlated equilibrium. If $p > 0$ and instructions are drawn independently for each player then p is a perfect correlated equilibrium.

<u>Proof</u>: In the game G(p) types are personal, since statistical independence is assumed and payoffs are independent of types and therefore are private. Since $p > 0$ the requirement that each type is drawn in some positive probability is also satisfied.

The corollary appears to be vacuous, because when instructions are drawn independently the correlated strategy p corresponds to an equilibrium in the game g. Each player can draw his instruction independently of the others and obey his instruction which seems to be tantamount to playing mixed strategy. But then if $p > 0$ we have a totally mixed strategy equilibrium in g which is perfect by definition.

This argument is flawed. Indeed our correlated equilibrium p corresponds to a totally mixed strategy equilibrium in the sense that both yield the same distribution over outcomes. But these two equilibria are two distinct equilibria in two different games. One is a real mixed strategy equilibrium in g; the other is a pure strategy equilibrium in G(p). This mathematical difference between playing the mixed strategy p in the game g and playing the pure obedient strategy in the game G(p) can be interpreted in terms of commitment. The question whether to obey the result of a random choice of action is left open in G(p) while in g playing the mixed strategy should be interpreted as a commitment made by the player before the choice has been made to obey it. This distinction is crucial, for example, in the minimax theory as is shown by Aumann and Maschler (1972); mixed minimax strategies cannot in general be supported if players are not precommited to obey the results of the random choice.

Corollary 5.1 should be understood, then, as asserting that perfection unlike minimax theory, is not affected by precommitment in games in normal form. A totally mixed strategy equilibrium remains perfect even if players can disobey the result of randomization.

In Example 2 of Section 7 we show that this perfection is due not only to total mixture but also to the independence of players' randomization. Indeed, correlated equilibria which are independent in this sense may fail to be perfect even when $p > 0$.

6. <u>General Type Decomposition</u>

If B is a partition of a set S and $s \in S$ then we let B(s) denote the element of B containing s. Given a Bayesian game we say that $L = (L_1, L_2, \ldots, L_n)$ is a <u>decomposition of</u>

<u>the type space</u> $T = \times_{i\in N} T_i$ if for every i

$$L_i = L_i^1, L_i^2, \ldots, L_i^{\ell(i)})$$

is a finite partition of T_i. For $t \in T$ we let

$$L(t) = \times_{i\in N} L_i(t)$$

and refer to it as the <u>cell</u> of L containing t. Obviously L generates only finitely many distinct cells. By a <u>personal type decomposition</u> we mean a decomposition L = $(L_1, L_2, \ldots, L_n)$ satisfying the following two conditions:

1. Types are <u>consistently independent relative to the decomposition</u> L, i.e., $p(t|L(t)) = \Pi_{i\in N}\, p(t_i|L_i(t_i))$ for every $t \in T$.
2. <u>Payoffs are private relative to decomposition</u> L, i.e., for every $t \in T$ and every $\bar{t} \in L(t)$ if $t_i = \bar{t}_i$ then $u_i(a,t) = u_i(a,\bar{t})$ for every $a \in A$.

Notice that personal type decompositions always exist because the trivial fine decomposition, $L_i = (\{t_i\})_{t_i \in T_i}$, trivially yields a personal decomposition. Also, if L is the coarse trivial decomposition, $L_i = (T_i)$, then we are back in the special cases discussed in the previous sections.

To illustrate a personal decomposition consider an example where two players are about to engage in a bidding game for a bottle of old wine. Each of the two players could be of three possible types, denoted by CH, CM, and D. The type CH is one who plans to consume the wine and has a high valuation for this act. The CM type is also a consumer but his desire to drink the wine is moderate. The D type, on the other hand, is a dealer who thinks that this wine has a high potential future value. Suppose that the prior distribution on type combinations is as follows:

	CH	CM	D
CH	.4 × .7 × .7	.4 × .7 × .3	.1 × .7
CM	.4 × .3 × .7	.4 × .3 × .3	.1 × .3
D	.1 × .7	.1 × .3	.4

This probability distribution satisfies the consistent independence condition relative to the partition illustrated in the table. The partition cells have the following probability distribution

	C	D
C	.4	.1
D	.1	.4

and within the cells the types are consistently independent with

$p(CH|C) = .7$ and $p(CM|C) = .3$

The condition of private payoff for this decomposition would require that a consumer of any type should not care against what type of consumer he won (or lost) but may care to know that he has won (or lost) against another dealer.

A strategy of player i, $s_i \in S_i$, is <u>completely mixed relative to a partition</u> L_i of player i types, T_i, if all the actions of player i are used by the members of every element of his partition, i.e., for every $L_i^j \in L_i$ and for every $a_i \in A$ there is a type $t_i \in L_i^j$ with $s_i(t_i)(a_i) > 0$.

<u>A strategy</u> $s \in S$ <u>is completely mixed relative to the decomposition</u> L if for every player i s_i is completely mixed relative to the partition L_i.

Consider, for example, the bidding game described above with the choice of bid high and bid low available to all types and the following strategy for every one of the two players. Bid high if you are CH and bid high with probability .50 if you are of type D. Clearly this is not an interior strategy but it is interior relative to the decomposition.

<u>Theorem 6.1</u>: A strategy tuple $s^* \in S$ is a perfect equilibrium if and only if for some (and equivalently for every) personal type decomposition L there exists a sequence of strategies $s^r \to s^*$ with every s^r being completely mixed relative to L and with s^* being best reply to each s^r.

<u>Proof</u>: The "only if" direction is obvious by the definition of perfect equilibrium and the fact that the trivial fine decomposition is personal. To prove the "if" direction w assume the existence of a sequence s^r as described in the theorem and will construct a sequence of strategy tuples $g^r \to s^*$, g^r being completely mixed and with the property tha for every player j:

$$u_j(g^r_{-j}: h_j) = u_j(s^r_{-j}: h_j) \qquad (*$$

for every $h_j \in S_j$. Thus, since s^* is a best reply to s^r it would also be a best reply t g^r showing that s^* is a perfect equilibrium. For $t_i \in T_i$ we let

$$a_i^r(t_i) = \Sigma_{c_i \in L_i(t_i)} s_i^r(c_i)/|L_i(t_i)|$$

Thus $a_i^r(t_i)$ is the average action of types of player i which are personally equivalent t t_i. Since the s_i^r's are completely mixed relative to L_i it follows immediately that the $a_i^r(t_i)$'s are completely mixed for every t_i. Now we define g_i^r by

$$g_i^r(t_i) = [1 - 1/rp(t_i|L_i(t_i))]s_i^r(t_i) + [1/rp(t_i|L_i(t_i))]a_i^r(t_i).$$

It is obvious that for sufficiently large r's, g_i^r is a well-defined completely mixed strategy with $g_i^r \to s_i^*$. Now, since for any strategy s, $u_j(s) = \Sigma_C\, u_j(s|C)p(C)$ (where we sum over all the cells generated by L) it suffices to show that (*) holds conditionally on C, for each cell C. Moreover, since the strategies g_i^r are defined on a cell C, only in terms of the restrictions of the strategies s_j^r to types in C, we may analyze the restriction of our game to types in C as a complete Bayesian game. Thus we assume without loss of generality that the game has only one such cell and that L is the coarsest decomposition, i.e., for i = 1,2,...,n,

$$L_t = (T_i),$$
$$L_i(t_i) = T_i,$$
$$|L_i(t_i)| = |T_i|,$$

$$a_i^r(t_i) = a_i^r = \Sigma_{t_i \in T_i}\, s_i^r(t_i)/|T_i|, \text{ and}$$

$$g_i^r(t_i) = [1 - 1/(rp(t_i))]s_i^r(t_i) + [1/rp(t_i))]a_i^r.$$

It would suffice to show (*) for this special circumstance. We show indeed that for each type $t_j \in T_j$ and $h_j \in S$; $u_j((g_{-j}^r,h_j)|t_j) = u_j((s_{-j}^r,h_j)|t_j)$. From this (*) follows because for any strategy s,

$$u_j(s) = \Sigma_{t_j \in T_j}\, u_j(s|t_j)p|t_j).$$

We observe now that if a game has only personal types then $u_j(a,t) = u_j(a,t_j)$ and for every strategy tuple f,

$$u_j(f|t_j)) = \Sigma_{a \in A}\, u_j(a,t_j)(f_j(t_j))(a_j)[\Pi_{i \in N\setminus j}\, \Sigma_{t_i \in T_i}\, (f_k i(t_i))(a_i)p_i(t_i)].$$

So, for our g^r strategies and any strategy h_j of player j

$$u_i((g_{-j}^r\colon h_j)|t_j) =$$
$$= \Sigma_{a \in N}\, u_j(a,t_j)(h_j(t_j))(a_j)[\Pi_{i \in N\setminus j}\, \Sigma_{t_i \in T_i}\, (g_i^r(t_i))(a_i)p_i(t_i)] =$$

$$= \Sigma_{a \in A}\, u_j(a,t_j)(h_j(t_j))(a_j)[\Pi_{i \in N\setminus j}\, \Sigma_{t_i \in T_i}\, (s_i^r(t_i))(a_i)p_i(t_i)] =$$

$$= u_j((s_{-j}^r\colon h_j)|t_j).$$ Q.E.D.

Clearly, the usefulness of Theorem 5.1 depends crucially upon the availability of "coarse" personal decompositions (for the trivially fine decomposition, Theorem 5.1 does no more than repeat the definition of a perfect equilibrium). It is therefore useful to know that a unique coarsest personal decomposition $C = (C_1,C_2,\ldots,C_n)$ exists. Also each one of its individual partitions C_i can be constructed independently without having to construct all the C_i's simultaneously.

For a subset of player i types, $D_i \subseteq T_i$, we say that D_i is a <u>set of personal types</u> if

1. $p(t_{-i}|d_i) = p(t_{-i}|D_i)$ for every $d_i \in D_i$ and $T_{-i} \in T_{-i}$, and

2. $u_j(a,(t_{-i}: d_i)) = u_j(a,(t_{-i}: d_i'))$ for every $j \neq i$, $t_{-i} \in T_{-i}$, d_i, $d_i' \in D_i$ and $a \in A$.

The following are two obvious facts.

1. If B_i and D_i are sets of personal types and $B_i \cap D_i \neq \emptyset$ then $B_i \cup D_i$ is a set of personal types.
2. $\{t_i\}$ is a set of personal types for every $t_i \in T_i$.

It follows immediately that there is a unique coarsest partition of T_i to sets of personal types. We denote this partition by $C_i = \{C_i^1, C_i^2, \ldots, C_i^{\ell(i)}\}$ and refer to it as the <u>coarsest personal type partition of player</u> i. Observe that it follows immediately that <u>a partition B_i is personal to player i if and only if it is a refinement of</u> C_i. We define $C = (C_1, C_2, \ldots, C_n)$ as <u>the coarsest personal type decomposition</u>. This terminology is justified by the following theorem.

<u>Theorem 6.2</u>: L is a personal decomposition of the type set T if and only if for every player i L_i is a refinement of C_i.

<u>Proof</u>: Let $L = (L_1, L_2, \ldots, L_n)$ be a decomposition of T, we want to show that it is a personal decomposition if and only if for every player i L_i is a personal partition of T_i. We first show that if L is personal then for every player i, for every $L_i^j \in L_i$, for every $a,b \in L_i^j$, and for every $t_{-i} \in T_{-i}$, $p(t_{-i}|a) = p(t_{-i}|b)$ (and hence $= p(t_{-i}|L_i^j)$).

Let $L = L(t_{-i}: a) = L(t_{-i}: b)$. Then from the definition of L being personal we obtain that

$$p((t_{-i}: a)|L)/p(a) = p((t_{-i}: b)|L)p(B)$$

it follows that

$$p((t_{-i}: a))/p(a) = p((t_{-i}: b))/p(b)$$

and hence

$$p((t_{-i}|a)) = p((t_{-i}|b)).$$

Next observe that if L is personal and $c,d \in L_i^j$ for some element of the partition L_i then for every $j \neq i$ for every $t_{-i} \in T_{-i}$, and for every $a \in A$

$$u_j(a,(t_{-i}: c)) = u_j((a,(t_{-i}: d)))$$

because $(t_{-i}: c)$ and $t_{-i}: d)$ belong to the same cell of L. Thus we conclude that if L is a personal decomposition then every L_i is a personal partition and must be a refinement of C_i.

Now, to prove the other direction of the theorem, assume that L is a decomposition with each of its L_i's being a personal partition (hence a refinement of C_i). We want to show that for every cell t, L(t) satisfies conditions (1) and (2) in the definition of a personal decomposition. To see that the second condition holds, assume that for player i $\bar{t} \in L(t)$ $t_i = \bar{t}_i$ and $a \in A$. Then from the fact that $L_j(t_j)$ $(= L_j(\bar{t}_j))$ is personal for all j, it follows that

$$\begin{aligned} u_i(\bar{t}) &= u_i(\bar{t}_1, \bar{t}_2, \ldots, \bar{t}_{i-1}, t_i, \bar{t}_{i+1}, \ldots, \bar{t}_n) \\ &= u_i(t_1, \bar{t}_2, \ldots, \bar{t}_{t-1}, t_i, \bar{t}_{i+1}, \ldots, \bar{t}_n) \\ &= u_i(t_1, t_2, \ldots, \bar{t}_{i-1}, t_i, \bar{t}_{i+1}, \ldots, \bar{t}_n) \end{aligned}$$

$= u_i(t_1, t_2, \ldots, t_{i-1}, t_i, t_{i+1}, \ldots, t_n).$

We can change one coordinate at a time because in each change we only change the coordinate of one player j without leaving his $L_j(t_j)$ and thus not affecting the payoff if i ($i \neq j$ and no change was necessary for i).

Condition 1 in the definition of a personal decomposition follows from the identity:

$$p(\times_{i=1}^{s} L_i(t_i) \times \{t_{-\{1,2,\ldots,s\}}\}) =$$

$$p(t_{s+1}|L_{s+1}(t_{s+1}))p(\times_{i=1}^{s+1} L_i(t_i) \times \{t_{-\{1,2,\ldots,s+1\}}\})$$

for $s = 0,1,2,\ldots,n-1$. Q.E.D.

7. Examples

The truth revealing equilibrium in the confession game described in the Introduction fails to be perfect. In terms of the conditional guaranteeing perfection in Corollary 3.2, this is due to the statistical dependence of types. But this game has also the special feature that although each type of player has a positive probability, there are type combinations which have probability zero. One might suspect that when $p > 0$ (i.e., when every type combination is possible) one of the conditions defining personal types is superfluous. The next two examples show that even when $p > 0$ neither statistical independence nor private payoffs can guarantee alone the perfection of truth revealing equilibrium.

Example 7.1: The Insufficiency of Statistical Independence Types Alone: In the following revelation game there are two players, I and II. The sets of types and actions are: $A_1 = T_1 = \{t_1, t_2\}$, $A_2 = T_2 = \{s_1, s_2\}$, and the probability distribution of types is $p(t_1) = p(t_2) = p(s_1) = p(s_2) = 1/2$.

The payoffs for I are given by

type t_1

	type s_1: s_1	type s_1: s_2	type s_2: s_1	type s_2: s_2
t_1	1	0	1	0
t_2	0	2	0	1

type t_2

	s_1	s_2	s_1	s_2
t_1	1	0	2	0
t_2	0	1	0	1

We assume for simplicity that player II's payoffs are constant in the game. Truth revealing is an equilibrium; player II is clearly indifferent between playing s_1 and s_2. As for player I, each one of his types is indifferent between playing t_1 and t_2; in either case, his expected payoff is 1/2. For example, as type t_1 when he plays t_1 he is paid 1 with probability 1/2 (the probability that player II is of type s_1 and therefore plays s_1) and he is paid 0 with probability 1/2 (the probability that player II is of type s_2 and therefore plays s_2). So in particular truth revealing for player I is a best response to the truth revealing of player II.

Now suppose that player II deviates from truth revealing and he plays $(1 - \varepsilon_1, \varepsilon_1)$ as type s_1 (i.e., he plays s_1 with probability $1 - \varepsilon_1$ rather than 1), and $(\varepsilon_2, 1 - \varepsilon_2)$ as type s_2 $(\varepsilon_1, \varepsilon_2 > 0)$. As type t_1, player I when playing t_1 is paid: $(1/2)(1 - \varepsilon_1) + (1/2)\varepsilon_2$ and when playing t_2: $(1/2)(2\varepsilon_1) + (1/2)(1 - \varepsilon_2)$. So I will not choose the truth (i.e., playing t_1) if $2\varepsilon_1 < 2\varepsilon_1$. Similar computation for type t_2 shows that player I will not choose the truth (i.e., t_2) if $2\varepsilon_1 < 3\varepsilon_2$. Since one of the inequalities always holds it follows that truth revealing for player I is not best response for any deviation of player II from truth revealing. Truth revealing is therefore not a perfect equilibrium.

<u>Example 7.2: The Insufficiency of Private Payoffs Alone</u>: In the revelation game G_2 there are two players, I and II. Types and action sets are $A_1 = T_1 = \{t_1, t_2, t_2\}$ and $A_2 = T_2 = \{s_1, s_2\}$. The probabilities $p(t_i, s_j)$ are given as entries in the following table:

$p(t_i, s_j)$

	s_1	s_2
t_1	2/9	1/9
t_2	1/9	2/9
t_3	1/6	2/6

The payoffs in the game are independent of any type.

II

I		s_1	s_2
	t_1	3,0	0,0
I	t_2	0,0	3,0
	t_3	2,0	2,0

Since the payoffs of each player are independent even of his own type a truth revealing equilibrium can be interpreted as a correlated equilibrium.

Truth revealing (or obedience) is an equilibrium in this game. Player II is always indifferent between his two actions. Player I is indifferent between actions t_1 and t_3 when he is of type t_1 (his expected payoff is 2 for either action) and he strictly prefers both actions to t_3. A similar result holds for player I as type t_2. As t_3, player I strictly prefers action t_3.

When player II trembles he plays $(1 - \epsilon_1, \epsilon_1)$ as type s_1, and $(\epsilon_2, 1 - \epsilon_2)$ as type s_2. When $\epsilon_2 < 2\epsilon$, player I prefers action t_3 to t_2 as type t_1 and when $\epsilon_1 < 2\epsilon_2$ player I prefers action t_3 to t_2 as player t_2. Under no tremble will player I reveal his type (or obey his instructions) both when he is t_1 and t_2. It follows that truth revealing is not a perfect equilibrium and as a result the correlated equilibrium p is not perfect.

References

Arrow, Kenneth J. (1951), Social Choice and Individual Values, New York: Wiley.

Aumann, Robert J. (1974), "Subjectivity and Correlation in Randomized Strategies," Journal of Mathematical Economics, 1, 67-96.

Aumann, Robert J. and M. Maschler (1972), "Some Thoughts on the Minimax Theorem," Management Science, 18, 54-63.

Bernheim, Douglas B. (1984), "Rationalizable Strategic Behavior," Econometrica, 52, 1007-1028.

Crawford, Vincent P. and Joel Sobel (1982), "Strategic Information Transmission," Econometrica, 50, 1431-51.

d'Aspremont, C. and L. A. Gerard-Varet (1979), "Incentives and Incomplete Information," Journal of Public Economics, 11, 25-45.

Forges, Francoise (1984), "An Approach to Communication Equilibria," CORE Discussion Paper No. 8435, Universite Catholique de Louvain.

Gibbard, Allan (1973), "Manipulation of Voting Schemes: A General Result," Econometrica 41, 587-601.

Groves, Theodore and John O. Ledyard (1977), "Optimal Allocation of Public Goods: A Solution to the Free Rider Problem," Econometrica, 45, 783-810.

Harris, M. and A. Raviv (1981), "Allocation Mechanisms and the Design of Auctions," Econometrica, 49, 1477-1499.

Harsanyi, John C. (1967-8), "Games with Incomplete Information Played by 'Bayesian' Players," Management Science, 14, 154-189, 320-334, 486-502.

Harsanyi, John C. and Reinhard Selten (1980), "A General Theory of Equilibrium Selection in Games" (Chapter 1, 2 and 3), Working Ppaers Nos. 91, 92 and 105, IME, University of Bielefeld, Bielefeld.

Hurwicz, L. (1972), "On Informationally Decentralized Systems," in Decision and Organization, McGuire and Radner (eds.), Amsterdam: North Holland.

Hurwicz, L. and David Schmeidler (1978), "Outcome Functions Which Guarantee the Existenc and Pareto Optimality of Nash Equilibria," Econometrica, 46, 144-174.

Kalai, E. and R. W. Rosenthal (1978), "Arbitration of Two-Party Disputes Under Ignorance," International Journal of Game Theory, 7, 65-72.

Kalai, E. and D. Samet (1984), "Persistent Equilibria in Strategic Games," International Journal of Game Theory, 13, 129-144.

Kalai, E. and D. Samet (1985), "Unanimity Games and Pareto Optimality," International Journal of Game Theory, 14, 41-50.

Kreps, David M. and Robert Wilson (1982), "Sequential Equilibria," Econometrica, 50, 863-894.

Maskin, Eric S. (1985), "The Theory and Implementation in Nash Equilibrium: A Survey," in Social Goals and Social Organization: Essays in Memory of Elisha Pazner, Hurwicz, Schmeidler and Sonnenschein (eds.), Cambridge University Press.

Maskin, Eric and J. Riley (1984), "Optimal Auctions with Risk Averse Buyers," Econometrica, 52, 1473-1518.

Matthews, Steven A. (1979), "Risk Aversion and the Efficiency of First and Second Price Auctions," Working Paper No. 586, University of Illinois.

McLennan, Andrew (1985), "Justifiable Beliefs in Sequential Equilibria," Econometrica, 54, 889-904.

Mertens, Jean-Francois and Elon Kohlberg (1982), "On the Strategic Stability of

Equilibria," CORE Discussion Paper No. 8248, Universite Catholique de Louvain.

Milgrom, P. R. and John Roberts (1982), "Limit Pricing and Entry Under Incomplete Information: An Equilibrium Analysis," Econometrica, 50, 443-459.

Milgrom, P. R. and R. J. Weber (1982), "A Theory of Auctions and Competitive Bidding," Econometrica, 50, 1089-1122.

Myerson, Roger B. (1978), "Refinement of the Nash Equilibrium Concept," International Journal of Game Theory, 7, 73-80.

Myerson, Roger B. (1981), "Optimal Auction Design," Mathematics of Operations Research, 6, 58-73.

Myerson, Roger B. (1984), "Acceptable Correlated Equilibria," Discussion Paper No. 591, CMSEMS, Northwestern University.

Myerson, Roger B. (1985), "Bayesian Equilibrium and Incentive Compatibility: An Introduction," in Social Goals and Social Organization: Essays in Memory of Elisha Pazner, Hurwicz, Schmeidler and Sonnenschein (eds.), Cambridge University Press.

Myerson, Roger, B. and Mark A. Satterthwaite (1983), "Efficient Mechanisms for Bilateral Trading," The Journal of Economic Theory, 29, 265-281.

Palfrey, Thomas R. and S. Srivastava (1986), "Private Information in Large Economies," The Journal of Economic Theory, 39, 34-58.

Pearce, David G. (1984), "Rationalizable Strategic Behavior and the Problem of Perfection," Econometrica, 52, 1029-1050.

Peleg, Bezalel (1977), "Consistent Voting Sytems," Econometrica, 46, 153-162.

Postlewaite, Andrew (1985), "Implementation via Nash Equilibria in Economic Environments," in Social Goals and Social Organization: Essays in Memory of Elisha Pazner, Hurwicz, Schmeidler and Sonnenschein (eds.), Cambridge University Press.

Postlewaite, Andrew and David Schmeidler (1986), "Implementation in Differential Information Economics," The Journal of Economic Theory, 39, 14-33.

Riley, J. and W. Samuelson (1981), "Optimal Auctions," American Economic Review, 71, 381-392.

Roberts, John D. (1986), "Incentives, Information and Iterative Planning," in Information, Incentives and Economic Mechanisms, Groves, Radner, and Reiter (eds.), University of Minnesota Press.

Satterthwaite, Mark A. (1975), "Strategy Proofness and Arrow's Conditions: Existence and Correspondence Theorems for Voting Procedures and Social Welfare Functions," Journal of Economic Theory, 10, 187-217.

Selten, Reinhard (1975), "Reexamination of the Perfectness Concept for Equilibrium Points in Extensive Games," International Journal of Game Theory, 4, 25-55.

Wilson, R. (1967), "Competitive Bidding with Proprietary Information," Management Science, 13, A816-A820.

van Damme, Eric (1983), Refinements of the Nash Equilibrium Concept, Berlin, Heidelberg, New York, Tokyo: Springer-Verlag.

Fair Division under Asymmetric Information

Eric van Damme

CentER, Tilburg University

P.O. Box 90153, 5000 LE Tilburg, The Netherlands

Abstract: This paper considers the situation in which a single indivisible object has to be allocated to one person out of a group whose members all have equal rights to it. Different persons value the object differently and each person only knows his own value exactly. The question is who should get the object and by how much this person should compensate the others in order to guarantee a fair and efficient allocation. After having shown that several well-known methods perform unsatisfactory, we derive an impossibility theorem showing that some classical fair division methods cannot be implemented when there is incomplete information. Finally, we give examples of mechanisms that do guarantee fair and efficient outcomes.

1 Introduction

The problem of how to divide an indivisible object among a group of persons who all have equal rights to it arises, for example, in divorce settlements and in inheritance situations in which there are equivalent heirs but there is no will. In a business setting, the problem arises in the dissolving of joint ventures. In this paper[1] we will assume that side payments between the parties are possible and we will try to answer the question of who should get the object and by how much each of the other players should be compensated in order to obtain (ex post) a fair and efficient allocation. Efficiency implies that the object should be allocated to the person who values is most. The fairness criterion we will employ is the one proposed in Foley (1967): It is required that the final allocation be envy free, i.e. no player should envy another, each player should prefer what he himself receives above what any other player receives.

There exists an extensive literature on the fair division problem[2]. In this literature various concepts of fairness have been proposed and several fair division methods, such as divide and

[1]Most of the material in this paper is taken from the unpublished working paper Van Damme (1985).

[2]For example Crawford (1977), Crawford and Heller (1979), Kuhn (1967), Luce and Raiffa (1957, Chapter 14), Pazner and Schmeidler (1978), Samuelson (1980), Steinhaus (1948) and Varian (1974).

choose, random allocation followed by bargaining, and auctioning the object followed by an equal division of the revenue, have been analyzed. In most of the literature, attention has been restricted to the case of complete information[3]. Frequently, however, it will be the case that each player, although he may know exactly how much he himself values the object, has only somewhat vague (probabilistic) information about his opponents' values. This opens up the possibility for strategic manipulation: A person might pretend that he values the object more (or less) than he actually does in order to obtain a more favorable outcome. Our aim in this paper is to study the consequences of such strategic behavior in fair division situations. Specifically we will investigate whether it is possible to obtain fair and efficient outcomes when players use their private information strategically.

Recently, considerable attention has been devoted to the study of bargaining under asymmetric information[4]. In this context the consequences of strategic behavior have been thoroughly investigated, and it has been shown that strategic behavior may prevent an ex post efficient outcome being reached. Although the insights generated by the bargaining literature are highly relevant for the problem at hand (indeed we will make extensive use of them), there are at least two novel aspects in the fair division problem. First of all we will see that division methods that treat the players asymmetrically aggravate the incentive problems. Methods that preserve the symmetry of the players are superior, they yield higher payoffs. As a consequence it is not desirable to reduce the division problem to a bargaining problem by first allocating the property rights. Secondly, in the bargaining literature, attention has been restricted to the efficiency aspect, questions of fairness have not been considered.

It should be noted that in the more abstract (cooperative) papers on games with incomplete information by Harsanyi and Selten (1972) and Myerson (1979, 1984) some of the axioms are based on equity considerations. However, these papers make the fundamental assumption that all that matters are the expected utilities at the interim stage, i.e. at the point in time where each player knows his own value but does not yet know those of his opponents. In contrast, in the present study our interest is also in the point in time where all values have been revealed as we want to guarantee that ex post there is no envy. Hence, the crucial parameters for our study are

[3]Exceptions are Güth (1986), Güth and Van Damme (1986) and Lyon (1986). In some older papers it is merely pointed out that the proposed methods are vulnerable to strategic manipulation, there is no analysis of its consequences. Dubins (1977) has shown that minmax strategies imply truthful revelation of values in the Steinhaus procedure.

[4]See Kennan and Wilson (1990) for a recent survey.

the actual, ex post, utilities.

The remainder of the paper is organized as follows. After having introduced some basic concepts in Section 2, we study, for the special case in which there are only two participants, some well-known division methods in Section 3. It is shown that methods that keep players in symmetric positions (such as auctions) outperform asymmetric division methods (such as divide and choose), but that also auctions may yield outcomes that are not envy free. In Section 4 we show that all ex post efficient mechanisms are equivalent at the interim stage (i.e. they generate the same expected utilities) and that random allocation is the worst possible mechanism. Section 5 shows that several 'classical' fair division methods cannot be implemented when there is incomplete information, and Section 6 gives an example of a mechanism that always generates fair and efficient outcomes. Section 7 investigates whether positive results can also be obtained if one requires that equilibria be in dominant strategies and Section 8 offers a brief conclusion.

2 The Division Problem

We consider a situation in which a single indivisible object has to be allocated to one person from a group of n. All players have equal rights to the object, side payments are possible, and each player is risk neutral and has a utility function that is additively separable in money and the object. Hence, if player i's value of the object is v_i and if t_i is the monetary transfer that this player pays, then his utility is $u_i = v_i - t_i$ if he gets the object while his utility is $u_i = -t_i$ if he doesn't get it. Each player's valuation is known privately, but it is common knowledge that all values are drawn[5] independently from a distribution F with support $[\underline{v}, \bar{v}]$. We assume that the density f is positive and continuous and without loss of generality we take $\underline{v} = 0$ and $\bar{v} = 1$. Hence, each player indeed values the object.

The problem is to determine which player should get the object and by how much he should compensate his partners so as to guarantee that the final allocation is both fair and efficient. A mechanism is a game form specifying the allocation rules. Formally, a *mechanism* is a tuple $\mu = < A, p, t >$, where

$$A = A_1 \times \ldots \times A_n \text{ with } A_i \text{ being the (nonempty) set of pure strategies of player } i, \quad (2.1)$$

[5]Hence, we follow the seminal idea from Harsanyi (1967-8) to convert a situation with incomplete information to one with asymmetric information. The assumption of independence simplifies the problem, we haven't analyzed the case where values are correlated.

$p = (p_1, \ldots, p_n)$ with $p_i : A \to [0,1]$ and $\sum_i p_i(a) = 1$ for all $a \in A$. ($p_i(a)$ is the probability that i receives the object if a is chosen), and (2.2)

$t = (t_{ij})_{i,j}$ with $t_{ij} : A \to \mathbf{R}$ and $\sum_j t_{ij}(a) = 0$ for all i and all $a \in A$. ($t_{ij}(a)$ is the monetary transfer that j has to pay (to a mediator) in case $a \in A$ is chosen and the object is allocated to i). (2.3)

Note that in (2.2) we require that the object be allocated under all circumstances and that in (2.3) it is required that the players' books always balance. In Section 7 the latter constraint is relaxed by allowing the mediator to act as a clearing house whose books have to balance only on average rather than for each value combination.

Given the mechanism μ, a (pure) strategy for player i is a map $\sigma_i : [0,1] \to A_i$, and a strategy combination σ is an n-tuple of strategies, one for each player. We now introduce some additional notation. For a value vector v we write v^h for the highest value and v^s for the second highest value. We write $a = (a_{-i}, a_i)$, $v = (v_{-i}, v_i)$, $\sigma = (\sigma_{-i}, \sigma_i)$ and $a \backslash a'_i = (a_{-i}, a'_i)$. Furthermore, $\sigma(v) = (\sigma_1(v_1), \ldots, \sigma_n(v_n))$ and $dF(v) = dF(v_1) \ldots dF(v_n)$. Player i's expected transfer when a is chosen is denoted by $t_i(a)$

$$t_i(a) = \sum_k p_k(a) t_{ki}(a) \tag{2.4}$$

If player i's opponents play according to σ while player i himself chooses action a_i, then his expected transfer is

$$T_i^{\sigma}(a_i) = \int t_i(\sigma(v) \backslash a_i)\, dF(v) \tag{2.5}$$

while the probability that he receives the object is

$$P_i^{\sigma}(a_i) = \int p_i(\sigma(v) \backslash a_i)\, dF(v) \tag{2.6}$$

In this case, if players i's value is v_i, then his expected utility is

$$U_i^\sigma(a_i; v_i) \quad = \quad v_i P_i^\sigma(a_i) - T_i^\sigma(a_i) \tag{2.7}$$

The strategy combination σ is a (Bayesian Nash) *equilibrium* of the mechanism μ if for all i and all v_i

$$\sigma_i(v_i) \in \arg\max_{a_i} U_i^\sigma(a_i; v_i) \tag{2.8}$$

A mechanism is said to be a direct mechanism if $A_i = [0, 1]$ for all i, i.e. players are asked to report their values. A direct mechanism is said to be *incentive compatible* if truthtelling (i.e. $\bar{\sigma}_i(v_i) \equiv v_i$) is an equilibrium. Note that if σ is an equilibrium of the mechanism μ, then the direct mechanism $\tilde{\mu}$ determined by $\tilde{p}_i(v) = p_i(\sigma(v))$ and $\tilde{t}_{ij}(v) = t_{ij}(\sigma(v))$ is incentive compatible and leads to the same allocation. Hence, the restriction to incentive compatible direct mechanisms is without loss of generality. (This is the so-called *revelation principle*, see e.g. Myerson (1979).) For an incentive compatible direct mechanism μ, we simplify notation by writing $P_i(v_i), T_i(v_i)$ and $U_i(v_i)$ instead of $P_i^{\bar{\sigma}}(v_i), T_i^{\bar{\sigma}}(v_i)$ and $U_i^{\bar{\sigma}}(v_i; v_i)$, resp. where $\bar{\sigma}$ denotes the strategy of truthtelling ($\bar{\sigma}_i(v_i) \equiv v_i$ for all i).

We conclude this section by specifying three additional conditions that we want mechanisms to satisfy. The requirements will be formulated only for direct mechanisms. An indirect mechanism μ is said to satisfy the requirements if it has an equilibrium σ which is such that the direct mechanism $\tilde{\mu} = \mu \circ \sigma$ satisfies them. First of all, since ex ante the players are in symmetrical positions we want the mechanism to be *symmetric*, i.e. the mechanism should be anonymous: The probability that a player gets the object should only depend on the vector of values and not on the player's name, and similarly for the transfers. In particular, symmetry implies that the functions P_i, T_i and U_i do not depend on the player index i, hence, from now on, we will drop this index. Secondly, we want the allocation to be ex post efficient, i.e. the object should end up with a player who values it most, hence

$$\text{if } p_i(v) > 0, \text{ then } v_i = v^h \tag{2.9}$$

A mechanism is *ex post efficient* if it satisfies (2.9) for all v. Finally, we want all players to be satisfied with the final allocation, i.e. the final allocation should be *envy free* (Foley (1967)): Ex post each player should (weakly) prefer what he himself receives above what is allocated to some other player. This requirement implies that any two players that do not receive the object get the same monetary transfer

$$t_{ij}(v) = t_{il}(v) = -t_{ii}(v)/(n-1) \qquad \text{for all } i, j, l \text{ with } j, l \neq i.$$

It also implies that if player i gets the object, he prefers to make the transfer, i.e.

$$v_i - t_{ii}(v) \geq t_{ii}(v)/(n-1) \qquad \text{for all } i, v \text{ with } p_i(v) > 0,$$

and that each player j not getting the object indeed prefers not to get it

$$t_{ii}(v)/(n-1) \geq v_j - t_{ii}(v) \qquad \text{for all } j \neq i, \text{ if } p_i(v) > 0.$$

Hence, we see that ex post efficiency is a necessary condition for ex post fairness. The final allocation is envy free if and only if the object is allocated efficiently and the winner pays each partner the same amount $\tau(v)$ with

$$v^s/n \leq \tau(v) \leq v^h/n \tag{2.10}$$

where v^h (resp. v^s) denotes the highest (resp. second highest) value. A mechanism that generates an envy free allocation for each possible value vector, will be called an *ex post fair mechanism*. The remainder of the paper is devoted to the question of whether such mechanisms exist and what their properties are.

3 Examples of Mechanisms

In this section several division methods that have been proposed in the literature will be illustrated. Attention will be confined to the case in which there are only two participants and in which F is the uniform distribution on $[0, 1]$.

A first procedure is random allocation: each player receives the object with probability $1/2$ and there are no side payments. Clearly, this mechanism is very inefficient, in fact, the results of the next section imply that this mechanism yields the lowest expected utility for each player, no matter what his value might be. One method for improving the performance of this mechanism readily suggests itself, viz. let the random allocation be followed by bargaining between the partners. Since a player cannot be forced to trade if he doesn't want to, each player can only gain by engaging in the bargaining and, therefore, the expected payoffs will be higher. The final allocation (and, hence, the expected payoffs) will depend on how the rules for the bargaining game are specified. For example, suppose that the rules are that simultaneously the buyer and the seller submit bids and that the object is transferred, for a price equal to the average of the bids, if and only if the buyer's bid exceeds that of the seller. Chatterjee and Samuelson (1982) have shown that this game has an equilibrium[6] given by

$$\sigma_s(v_s) = 2/3\ v_s + 1/4 \qquad \sigma_b(v_b) = 2/3\ v_b + 1/12, \tag{3.1}$$

hence, in the range where trade is possible, the seller overstates his value while the buyer understates his, and this has the consequence that the outcome is not always ex post efficient. Straightforward computations show that the strategies from (3.1) yield the expected utility function U given by

$$U(v_i) = \begin{cases} 1/4\ v_i^2 + 1/8\ v_i + 9/64 & \text{if} \quad v_i \leq 1/4 \\ 1/2\ v_i^2 + 10/64 & \text{if} \quad 1/4 \leq v_i \leq 3/4 \\ 1/2\ v_i^2 + 3/8\ v_i + 1/64 & \text{if} \quad v_i \geq 3/4 \end{cases} \tag{3.2}$$

The reader might think that by using a different bargaining procedure, the performance of this type of mechanism can be improved. However, it follows from the results of Myerson and Satterthwaite (1983) that, no matter what the rules are, there will always be combinations of values for which the object ends up with the person who values it least. Random allocation followed by bargaining performs badly because players are treated asymmetrically: Once the initial property rights have been assigned, the partners have conflicting interests. In order to get a better price, the seller pretends that the object is worth more to him than it actually is while the buyer understates his value; strategic behavior which implies that the players may fail to strike an efficient bargain. These incentives for strategic manipulation can be reduced by forcing the players to announce their bids before the object is allocated; if a player doesn't know whether he will be the buyer or

[6] There exist other equilibria as well.

the seller, then his bid will be closer to the actual value.[7] Formally, one may proceed as follows. The players are asked to simultaneously submit bids b_1 and b_2, and then the object is randomly allocated. If the random move assigns the object to player i, then player i may retain the object if $b_i \geq b_j$, otherwise he sells it to player j for the price $p = (b_i + b_j)/2$. Because of risk neutrality, this mechanism is equivalent to the auction mechanism in which the players bid and the object is allocated to the highest bidder who pays his partner a compensation of $p = (b_1 + b_2)/4$. This auction mechanism will be analyzed at the end of this section (it corresponds to the auction with $\lambda = 1/2$) and we will see that the expected utilities generated by this mechanism dominate those given by (3.2).

Another mechanism that may be used is the divide and choose method, specifically the variant of this method that has been proposed in Luce and Raiffa (1957). Each person adds 1 (perfectly divisible) money unit to the pot, then a random move determines who will be the divider, this divider transfers a certain amount x from the pot to the object and his partner has the choice between the object plus x or the remainder, $2 - x$, of the money. If player d is the divider and he transfers $x \in [1/2, 1]$, then player c chooses the object if $v_c \geq 2 - 2x$, so that player d's expected payoff is

$$(2 - x)(2x - 1) + (v_d + x)(2 - 2x).$$

Hence, player d's optimal choice is

$$x_d = 7/8 - v_d/4$$

and the divider's equilibrium payoffs are

$$U_d(v_d) = (v_d/2 + 1/4)^2 \tag{3.3}$$

while those of the chooser are given by

$$U_c(v_c) = \begin{cases} 1/4 & \text{if } v_c \leq 1/4 \\ 1/4 + (v_c - 1/4)^2 & \text{if } 1/4 \leq v_c \leq 3/4 \\ v_c - 1/4 & \text{if } v_c \geq 3/4 \end{cases} \tag{3.4}$$

[7] This argument has also been made in Samuelson (1985). That paper also discussed the λ-auction mechanism with $\lambda = 1/2$.

It is interesting to note that $U_c(v_i) > U_d(v_i)$ for all v_i, hence, the method favors the chooser. This is in sharp contrast to the case in which the values are perfectly known. In the latter case, the method favors the divider, he can extract all the surplus. Also note that, under incomplete information, the object ends up with the chooser if and only if $v_c \geq 1/4 + v_d/2$, hence, also this method may lead to an inefficient allocation. Again the reason is strategic manipulation: If v_d is small (resp. large), then the divider is tempted to transfer relatively little (resp. relatively much) to the object, and if the chooser's value is below (resp. above) average, then the chooser takes the alternative that the divider doesn't "expect" and an inefficient outcome results.

The cause of the inefficiency associated with the divide and choose method is again the fact that players are in asymmetric roles. It is better not to introduce roles and keep the symmetry. There are various possibilities for modifying the method in this way. We now discuss two of these, both based on ideas of Banach and Knaster as reported in Steinhaus (1948). The essential idea is to let the division be performed by a mediator. For example, the mediator continuously transfers money from the pot to the object until one of the players shouts 'stop'. This player then receives the object plus the money that has been transferred, his partner receives the remainder of the money. It is clear that this mechanism is actually an auction mechanism. If we write $s_i = 1 + b_i/2$ for the value of the pot at which player i shouts 'stop', and interpret b_i as player i's bid, then the highest bidder gets the object and he pays his partner a price equal to half of his bid. (Hence, this is the special case of the auction mechanism introduced below with $\lambda = 1$.)

Alternatively, the mediator may first add all money to the object and then continuously transfer money from the object to the pot. In this case the person who first shouts 'stop' receives the pot, his partner receives the object plus the remainder of the money. Again this mechanism is actually an auction mechanism. If player i shouts 'stop' when the amount that has been transferred to the pot is s_i and we write $s_i = 1 + b_i/2$ then (with b_i being the bid of player i) the highest bidder gets the object and he pays this partner a price equal to half of the bid that the partner made. (Hence, this is the special case of the auction mechanism introduced next with $\lambda = 0$.)

Generally, an auction mechanism may be described as follows. Simultaneously the players bid and the object is allocated to the highest bidder for a price of λ times the highest bid, b_h, plus $1-\lambda$ times the second highest bid, b_s. The revenues of the auction are shared equally by the partners so that, effectively, the winner pays his partner a compensation equal to $(\lambda b_h + (1-\lambda)b_s)/2$. Above we encountered the Dutch auction ($\lambda = 1$), the Vickrey auction ($\lambda = 0$) and the case with $\lambda = 1/2$

as special cases. It is straightforward to verify that the strategy combination σ given by

$$\sigma_i(v_i) = 2/3v_i + 1/3(1-\lambda) \tag{3.5}$$

is a symmetric equilibrium of the auction mechanism. In Van Damme (1984) it was shown that this strategy pair is actually the unique symmetric equilibrium. Note that the equilibrium strategy is monotonic, hence, the person with the highest value bids highest so that the object always ends up with the person who values it most. Hence, any auction mechanism is ex post efficient[8]. The reader can easily verify that the expected payoffs associated with the auction mechanism are given by

$$U(v_i) = v_i^2/2 + 1/6. \tag{3.6}$$

Hence, the expected payoffs are independent of λ, a fact that is explained by the results of Sect. 4. The reader also notes that in terms of expected utilities the auction mechanisms indeed dominate random allocation followed by bargaining (the RHS from (3.6) is always strictly larger than that of (3.2)) as well as the divide and choose method (if U is as in (3.6), then $U(v_i) \geq U_d(v_i)/2 + U_c(v_i)/2$ with equality only for $v_i \in \{1/6, 5/6\}$). Although the different auction methods are equivalent at the interim stage (i.e. when a player only knows his own value), they are not equivalent ex post: the transfer payments depend on λ. If $v_1 > v_2$, then player 1 pays player 2 the amount $\lambda v_1/3 + (1-\lambda)(v_2/3 + 1/2)$. One notes that this transfer cannot always lie between $v_2/2$ and $v_1/2$ so that no auction mechanism is ex post fair.

The following two conclusions emerge from these examples. Mechanisms that have players play different roles typically lead to inefficient outcomes. Symmetric mechanisms have better efficiency properties, but also these mechanisms may yield outcomes that are not ex post fair. In the Sections 6 and 7 we will describe mechanisms that are ex post fair. In Sect. 4 we explain why all auction mechanisms are equivalent at the interim stage.

[8]In Van Damme (1984) it was shown that these properties hold for any number of players, n, and for any distribution F.

4 Revenue Equivalence

In this section we derive a 'revenue equivalence theorem' that explains why all auction mechanisms are equivalent at the interim stage. From now on attention will be restricted to direct mechanisms. By the revelation principle, this is without loss of generality.

We start by giving a characterization of incentive compatibility. Recall that a direct mechanism is said to be incentive compatible if truthtelling is a Nash equilibrium, i.e. if for all i and all $v_i, w_i \in [0,1]$

$$v_i P(v_i) - T(v_i) \geq v_i P(w_i) - T(w_i) \tag{4.1}$$

where P and T are defined by (2.5) and (2.6) and by the remarks concerning notation that were made after (2.8). The proof of Lemma 1 is standard and follows ideas outlined in Myerson and Satterthwaite (1983). (See also Cramton et al. (1987).) To make the paper self-contained, however, the proof is given in the appendix.

Lemma 1 . *A direct mechanism μ is incentive compatible if and only if P is nondecreasing and T is related to P according to*

$$T(v_i) = T(0) + \int_0^{v_i} x dP(x) \qquad (v_i \in [0,1]) \tag{4.2}$$

Since transfers sum to zero for each realized vector of values (Eq. (2.3)) and since players are treated symmetrically, we have that

$$\int T(v_i) dF(v_i) = 0. \tag{4.3}$$

Together with (4.2) this boundary condition implies that, for incentive compatible mechanisms, T and therefore U is completely determined by P. In particular it follows that all ex post efficient mechanisms are equivalent at the interim stage. (All such mechanisms have $P(v_i) = F(v_i)^{n-1}$.) Straightforward calculations show that

$$T(0) = \int_0^1 (1 - F(x) - xf(x))P(x)dx,$$

hence that

$$U(v_i) = \int_0^1 (F(x) + xf(x) - 1)P(x)dx + \int_0^{v_i} P(x)dx \tag{4.4}$$

For ex post efficient mechanisms, this expression can be rewritten as

$$U(v_i) = \frac{1}{n} + \frac{n-1}{n} \int_0^1 F^n(x)dx - \int_0^1 F^{n-1}(x)dx + \int_0^{v_i} F^{n-1}(x)dx \tag{4.5}$$

Let us remark that the constant term $U(0)$ in Eq. (4.5) is equal to $(1/n)$th of the expected revenue generated when the object is auctioned, say by the Vickrey procedure (or indeed by any auction procedure that always allocates the object to the person who values it most). We return to this property in Section 7. The following proposition summarizes the results obtained thus far

Proposition 1 . *The expected utility function U associated with an incentive compatible mechanism depends only on the probability function P and is given by (4.4). For ex post efficient mechanisms, $P(x) = F(x)^{n-1}$ and then the expected utility function is given by (4.5).*

Note that Lemma 1 and Proposition 1 imply that the expected utility function U is nondecreasing and convex. Another interesting property is that $U(v_i) \geq v_i/n$ for each mechanism and each value v_i, hence, each mechanism yields at least as much utility as random allocation does. Note that Proposition 2 implies that the individual rationality constraints are not binding, each player is, no matter what his value might be, willing to participate in any mechanism.

Proposition 2 . *Every mechanism is (weakly) preferred to random allocation, i.e. $U(v_i) \geq v_i/n$ for any incentive compatible mechanism.*

Proof. See the appendix.

From now on attention will be restricted to ex post efficient mechanisms. The following Proposition, which states that ex ante efficiency is equivalent to ex post efficiency, provides an additional justification for this restriction. Namely the proposition implies that there does not exist a mechanism that is uniformly preferred above an ex post efficient mechanism. A mechanism is said to be ex ante efficient if it maximizes the ex ante expected payoffs over the set of all incentive compatible mechanisms, i.e. if M is the set of all incentive compatible mechanisms satisfying (2.1) - (2.3), then $\nu \in M$ is *ex ante efficient* if

$$\int U_\mu(v_i)dF(v_i) \leq \int U_\nu(v_i)dF(v_i)$$

for all $\mu \in M$, where $U_\mu(v_i)$ is player i's expected payoff associated with μ if his value is v_i.

Proposition 3 . *The mechanism μ is ex ante efficient if and only if it yields an ex post efficient outcome for almost all value combinations, i.e. if (2.9) is satisfied for almost all v.*

Proof. Let ν be any ex post efficient mechanism[9]. Then for any mechanism μ we have in view of (4.3)

$$\int U_\mu(v_i)dF(v_i) = \int v_i P(v_i)dF(v_i)$$

Furthermore, with v^h being the highest component of v, we have

$$n\int v_i P(v_i)dF(v_i) = n\int v_i p_i(v)dF(v)$$

$$= \int \sum_i v_i p_i(v)dF(v) \leq \int v^h \sum_i p_i(v)dF(v)$$

$$= \int v^h dF(v) = n\int U_\nu(v_i)dF(v_i)$$

Since the inequality in the above chain is strict unless (2.9) is satisfied for almost all v, the proof is complete. □

5 An Impossibility Theorem

In this section we turn to some resolutions of the fair division problem that have been proposed in the literature for the case in which the value vector is commonly known, and we will show that the

[9]Such a mechanism indeed exists: In Van Damme (1984) it was shown that the λ-auction methods discussed in Section 3 have this property. Cramton et al. (1987) have shown that, more generally, ex post efficient mechanisms exist if and only if the initial distribution of ownership shares is not "too asymmetric".

allocations corresponding to these procedures cannot be implemented when there is incomplete information about the players' values.

A fair division method is a mapping m that assigns to each value vector v a utility $u_i^m(v)$ for each player i. For an incentive compatible direct mechanism μ let us write $u_i^\mu(v)$ for the utility that player i gets from μ when the value vector is v (assuming, of course, that the truthtelling equilibrium is played in μ). Hence $u_i^\mu(v) = v_i p_i(v) - t_i(v)$. We say that the fair division method m can be *implemented* if there exists an incentive compatible mechanism μ with $u_i^\mu(v) = u_i^m(v)$ for all i and v. We will show that neither the Steinhaus division method (Steinhaus (1948)), nor the egalitarian division can be implemented in case there is incomplete information. Since both methods can be obtained by applying the Nash bargaining solution (Nash (1950)) resulting from an appropriately chosen threat point, we turn to the Nash bargaining solution first.

Assume that the vector of values v is common knowledge. Since side payments are possible, the Pareto efficient frontier in utility space is given by

$$\{u \in \mathbf{R}^n;\ \sum u_i = v^h\}$$

If d is the utility vector in case of disagreement and $\sum_i d_i < v^h$, then the Nash bargaining solution of the problem is the utility vector u^* determined by

$$\sum u_i^* = v^h \quad \text{and} \quad u_i^* - d_i = u_1^* - d_1 \qquad \text{all } i$$

hence

$$u_i^* = d_i + (v^h - \sum d_j)/n \tag{5.1}$$

Two choices of d appear natural. If in case of conflict the object is destroyed then $d_i = 0$ for all i and $u_i^* = v^h/n$. This allocation corresponds to the egalitarian solution, the side payments are arranged in such a way that all players have equal payoffs. A second possibility is to resort to random allocation in case of conflict. Then $d_i = v_i/n$ for all i and the resulting allocation is the one proposed in Steinhaus (1948). Steinhaus calls v_i/n the 'fair share' of player i and he proposes to divide equally the surplus that remains after each player has received his fair share. Note that the Steinhaus allocation is not envy free.

The main result of this section is

Proposition 4 . *Neither the egalitarian allocation nor the Steinhaus allocation can be implemented when there is asymmetric information about the value vector v.*

Proof. For $\lambda \in [0,1]$ write

$$u_i^\lambda(v) = \lambda v_i/n + \left(v^h - \lambda \sum v_j/n\right)/n \tag{5.2}$$

Then $\lambda = 0$ corresponds to the egalitarian solution while $\lambda = 1$ yields the Steinhaus allocation. We have to show that there does not exist an incentive compatible mechanism μ with $u_i^\mu(v) = v_i p_i(v) - t_i(v) = u_i^\lambda(v)$ for all v. Assume such a mechanism μ does exist. Then μ is ex post efficient so that, from Proposition 1, we get $U'(v_i) = P(v_i) = F(v_i)^{n-1}$. On the other hand, Eq. (5.2) yields

$$\begin{aligned} nU(v_i) &= n\int u_i^\lambda(v_{-i}, v_i)dF_{-i}(v_{-i}) \\ &= \frac{n-1}{n}\lambda v_i + v_i F(v_i)^{n-1} + \int_{v_i}^1 x dF(x)^{n-1} - (n-1)\lambda \int x dF(x)/n \end{aligned}$$

so that

$$nU'(v_i) = \frac{n-1}{n}\lambda + F(v_i)^{n-1}$$

Combining this expression with $U'(v_i) = F(v_i)^{n-1}$ yields

$$F^{n-1}(v_i) = \lambda/n,$$

hence, F is constant. But this contradicts the assumption that we have a situation with incomplete information. □

It is interesting to investigate which consequences strategic behavior has on the final outcome if the Steinhaus procedure or the egalitarian procedure is used to determine the allocation. Identifying a stated value with a bid, we see that the rules of the latter procedure correspond to those of an auction in which the object is allocated to the highest bidder who pays each of his partners $1/n$th of his bid. In general this method yields an ex post efficient outcome, but as we know from Section 3 the outcome need not be ex post fair. (The egalitarian method corresponds to the λ-auction from Section 3 with $\lambda = 1$.) The Steinhaus procedure corresponds to an auction in which the highest bidder (say this is player i) receives the object and in which this person pays each player j an amount equal to $b_j/n + (b_i - \sum_j b_j/n)/n$. Lyon (1986) has shown that this auction mechanism has a symmetric, increasing equilibrium. Hence, this mechanism is also ex post efficient and the interim expected payoffs are given by (4.5). In the case of 2 players these rules correspond to those of the auction mechanism from Section 3 with $\lambda = 1/2$, hence, the final outcome need not be fair.

6 A Possibility Theorem

In this section we show that it is possible to ensure that the final allocation will be envy free by giving an example of a mechanism leading to fair allocations.

Proposition 5 . *There exists an incentive compatible, ex post fair mechanism.*

Proof: For a value vector $v \in \mathbf{R}_n$ let v^h be the highest component of v and let v^s be the second highest component of v. Consider the following direct mechanism: The object is allocated to the person with the highest value, he pays the amount

$$t(v^h, v^s) = \frac{1}{n}\left\{v^h - \int_{v^s}^{v^h} F(x)dx\right\} \tag{6.1}$$

to each of his partners. In case there are multiple, say k, players with the highest value then each of them receives the object with probability $1/k$ and the person getting the object pays each of his partners v^h/n.

Clearly, this mechanism treats the players symmetrically. Furthermore, the amount that the winner pays to each of his partners lies inbetween v^s/n and v^h/n so that the mechanism is ex post fair. It remains to verify that the mechanism is incentive compatible, i.e. that truthtelling is a Nash equilibrium. Since the mechanism is ex post efficient, it suffices, by Lemma 1, to show that the expected transfer payments satisfy $T'(v_i) = v_i P'(v_i) = (n-1)v_i f(v_i)F(v_i)^{n-2}$. The verification of this identity involves straightforward calculations which are carried out in the appendix. □

Note that, since the mechanism from Proposition 3 is a direct mechanism, it is context dependent, i.e. the rules of the mechanism directly depend on the characteristics of the underlying uncertainty, i.e. the mechanism depends on the distribution function F. The author has not succeeded in finding a context independent mechanism of which the equilibrium gives rise to the direct mechanism from Proposition 3. It should also be noted that the mechanism described in (6.1) is probably not the unique ex post fair mechanism, however, in a certain sense it is the simplest one possible. Namely, it is the unique ex post fair mechanism for which the transfers depend only on v^h and v^s and for which

$$\frac{\partial t(v^h, v^s)}{\partial v^h} \text{ is independent of } v^s, \text{ and}$$
$$\frac{\partial t(v^h, v^s)}{\partial v^s} \text{ is independent of } v^h.$$

7 Dominant Strategy Mechanisms

Up to now we have focused on Bayesian Nash equilibria of mechanisms. It is easily seen that, without relaxing some of our requirements, no positive results can be obtained for the stronger notion of dominant strategy equilibria. Truthtelling is a dominant strategy for each player in the direct mechanism μ if

$$v_i p_i(v) - t_i(v) \geq v_i p_i(v \backslash w_i) - t_i(v \backslash w_i) \quad \text{for all } i, v, w_i \tag{7.1}$$

with the inequality being strict for at least one value vector v. For ex post efficient mechanisms, condition (7.1) implies that the winner's transfer payment should be independent of his value, and that also each looser's transfer must be independent of this player's value. In ex post fair mechanisms all loosers receive the same payoff, which by (7.1) can, therefore, only depend on the winner's value. Hence, if the transfers have to balance for each combination of values (condition (2.3)), then the transfers must be constant, but surely a mechanism in which the transfers do not depend on the value vector cannot be ex post fair. Hence,

Proposition 6 . *There does not exist an ex post fair mechanism for which truthtelling is an equilibrium in dominant strategies.*

If one insists on equilibria in dominant strategies, then positive results can be obtained by relaxing condition (2.3): One may be satisfied if the transfers balance on average, i.e. if condition (4.3) is satisfied instead of condition (2.3). Condition (4.3) may be established by allowing the mediator to act as a clearing house who balances the payments. (If mediators are risk neutral and if the market for mediators is competitive, the players will indeed be able to find a mediator who is willing to play this role.) By reviewing the proof of Proposition 1, one sees that condition (2.3) is not essential for this result to hold, the proof just uses (4.3). Hence, if the mediator has zero expected profits, then (in an incentive compatible mechanism) the player's expected utilities are still given by (4.4). To have an ex post fair mechanism with an equilibrium in dominant strategies, the winner should pay an amount $w(v)$ that does not depend on his own value, while each looser should get an amount $l(v)$ that only depends on the winner's value. Furthermore, it should be the case that $v^h - w(v) \geq l(v) \geq v^s - w(v)$. One possibility immediately suggests itself: The mediator first buys the object for a price B of which each player receives B/n; after the mediator has acquired it, he sells is again to one of the partners by using the Vickrey (1962) procedure, i.e. the object is allocated to the highest bidder who pays the second highest bid. Hence, using the above notation, $w(v) = -B/n + v^s$ and $l(v) = B/n$. Clearly, truthtelling is a dominant strategy and the resulting allocation is always fair. Furthermore, in light of the remarks made above, the mediator's expected payoffs are zero if his bid B is equal to $nU(0)$ where $U(0)$ is given by (4.5).

An alternative possibility is that the mediator uses the information revealed by the auction to determine the compensation of the loosers rather than to determine the price that the winner should pay. Specifically, the mediator may use the following procedure. First he asks each player to contribute an amount C equal to $(n-1)/n$ times the expectation of the highest value (i.e. $C = (n-1)\int v^h dF(v)/n$). After these contributions have been made he auctions the object. This time, however, the winner does not have to pay anything, rather it is the case that each looser receives an amount equal to the winner's value. Hence, in terms of the above notation, $w(v) = C$ and $l(v) = v^h - C$. Again truthtelling is a dominant strategy and, no matter what the values are, all players have the same net payoffs. Hence we have shown

Proposition 7 . *With an active mediator, who is used to balance the budgets on average, there exists an ex post fair mechanism for which truthtelling is an equilibrium in dominant strategies. In fact there exists such a mechanism that leads to the egalitarian outcome, i.e. all players have the same net payoff.*

Clearly, the mechanisms discussed in this section are viable only if the mediator is able to make an accurate assessment of the value of the object, i.e. if he knows the distribution F. It will also be clear that even if he knows F, he will be reluctant to use the second procedure discussed above. Namely, although truthtelling is a dominant strategy in the game, the mediator should fear that the agents will try to increase their payoffs by more sophisticated ways of manipulating the outcome. For example, the players could make a secret contract that each player will bid his value plus an amount x and that each looser will pay the winner x/n. With this contract in place it is a dominant strategy for each player i to bid $v_i + x$ in the mediator's auction game. Compared with the original situation in which there is no contract each player increases his payoff by $(n-1)x/n$, at the expense of the mediator who makes an expected loss of $(n-1)x$. Hence, it is very unlikely that we will observe such a mechanism in practice.

8 Conclusion

In this paper we have investigated a simple fair division situation with incomplete information. The analysis was simplified by the assumptions of symmetry (all values are drawn from the same distribution) and independence (all values are drawn independently). Further research should be devoted to relaxing these assumption. The assumption of independence seems especially inappropriate in the case of the dissolving of joint ventures, one of the examples that was mentioned in the Introduction. In that case the value of each partner depends on the future prospects for the business about which the partners may have different information.

The paper was motivated by the idea that it is desirable to guarantee an ex post fair outcome. It may be questioned whether ex post fairness is indeed desirable, especially since some types of a player may, at the interim stage, prefer non-fair mechanisms above fair ones[10]. In Van Damme (1985) it was shown that, for the example discussed in Section 3, the expected utility for a player with value 1/2 is maximized by a mechanism that assigns the object to the person with the highest value when this value is not in the interval [1/4, 3/4] and that allocates the object randomly if the highest value is in this interval. In that paper it was also shown that only a type with value $v_i = 0$ or $v_i = 1$ prefers an ex post fair mechanism above any other mechanism. If one does

[10]Why this is so can easily be seen in an asymmetric example. Suppose that player 1 values the object at $v_1 = 0$ and that player 2 values the object either at $v_2 = 0$ (with probability p) or at $v_2 = 100$ (with probability $1-p$). In an ex post fair incentive compatible mechanism, player 2 always gets the object and he pays at most 5 for it. If p is sufficiently small player 1 clearly prefers a mechanism in which the object is allocated only to player 2 if this player is willing to pay a sufficiently high price for it.

not insist on ex post efficiency, then the allocation mechanism should reflect a fair compromise between the alternative types of a player[11] and the solutions of Harsanyi and Selten (1972) and Myerson (1984) specify axioms for determining such fair compromises. If, however, one insists on ex post efficiency, then Proposition 1 shows that the interim utilities are completely determined so that according to Harsanyi/Selten and Myerson all mechanisms are equivalent. At the ex post stage, however, such mechanisms are not necessarily equivalent: some of these guarantee ex post fair mechanisms while others do not.

References

Chatterjee, K. and W. Samuelson (1983). "Bargaining under Incomplete Information", *Operations Research*, **31**, 838-851.

Cramton, P., R. Gibbons and P. Klemperer (1987). "Dissolving a Partnership Efficiently", *Econometrica*, **55**, 615-632.

Crawford, V. (1977). "A Game of Fair Division", *Review of Economic Studies*, **44**, 235-247.

Crawford, V. and W. Heller (1979). "Fair Division with Indivisible Commodities", *Journal of Economic Theory*, **21**, 10-27.

Dubins, L. (1977). "Group Decision Devices", *American Mathematical Monthly*, **84**, 350-356.

Foley, D. (1967). "Resource Allocation and the Public Sector", *Yale Economic Essays*, **7**.

Güth, W. (1986). " Auctions, Public Tenders, and Fair Division Games: An Axiomatic Approach", *Mathematical Social Sciences*, **11**, 283-294.

Güth, W. and E. van Damme (1986). "A Comparison of Pricing Rules for Auctions and Fair Division Games", *Social Choice and Welfare*, **3**, 177-198.

Harsanyi, J. (1967-8). "Games with Incomplete Information Played by 'Bayesian' players", *Management Science*, **14**, 159-189, 320-344, 486-504.

Harsanyi, J. and R. Selten (1972). "A Generalized Nash Solution for Two-Person Bargaining Games with Incomplete Information", *Management Science*, **18**, 80-105.

Kennan, J. and R. Wilson (1990). "Bargaining with Private Information", *SITE Technical report*, **5**, Stanford University.

Kuhn, H. (1967). "On Games of Fair Division", in *Essays in Mathematical Economics*

[11]In our symmetric setup there is no conflict between different players.

in Honor of Oskar Morgenstern, ed. M. Shubik, Princeton University Press, Princeton.

Luce, R. and H. Raiffa (1957). *Games and Decisions*, Wiley, New York.

Lyon, R. (1986). "Equilibrium Properties of Auctions and Alternative Procedures for Allocating Transferable Permits", *Journal of Environmental Economics and Management*, **13**, 129-152.

Myerson, R. (1979). "Incentive Compatibility and the Bargaining Problem", *Econometrica*, **47**, 61-73.

Myerson, R. (1984a). "Two-Person Bargaining Problems with Incomplete Information", *Econometrica*, **52**, 461-487.

Myerson, R. (1984b). "Cooperative Games with Incomplete Information", *International Journal of Game Theory*, **14**, 69-97.

Myerson R. and M. Satterthwaite (1983). " Efficient Mechanisms for Bilateral Trading", *Journal of Economic Theory*, **29**, 265-281.

Nash, J. (1950). "The Bargaining Problem", *Econometrica*, **18**, 155-162.

Pazner, E. and D. Schmeidler (1978). "Egalitarian Equivalent Allocations: A New Concept of Economic Equity", *Quarterly Journal of Economics*, **92**, 671-687.

Samuelson, W. (1980). "The Object Distribution Problem Revisited", *Quarterly Journal of Economics*, **94**, 85-98.

Samuelson, W. (1985). "A comment on the Coase Theorem". In A.E. Roth (Ed.) *Game-theoretic Models of Bargaining*. Cambridge University Press, Cambridge, 321-340.

Steinhaus, H. (1948). "The Problem of Fair Division", *Econometrica*, **16**, 101-104.

Van Damme, E. (1984). "Auctions and Fair Division", Mimeo, Dept. of Mathematics and Computer Science, University of Technology, Delft.

Van Damme, E. (1985). "Fair Allocation of an Indivisible Commodity", Mimeo, Dept. of Mathematics and Computer Science, University of Technology, Delft.

Varian, H. (1974). "Equity, Envy and Efficiency", *Journal of Economic Theory*, **9**, 63-90.

Vickrey, W. (1961). "Counterspeculation, Auctions and Competitive Sealed Tenders", *Journal of Finance*, **16**, 8-37.

Appendix

Proof of Lemma 1. If μ is incentive compatible, then (4.1) holds and by rearranging and interchanging the roles of v_i and w_i we obtain

$$w_i(P(v_i) - P(w_i)) \leq T(v_i) - T(w_i) \leq v_i(P(v_i) - P(w_i)),$$

so that P must be nondecreasing. Furthermore, T is differentiable whenever P is, and at such points of differentiability $T'(v_i) = v_iP'(v_i)$. At other points, a jump in $v_iP(v_i)$ is matched by a jump in $T(v_i)$ and such discontinuities are accounted for by in the integral in (4.2), which is understood to be a Lebesgue-Stieltjes integral. Hence, (4.2) is indeed correct.

Conversely, if (4.2) holds, then for any $v_i, w_i \in [0,1]$

$$T(v_i) - T(w_i) = \int_{w_i}^{v_i} x dP(x) \leq \int_{w_i}^{v_i} v_i dP(x) = v_i(P(v_i) - P(w_i)),$$

which, after rearranging, yields the incentive compatibility condition (4.1). □

Proof of Proposition 2. Because of symmetry, the ex ante probability that a player receives the object is $1/n$, hence $\int P(x)dF(x) = 1/n$. In view of (4.4) we therefore have to show that

$$v_i \int_0^1 f(x)P(x)dx \leq \int_0^1 (F(x) + xf(x) - 1)P(x)dx + \int_0^{v_i} P(x)dx,$$

or, equivalently

$$\int_0^{v_i} v_i f(x)P(x)dx + \int_{v_i}^1 (v_i f(x) + 1)P(x)dx \leq \int_0^1 (F(x) + xf(x))P(x)dx$$

Now by using partial integration it is easily seen that

$$\int_0^{v_i} (v_i - x)f(x)P(x)dx \leq \int_0^{v_i} F(x)P(x)dx,$$

so that it suffices to show that

$$\int_{v_i}^1 (v_i f(x) + 1)P(x)dx \leq \int_{v_i}^1 (F(x) + xf(x))P(x)dx,$$

or equivalently

$$\int_{v_i}^{1}(1-F(x))P(x)dx \leq \int_{v_i}^{1}(x-v_i)f(x)P(x)dx.$$

Integrating the integral of the RHS by parts we see that we have to show that

$$\int_{v_i}^{1}(1-F(x))P(x)dx \leq (1-v_i)P(1) - \int_{v_i}^{1}F(x)P(x)dx - \int_{v_i}^{1}(x-v_i)F(x)dP(x)$$

which is equivalent to

$$\int_{v_i}^{1}P(x)dx \leq (1-v_i)P(1) - \int_{v_i}^{1}(x-v_i)F(x)dP(x)$$

By integrating the integral of the LHS by parts it is seen that this inequality is equivalent to

$$v_i(P(1)-P(v_i)) - \int_{v_i}^{1}xdP(x) \leq -\int_{v_i}^{1}(x-v_i)F(x)dp(x),$$

which in turn is equivalent to

$$\int_{v_i}^{1}(x-v_i)(1-F(x))dP(x) \geq 0,$$

an inequality which is clearly satisfied. □

Proof of Proposition 5: It remains to be shown that the mechanism determined by (6.1) is incentive compatible, i.e. that $T'(v_i) = v_iP'(v_i) = (n-1)v_if(v_i)F(v_i)^{n-2}$. Write $x = v_{-i}^h$ (resp. $y = v_{-i}^s$) for the highest (resp. second highest) component of v_{-i}. Then the joint distribution Φ of x and y is given by

$$\Phi(x,y) = \begin{cases} F^{n-1}(x) + (n-1)F^{n-2}(y)(F(x)-F(y)) & \text{if } x > y \\ F^{n-1}(y) & \text{if } x \leq y, \end{cases}$$

and the associated joint density is equal to

$$\varphi(x,y) = \begin{cases} (n-1)(n-2)f(x)f(y)F(y)^{n-3} & \text{if } x > y \\ 0 & \text{if } x \leq y \end{cases}$$

With t given by (5.1), the transfer that player i makes if he has the value v_i is equal to

$$t_i(v_i) = \begin{cases} (n-1)\, t(v_i, x) & \text{if } x < v_i \\ -t(x, v_i) & \text{if } y < v_i < x \\ -t(x, y) & \text{if } v_i > y, \end{cases}$$

and the expected transfer is

$$T(v_i) = \int\int t_i(v_i) d\Phi(x, y).$$

Differentiating with respect to v_i we find that

$$\begin{aligned} T'(v_i) &= (n-1)\int_0^{v_i}\int_0^{v_i} \frac{\partial}{\partial v_i} t(v_i, x) d\Phi(x, y) + (n+1)^2\, t(v_i, v_i) f(v_i) F(v_i)^{n-2} \\ &- \int_{v_i}^1 \int_0^{v_i} \frac{\partial}{\partial v_i} t(x, v_i) d\Phi(x, y) - \int_{v_i}^1 t(x, v_i) d\Phi(x, v_i) \\ &+ (n-1)\, t(v_i, v_i) f(v_i) F(v_i)^{n-2} + \int_{v_i}^1 t(x, v_i) d\Phi(x, v_i). \end{aligned}$$

Rearranging and collecting terms we see that

$$\begin{aligned} T'(v_i) &= (n-1) v_i\, f(v_i) F(v_i)^{n-2} + \frac{(n-1)}{n}(1 - F(v_i)) F(v_i)^{n-1} \\ &= \frac{n-1}{n} \int_{v_i}^1 \int_0^{v_i} F(v_i) d\Phi(x, y) \\ &= (n-1) v_i\, f(v_i) F(v_i)^{n-2}, \end{aligned}$$

and this is what had to be proved. □

Non-Transferable Utility and Bayesian Incentives

Claude d'Aspremont
CORE, Université Catholique de Louvain,
Voie du Roman Pays 34, B-1348 Louvain-la-Neuve

Louis-André Gérard-Varet
GREQE, Ecole des Hautes Etudes en Sciences Sociales,
Hospice de la Vieille Charité, Rue de la Charité, F-13002 Marseille

I. Introduction

Whenever information is decentralized, efficient collective decisions have to rely on the pieces of information that each individual privately controls. Incentives have to be provided to guarantee proper revelation of these private informations. The literature on mechanism design under incomplete information addresses the question.[1]

In a multiperson decision problem, every individual evaluates the outcomes on the basis of a utility function which is conditional on his personal characteristics, the vector of which defines his type. The selection of an outcome is accordingly defined as a function of the configuration of possible types in society. To any such "outcome function" is associated a game of incomplete information in which individuals can strategically report their types. "Incentive compatibility" means that "telling the truth about his type" has to be a non cooperative equilibrium of that game. Most of the results in the literature have however been obtained assuming full transferability of the utilities through some "money", i.e. assuming that individuals' utilities are quasi-linear with respect to that money. The aim of the present paper is to discuss issues arising when monetary transfers are not available.

Following Harsanyi (1967-68), we model incomplete information from a Bayesian point of view leading to a Bayesian equilibrium concept.[2] In a Bayesian framework individuals are described by their utilities but also by their "beliefs" about what may be the others' types. When the utilities are fully transferable and the individual types uncorrelated, one may design a mechanism which associates a Pareto-optimal outcome to any configuration of individual types – an ex post collective efficiency property – and which is such that telling the truth is a Bayesian equilibrium – the Bayesian incentive compatibility property (d'Aspremont, Gérard-Varet (1979a)). The two properties remain

[1]For a recent survey, see Groves-Ledyard (1987).
[2]The use of this concept for mechanism design was motivated by the difficulties in using the stronger concept of equilibrium in dominant strategies, see Green-Laffont (1979).

consistent for much more general families of beliefs. But then the explicit construction of the mechanism may be difficult.[3]

Here we shall maintain the "uncorrelated type" hypothesis. Our aim is to relax instead the full-transferability assumption. We shall adapt the general procedure introduced by Shapley (1969) to extend a solution concept from the transferable to the non-transferable utility case. A utilitarian social welfare index under incomplete information will replace here the Shapley value under complete information and incentive constraints will be substituted to the standard "individual rationality conditions".

The idea is to find individual weighing factors, and to rescale the Neuman-Morgenstern (NM) utilities accordingly, in such a way that the maximization of the weighted sum of utilities is incentive compatible without requiring transfers. Moreover in this procedure one wants to exclude "dictatorship" which would arise if one individual had positive weights in every type with all other individuals having weight zero in all their types. As a matter of fact, under full transferability the non-dictatorship condition is satisfied since monetary transfers cannot be infinite. Here, for the non-transferable utility case we have to impose that any individual of any type receives a positive but finite weight.

In section II we formally posit the problem. Our no-transfer solution will be established in section III and although avoiding dictatorship, we still need that some individual plays the role of an "arbitrator". We show in section IV that the arbitrator is no longer necessary in a "symmetric" situation where there is a common utility function defined on a common set of possible types as in the fundamental utility approach of Harsanyi (1955,1977).

II. Multiperson decision problems under incomplete information

Let X be a compact set of alternative social states concerning a (finite) set $N = \{1, \cdots, i, \cdots, n\}$ of individuals. Each individual $i \in N$ is described by a list α_i of personal characteristics (e.g. natural talents, private endowments, ...) and by a continuous NM utility function $u_i(\cdot, \alpha_i)$ defined on the set X. Incomplete information means that α_i is private information for every i : any $j \neq i$ or any outside observer only knows that the actual listing of characteristics of i belongs to a given set A_i also called the set of *possible types* of i. We assume A_i is finite. Thus, even though all functions $u_i : X \times A_i \to \mathbb{R}$ are common knowledge, the "true" utility function $u_i(\cdot, \alpha_i)$ remains i's private information. Moreover, each individual i being uncertain about the types $\alpha_{-i} = (\alpha_j)_{j \neq i}$ in $A_{-i} = \times_{j \neq i} A_j$ describing the actual characteristics of others, has beliefs concerning their likelihood. The beliefs of i are a probability represented by a transition p_i from A_i to A_{-i} giving for every possible α_i in A_i a probability distribution $p_i(\cdot \mid \alpha_i)$ over A_{-i} (i.e. $p_i(\alpha_{-i} \mid \alpha_i) \geq 0$, $\alpha_{-i} \in A_{-i}$ and $\sum_{\alpha_{-i}} p_i(\alpha_{-i} \mid \alpha_i) = 1$). Again, even though all transitions p_i are common knowledge, the "true"

[3]For a discussion and recent results, see d'Aspremont, Crémer and Gérard-Varet (1990) and Johnson, Pratt, Zeckhauser (1990).

beliefs $p_i(\cdot \mid \alpha_i)$ remain i's private information.

In this incomplete information set-up, a "decentralized mechanism" is needed to select an alternative in the set X. An *outcome function* is a function $x(\cdot) \ : A \rightarrow X$ where $x(\alpha)$ is the alternative selected in X on the basis of a listing of types $\alpha = (\alpha_1, \cdots, \alpha_n) \in A = \times A_i$ for all individuals. We say that an outcome function $x(\cdot)$ is *(weakly) Pareto-optimal* if for every α in A one cannot find an alternative y in X such that, for every $i \in N$, $u_i(y, \alpha_i) > u_i(x(\alpha), \alpha_i)$. This is an ex-post efficiency criterion in the sense that it does not apply to the a priori choice of an outcome function but, for a given function, to the choice of an eventually feasible outcome.

Given an outcome function $x(\cdot)$, one needs some procedure by which information on the actual (or "true") values of the individual characteristics is collected. We adopt the following simple procedure : individuals are supposed to announce their characteristics, say $a = (a_1, \cdots, a_n)$ in A to some processing center which selects accordingly the alternative $x(a)$ in X. This procedure, called a "direct revelation mechanism"[4], defines a game of incomplete information where every individual is a player whose strategic choice consists in announcing an element a_i of A_i. A strategy for a player is a function from A_i to A_i. The "truth-telling" strategy is the identity function. The outcome function $x(\cdot)$ is said to be a *Bayesian Incentive Compatible* (BIC) mechanism if the truth-telling strategy is a Bayesian equilibrium (in the sense of Harsanyi (1967-1968)), which amounts to require :

$$\sum_{\alpha_{-i}} u_i(x(\alpha_i, \alpha_{-i}), \alpha_i) p_i(\alpha_{-i} \mid \alpha_i) \geq \sum_{\alpha_{-i}} u_i(x(a_i, \alpha_{-i}), \alpha_i) p_i(\alpha_{-i} \mid \alpha_i)$$

$$\text{for all } i \in N, \alpha_i, a_i \in A_i.$$

We consider selections of alternatives based on some possible rescaling of the individual utilities and determined by *positive scaling factors* $\lambda_i(\alpha_i)$ multiplying the utility $u_i(\cdot, \alpha_i)$ for every i and every α_i. Individual utilities are thus calibrated in the common unit in which the aggregate social welfare is to be evaluated. For formal reasons we temporarily admit that some scaling factors may vanish and define

$$\Lambda = \{\lambda \in \times_{i=1}^{n} \mathbb{R}_+^{A_i} \ : \text{ for every } \alpha \text{ there exists some } i \text{ for which } \lambda_i(\alpha_i) > 0\}.$$

For a given $\lambda \in \Lambda$ and a vector $\alpha \in A$ the aggregate social welfare is

$$\sum_{i=1}^{n} \lambda_i(\alpha_i) u_i(\cdot, \alpha_i).$$

Clearly, if the outcome function $x(\cdot)$ is such that, for every $\alpha \in A$

$$\sum_{i=1}^{n} \lambda_i(\alpha_i) u_i(x(\alpha), \alpha_i) = \sup_{y \in X} \sum_{i=1}^{n} \lambda_i(\alpha_i) u_i(y, \alpha_i)$$

then it is Pareto-optimal. For the same formal reasons we shall temporarily use a weaker notion of Bayesian incentive compatibility which is relative to $\lambda \in \Lambda$. The outcome function $x(\cdot)$ is λ-BIC if for every $i \in N$, α_i, $a_i \in A_i$,

[4]See the survey of Myerson (1985).

$$\sum_{\alpha_{-i}} \lambda_i(\alpha_i) u_i(x(\alpha_i, \alpha_{-i}), \alpha_i) p_i(\alpha_{-i} \mid \alpha_i) \geq \sum_{\alpha_{-i}} \lambda_i(\alpha_i) u_i(x(a_i, \alpha_{-i}), \alpha_i) p_i(\alpha_{-i} \mid \alpha_i).$$

Clearly when the scaling factors $\lambda_i(\alpha_i)$ are all positive the BIC property is equivalent to the λ-BIC property.

As already mentioned, most of the results in the literature are obtained under the assumption of unrestricted side-payments with full transferability. This amounts to introduce a good, say "money", perfectly divisible, freely available and transferable, and measured in the common utility unit fixed by the chosen scaling factors. Then for some chosen vector $\lambda \in \Lambda$, for any $\alpha \in A$ and any $x \in X$, if the utility vector $(\lambda_1(\alpha_1)u_1(x, \alpha_1), \lambda_2(\alpha_2)u_2(x, \alpha_2), \cdots, \lambda_n(\alpha_n)u_n(x, \alpha_n))$ is feasible, then any vector of the form $(\lambda_1(\alpha_1)u_1(x, \alpha_1) + t_1, \lambda_2(\alpha_2)u_2(x, \alpha_2) + t_2, \cdots, \lambda_n(\alpha_n)u_n(x, \alpha_n) + t_n)$ where $(t_1, t_2, \cdots, t_n) \in I\!R^n$ and $\sum_{i=1}^n t_i = 0$ is also feasible. In this case, a *(direct revelation) mechanism* is determined by a pair of functions $(x(\cdot), t(\cdot)), x(\cdot) : A \to X$, $t(\cdot) : A \to I\!R^N$, such that $\sum_{i=1}^n t_i = 0$. The function $t(\cdot)$ is called a *transfer function*. The mechanism is Pareto-optimal when the outcome function $x(\cdot)$ is itself Pareto-optimal and $(x(\cdot), t(\cdot))$ has the λ-BIC property when for every $i \in N$, $\alpha_i \in A_i$ and $a_i \in A_i$

$$\begin{aligned} &\sum_{\alpha_{-i}} [(\lambda_i(\alpha_i) u_i(x(\alpha_i, \alpha_{-i}), \alpha_i) + t_i(\alpha_i, \alpha_{-i})] p_i(\alpha_{-i} \mid \alpha_i) \\ \geq \; &\sum_{\alpha_{-i}} ([\lambda_i(\alpha_i) u_i(x(a_i, \alpha_{-i}), \alpha_i) + t_i(a_i, \alpha_{-i})](p_i(\alpha_{-i} \mid \alpha_i). \end{aligned}$$

In the following, as in Shapley (1969), we shall, for every $\lambda \in \Lambda$, consider the possibility of treating the rescaled utilities as fully transferable and, consequently of using transferable utility mechanisms. We can then apply for every $\lambda \in \Lambda$ some of the results obtained in the transferable case.

Let us introduce the following assumptions.

H1. *For every $\alpha \in A$, there exists a selection*

$$x(\alpha, \lambda) \in \text{Arg} \sup_{y \in X} \sum_{i=1}^{n} \lambda_i(\alpha_i) u_i(y, \alpha_i)$$

which is continuous as a function of $\lambda \in \Lambda$.

H2. *For every $i \in N$, every $\alpha_i \in A_i$ and every $a_i \in A_i$* $: p_i(\cdot \mid \dot{a}_i) = p_i(\cdot \mid \alpha_i) = p_i(\cdot)$.

The first assumption is a standard regularity condition. The second assumption states that individual types are uncorrelated and is quite restrictive : it implies that individual beliefs are common knowledge. Our reason for introducing such a restriction is our need of the following lemma derived from theorem 6 of d'Aspremont and Gérard-Varet (1979a).

LEMMA 1. *Under assumptions* **H1** *and* **H2**, *for every $\lambda \in \Lambda$, there exists $(x(\cdot,\lambda), t(\cdot,\lambda))$, a Pareto-optimal λ-BIC transferable utility-mechanism, with transfers given by :*

$$t_i(a,\lambda) = g_i(a_i,\lambda) - \frac{1}{n-1}\sum_{j\neq i} g_j(a_j,\lambda);$$

$$g_i(a_i,\lambda) \stackrel{\text{def}}{=} \sum_{\alpha_{-i}}\sum_{j\neq i}\lambda_j(\alpha_j)u_j(x(a_i,\alpha_{-i},\lambda),\alpha_j)p_j(\alpha_{-i}).$$

Every function $(x(a,\cdot), t(a,\cdot))$ is continuous in $\lambda \in \Lambda$.

III. Pareto-optimal Bayesian incentive compatible outcome functions with an arbitrator

Shapley (1969) proposes a method to extend the application of the "value solution" to bargaining problems involving non-transferable utilities. Complete information and individual rationality conditions were required. We shall now use an analogous method to extend the application of the utilitarian solution to multiperson decision problems involving non-transferable utilities. But we allow for incomplete information and replace the individual rationality condition by incentive compatibility.

By lemma 1 we know that to every vector $\lambda \in \Lambda$ of scaling factors is associated a λ-BIC mechanism with transfers $(x(\cdot,\lambda), t(\cdot,\lambda))$ which is Pareto-optimal (i.e. in particular $\sum_{i-1}^{n} t_i(\cdot,\lambda) \equiv 0$). The idea is to find a vector λ^* in Λ^0, the interior of Λ, such that $t_i(\cdot,\lambda^*)$ be identically zero. This will show that there exists a Pareto-optimal BIC outcome function $x(\cdot,\lambda^*)$ with respect to scaling factors λ^* which are endogenously determined.

Non-dictatorship is a minimal requirement that we want to impose on the scaling factors. As a matter of fact, the BIC property is trivially obtained whenever one individual – the dictator – gets positive weights in every type and all other individuals have zero weights for all their types : the dictator cannot gain by deviating from his truth-telling strategy since by Pareto-optimality he gets his preferred outcome and all other individuals having utilities identically zero cannot gain either by distorting their characteristics. The condition that λ^* belongs to the interior of Λ excludes dictatorship. We shall, at least monentarily, stick to the weaker non-dictatorship condition that λ^* belongs to Λ and is such that for every individual $i \in N$, $\sum_{\alpha i} \lambda_i^*(\alpha_i) > 0$. It does not exclude that an individual receives an arbitrarily large weight similarly to the possibility, in the transferable utility case, that an individual receives an arbitrarily large positive transfer. To obtain weak non-dictatorship though, the next lemma has to rely on an additional condition : at least one individual i (say $i = 1$) has only a single type so that no incentive constraints are required for him.

Lemma 2. *Under* **H1** *and* **H2**, *if $|A_1| = 1$ there exists a Pareto-optimal λ-BIC outcome function $x(\cdot,\lambda^*)$ for some $\lambda^* \in \Lambda$ such that, for every $i \in N$, $\sum_{\alpha_i} \lambda_i^*(\alpha_i) > 0$.*

Proof. Under **H1** and **H2**, one may associate to any $\lambda \in \Lambda$ a Pareto-optimal λ-BIC mechanism with transfers $(x(\cdot,\lambda), t(\cdot,\lambda))$, which is continuous in $\lambda \in \Lambda$. Notice that the λ-BIC property becomes equivalent to

$$\lambda_i(\alpha_i)\overline{u}_i(\alpha_i, a_i, \lambda) + g_i(\alpha_i, \lambda) - g_i(a_i, \lambda) \geq 0, \ \forall\, i \in N, \alpha_i \in A_i, a_i \in A_i,$$

where $\overline{u}_i(\alpha_i, a_i, \lambda) \stackrel{\text{def}}{=} \sum_{\alpha_{-i}} [u_i(x(\alpha_i, \alpha_{-i}, \lambda), \alpha_i) - u_i(x(a_i, \alpha_{-i}, \lambda), \alpha_i)] p_i(\alpha_{-i})$ and

$$g_i(a_i, \lambda) \stackrel{\text{def}}{=} \sum_{\alpha_{-j}} \sum_{j \neq i} \lambda_j(\alpha_j) u_j(x(a_i, \alpha_{-i}, \lambda), \alpha_j) p_j(\alpha_{-j}).$$

Step 1. Since $|A_1| = 1$, one may fix $\lambda_1^* = \varepsilon > 0$. Define

$$\Lambda^1 \stackrel{\text{def}}{=} \{\lambda \in \mathop{\mathsf{X}}_{j \neq 1} I\!R_+^{|A_i|} : \sum_{\alpha_i} \lambda_i(\alpha_i) = 1, i \neq 1\}$$

which is a non empty convex compact subset of $\mathsf{X}_{j \neq 1} I\!R_+^{|A_i|}$. Define also the function $f : \Lambda^1 \to \mathsf{X}_{i \neq 1} I\!R_+^{|A_i|}$ by letting for every $i \neq 1$ and $\alpha_i \in A_i$

$$f_{i\alpha_i} \stackrel{\text{def}}{=} \lambda_i(\alpha_i) + |A_i| g_i(\alpha_i, \lambda) - \sum_{a_i} g_i(a_i, \lambda).$$

Clearly for every $i \neq 1$, we have $\sum_{\alpha_i} f_{i\alpha_i}(\lambda) = \sum_{\alpha_i} \lambda_i(\alpha_i) = 1$. Let now :

$$L = \{\ell \in \mathop{\mathsf{X}}_{i \neq 1} I\!R^{|A_i|} : \ell_i(\alpha_i) \geq -\delta, \alpha_i \in A_i \text{ and } \sum_{\alpha_i} \ell_i(\alpha_i) = 1, i \neq 1\}$$

with $\delta > 0$ large enough so that $f(\Lambda^1) \subset \Lambda$. (Such a δ exists since Λ^1 is compact and each $g_i(\alpha_i, \lambda)$ is continuous in λ). Then extend f to L by letting for every $i \neq 1$ and every $\alpha_i \in A_i$

$$\tilde{f}_{i\alpha_i}(\ell) = f_{i\alpha_i}\left(\frac{\max\{0, \ell_2(\alpha_2)\}}{\sum_{a_2} \max\{0, \ell_2(a_2)\}}, \ldots, \frac{\max\{0, \ell_n(\alpha_n)\}}{\sum_{a_n} \max\{0, \ell_n(a_n)\}} \right).$$

We obtain a continuous function $\tilde{f}$ from a nonempty compact convex subset L of $\mathsf{X}_{i \neq 1} I\!R^{|A_i|}$ into itself. Hence by Brouwer's theorem there is a $\ell^* \in L$ such that $\tilde{f}(\ell^*) = \ell^*$.

Step 2. Suppose now that there is some $(j, \alpha_j), j \neq 1$, $\alpha_j \in A_j$ such that $\ell_j^*(\alpha_j) \neq \lambda_j^*(\alpha_j)$, with for every $i \neq 1$ and $\alpha_i \in A_i$

$$\lambda_i^*(\alpha_i) = \frac{\max\{0, \ell_i^*(a_i)\}}{\sum_{a_i} \max\{0, \ell_i^*(a_i)\}}.$$

Then at some $\hat{\alpha}_j$, we have $\ell_j^*(\hat{\alpha}_j) < \lambda_j^*(\hat{\alpha}_j) = 0$ so that by the definition of $\tilde{f}$,

$$|A_j| g_j(\hat{\alpha}_j, \lambda^*) - \sum_{a_j} g_j(a_j, \lambda^*) = \ell_j^*(\hat{\alpha}_j) < 0 \text{ and } \lambda_j^*(\hat{\alpha}_j) \sum_{a_j} \overline{u}_j(\hat{\alpha}_j, a_j, \lambda^*) = 0.$$

But this is a contradiction to the λ-BIC property. Hence we must have for every $i \neq 1$ and every $\alpha_i \in A_i$, $\ell_i^*(\alpha_i) = \lambda_i^*(\alpha_i)$. On the other hand, by definition of f, we also have, for every $i \neq 1$ and $\alpha_i \in A_i$

$$\ell_i^*(\alpha_i) = \lambda_i^*(\alpha_i) + |A_i| g_i(\alpha_i, \lambda^*) - \sum_{a_i} g_i(a_i, \lambda^*).$$

This implies that for all $i \neq 1$, $g_i(\cdot, \lambda^*)$ is constant. Since $|A_1| = 1$, $g_1(\cdot, \lambda^*)$ is also constant. Therefore we obtain :

$$\forall\, i \in N,\ \forall\, \alpha_i \in A_i,\ \forall\, a_i \in A_i,\ \lambda_i^*(\alpha_i)\overline{u}_i(\alpha_i, a_i, \lambda^*) \geq 0$$

i.e. the λ-BIC property holds without transfers. ∎

The result of this lemma is unsatisfactory because it allows some $\lambda_i(\alpha_i)$ to be zero and thus to the corresponding incentive constraint to vanish : we don't get the BIC property. However, as is known from the study of the Shapley value, it is not possible to ensure all scaling factors to be positive without an additional assumption.[5] The one we shall use is the following :

H1'. *For every* $\alpha \in A$ *and every* $\lambda \in \Lambda$, $x(\alpha, \lambda) \stackrel{\text{def}}{=} \operatorname{Arg\,max}_{y \in X} \sum_{i=1}^{n} \lambda_i(\alpha_i) u_i(y, \alpha_i)$ *defines a function, continuous in* $\lambda \in \Lambda$, *and such that :*

i) for any $i \in N$, $u_i(x(\alpha, \lambda), \alpha_i) = 0$ *if and only if* $\lambda_i(\alpha_i) = 0$ *at any* $\alpha \in A$, *and*

ii) if for some $i \in N$, $\alpha_i \in A_i$, *and* $a_i \in A_i$, $\lambda_i(\alpha_i) = 0 < \lambda_i(a_i)$ *then for any* $\alpha_{-i} \in A_{-i}$ *such that* $\lambda_j(\alpha_j) > 0$ *for some* j *and* α_j, $j \neq i$,

$$\sum_{j \neq i} \lambda_j(\alpha_j) u_j(x(a_i, \alpha_{-i}, \lambda), \alpha_j) < \sum_{j \neq i} \lambda_j(\alpha_j) u_j(x(\alpha_i, \alpha_{-i}, \lambda), \alpha_j).$$

One may observe that : (i) is a 0-normalization property implying a weak individual rationality condition at any Pareto-optimal outcome; (ii) is a "non-levelness" condition in the spirit of Aumann (1985). It amounts to require that choosing a Pareto-optimal outcome be a "strictly competitive" problem : when the weight of an individual is increased, ceteris paribus the utility levels obtained by the others must decrease. It is illustrated by figure 1.

[5] See Aumann, Kurz (1977), and Aumann (1985). To quote : "Shapley's treatment permits some of the comparison weights λ_i to vanish. Ours does not. Vanishing comparisons weights are undesirable for several reasons : in most specific cases it can be shown that the λ_i must be positive though the definition allows them to vanish. Our definition of 'Shapley Value' explicitly takes λ positive, and the nonlevelness condition (3.2) assures that whatever emerges from the axioms will be associated with a positive λ. A verbal statement of (3.2) is that weak and strong Pareto-optimality are equivalent" (Aumann (1985) p. 609).

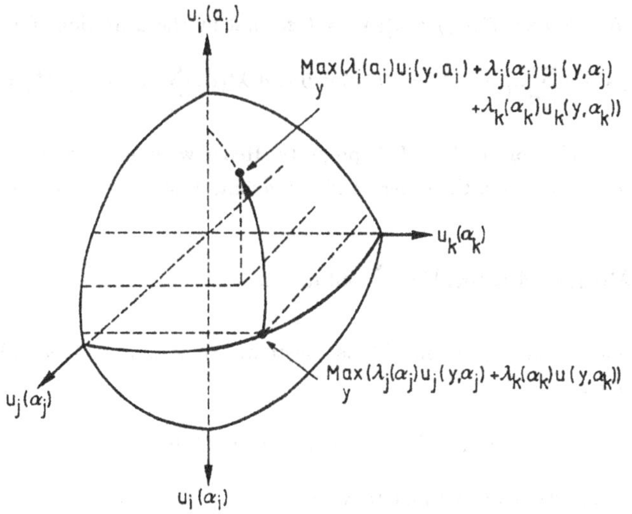

Figure 1

Since **H1'** implies **H1**, we can, assuming **H1'** and **H2** rely on the argument of lemma 2. It shows that there exists some $\lambda^* \in \Lambda$ such that for every $i \in N$, $\alpha_i \in A_i$ and $a_i \in A_i$

$$\lambda_i^*(\alpha_i)\overline{u}_i(\alpha_i, a_i, \lambda^*) + g_i(\alpha_i, \lambda^*) - g_i(a_i, \lambda^*) \geq 0$$

and where $g_i(\cdot, \lambda^*)$ is constant. Let us now show that, at such $\lambda^* \in \Lambda$, we must have $\lambda_i^*(\alpha_i) > 0$ for every $i \in N$ and $\alpha_i \in A_i$. If not, one could find $i \in N$, $i \neq 1$, and $\alpha_i \in A_i$ such that $\lambda_i^*(\alpha_i) = 0 < \lambda_i^*(a_i)$. Thus, using **H1'**, we could have

$$g_i(\alpha_i, \lambda^*) - g_i(a_i, \lambda^*)$$

$$= \sum_{\alpha_{-i}} \sum_{j \neq i} \lambda_j^*(\alpha_j) u_j(x(\alpha_i, \alpha_{-i}, \lambda^*), \alpha_j) p_i(\alpha_{-i}) - \sum_{\alpha_{-i}} \sum_{j \neq i} \lambda_j^*(\alpha_j) u_j(x(a_i, \alpha_{-i}, \lambda^*), \alpha_j) p_i(\alpha_{-i}) > 0$$

which is impossible since $g_i(\cdot, \lambda^*)$ is constant. This actually proves the following :

Proposition 1. *Under* **H1'** *and* **H2**, *if* $|A_1| = 1$ *then there exists a Pareto-optimal BIC outcome function* $x(\cdot, \lambda^*)$ *with* $\lambda^* \in \Lambda$.

It is easy to see that the same arguments hold under a weakening of the condition imposed on individual 1, namely that there exists a continuous function $\lambda_1 : \mathsf{X}_{j \neq 1} \mathbb{R}_+^{A_i} \to \mathbb{R}_{++}^{A_i}$ such that, for every $\lambda_{-1} \in \mathsf{X}_{j \neq 1} \mathbb{R}_+^{A_i}$, every $\alpha_1 \in A$ and $a_1 \in A_1$,

$$\lambda_1(\lambda_{-1})(\alpha_1)\{\sum_{\alpha_{-1}} [u_1(x(\alpha, \lambda_1(\lambda_{-1}), \lambda_{-1}), \alpha_1) - u_1(x(a_1, \alpha_{-1}, \lambda_1(\lambda_{-1}), \lambda_{-1}), \alpha_1)] p_1(\alpha_{-1})\} \geq 0.$$

It says that individual 1 would never need to receive transfers to satisfy the incentive constraints. This means that individual 1 is not really concerned with the incentive problem and thus can play the role of an "arbitrator". We will see now that introducing a symmetry condition in the spirit of Harsanyi's (1955, 1977)[6] allows to get rid of the need for an arbitrator.

IV. Pareto-optimal Bayesian incentive compatible outcome function with symmetry

Let us assume that the set of all types for every individual i coincides with a common set of possible types in the society, say $A_i = B$, and that there is a common NM utility function $u : X \times B \rightarrow I\!R$. A vector α of individual characteristics can be seen as a function from N to B, i.e. as an element of $A = B^N$. A personal utility function is denoted $u(\cdot, \alpha_i)$.

Besides a common utility function, we assume that all individuals "share the same statistical information".[7] Let ν be a probability distribution over B (i.e., $\nu(b) \geq 0$, $b \in B$ and $\sum_b \nu(b) = 1$). All individual beliefs are now assumed to be derived from the product probability $\mathsf{X}_{i=1}^{n} \nu(\alpha_i)$ over A. In particular, player i beliefs are given by $p(\alpha_{-i}) = \mathsf{X}_{j \neq i} \nu(\alpha_j), \alpha_{-i} \in A_{-i} = B^{N-i}$, thus free from any dependence on his own type in A_i, implying **H2**. This means that each individual believes that each one of the others is independently drawn from the given distribution.

Consider now the set Σ of all permutations $\sigma : N \rightarrow N$. We denote α_σ the vector in A of permuted elements of α, i.e. $\alpha_\sigma = (\alpha_{\sigma(i)})_{i \in N}$ where $\alpha_{\sigma(i)} \in B$ is individual's i permuted type. Hence, given any permutation σ and any listing $\alpha \in A$, individuals $i \in N$ and $\sigma(i) \in N$ have the same utility function

$$u(\cdot, \alpha_i) = u(\cdot, \alpha_{\sigma(i)}).$$

Also, since for any $\alpha \in A$ and any $\sigma \in \Sigma$, $\mathsf{X}_{i=1}^{n} \nu(\alpha_i) = \mathsf{X}_{i=1}^{n} \nu(\alpha_{\sigma(i)})$, all individual beliefs are also invariant under a permutation of the personal characteristics. This defines a *symmetric multiperson decision problem.*

We now want to introduce the notion of a "symmetric" outcome function. Given $x(\cdot) : A \rightarrow X$, such a function is *symmetric* if for any $\alpha \in A$, any $a \in A$ and any $i \in N$:

$$u(x(a), \alpha_i) = u(x(a_\sigma), \alpha_{\sigma(i)}).$$

Thus, individual payoffs obtained at a symmetric outcome function are invariant under permutations of the individual types and possible strategic choices. Also define $\overset{\circ}{\Lambda}_B = \{\ell \in I\!R_+^B : \sum_b \ell(b) < 1, \ell(b) > 0, \text{ for all } b \in B\}$. A personal scaling factor is denoted $\ell(\alpha_i)$ and we have, for any $\sigma \in \Sigma$, $\ell(\alpha_i) = \ell(\alpha_{\sigma(i)})$. To simplify notation, we define, for every $\ell \in \overset{\circ}{\Lambda}_B$ and $\alpha \in A$, $\tilde{x}(a, \ell) = x(\alpha, \mathbf{1}_n \ell)$ with $\mathbf{1}_N = (1, 1, \cdots, 1)$.

---

[6]A discussion of this symmetry condition in relation to Harsanyi's fundamental theory of impersonal choice utilitarianism is given in d'Aspremont and Gérard-Varet (1990).

[7]The assumption is used in the incentive literature for large economies. See Hammond (1979) or Champsaur (1989). Harsanyi goes further by assuming a uniform distribution.

Notice that for every $\ell \in \mathring{\Lambda}_B$, $\tilde{x}(\cdot,\ell)$ is a symmetric outcome function whenever $\ell(\alpha_i)u(\tilde{x}(a,\ell),\alpha_i) = \ell(\alpha_{\sigma(i)})u(\tilde{x}(a_\sigma,\ell),\alpha_{\sigma(i)})$.

By restricting to symmetric scaling factors and symmetric outcome functions, we can avoid having an individual playing a special role, as in the previous section where we needed an "arbitrator". We actually may prove the following proposition :

Proposition 2. *Given a symmetric multiperson decision problem satisfying* **H1'**, *there exists a symmetric Pareto-optimal BIC outcome function* $x(\cdot,\ell^*)$ *with* $\ell^* \in \mathring{\Lambda}_B$.

Proof.

Step 1. Take any $\ell \in \mathring{\Lambda}_B$ and let, for every $a \in A$,

$$\tilde{x}(a,\ell) = \text{Arg}\max_{y\in X} \sum_{i=1}^{n} \ell(a_i)u(y,a_i) \text{ resulting from } \mathbf{H1'}.$$

Since, for every $\sigma \in \Sigma$ and every $a \in A$, $\tilde{x}(a,\ell) = \tilde{x}(a_\sigma,\ell)$, we have, for every $i \in N$ and every $\alpha \in A$,

$$\ell(\alpha_i)u(\tilde{x}(a,\ell),\alpha_i) = \ell(\alpha_{\sigma(i)})u(\tilde{x}(a_\sigma,\ell),\alpha_{\sigma(i)}).$$

Define

$$t_i(a,\ell) = g(a_i,\ell) - \tfrac{1}{n-1}\sum_{j\neq i} g(a_j,\ell)$$

where $g(a_i,\ell) = \sum_{a_{-i}}\sum_{j\neq i} \ell(a_j)u(\tilde{x}(a_i,a_{-i},\ell),a_j)p(a_{-i})$. Clearly, for every $\sigma \in \Sigma$, and every $i \in N$, $t_i(a,\ell) = t_{\sigma(i)}(a_\sigma,\ell)$. Thus, to any $\ell \in \mathring{\Lambda}_B$ is associated by lemma 1, a Pareto-optimal BIC mechanism with transfers $(\tilde{x}(\cdot,\ell),t(\cdot,\ell))$ which is symmetric, namely

$$\ell(\alpha_i)u(\tilde{x}(a,\ell),\alpha_i) + t_i(a,\ell) = \ell(\alpha_{\sigma(i)})u(\tilde{x}(a_\sigma,\ell),\alpha_{\sigma(i)}) + t_{\sigma(i)}(a_\sigma,\ell).$$

By **H1'** it is a continuous function of $\ell \in \mathring{\Lambda}_B$.

Step 2. Considering for any $\ell \in \mathring{\Lambda}_B$ the mechanism $(\tilde{x}(\cdot,\ell),t(\cdot,\ell))$, resulting from step 1, define $f : \mathring{\Lambda}_B \to \mathbb{R}^B$ the continuous function such that

$$f_\beta(\ell) = |B|g(\beta,\ell) - \textstyle\sum_{b\in B} g(b,\ell), \beta \in B.$$

If there exists some $\ell^* \in \mathring{\Lambda}_B$ at which $f(\ell^*) = 0$ then, for every $\beta \in B$,

$$g(\beta,\ell^*) = \frac{1}{|B|}\sum_{b\in B} g(b,\ell^*) \stackrel{\text{def}}{=} \overline{g}(\ell^*).$$

This implies that for every $\alpha \in A$, and every $i \in N$, $t_i(\alpha, \ell^*) = \overline{g}(\ell^*) - \frac{n-1}{n-1}\overline{g}(\ell^*) = 0$. Thus we obtain at $\ell^* \in \mathring{\Lambda}_B$ a BIC Pareto-optimal outcome function $x(\cdot, \ell^*)$.

Step 3. It remains to be shown that there is some $\ell^* \in \mathring{\Lambda}_B$ such that $f(\ell^*) = 0$. For that purpose we may apply a theorem of Schulz (1985), using degree theory, and prove the existence of such a ℓ^* by verifying the boundary behavior of the continuous function f. These boundary conditions[8] are that, for any sequence $(\ell^k)_{k\geq 1}$ in $\mathring{\Lambda}_B$ converging to ℓ^0.

a) if $\sum_{b\in B} \ell^0(b) = 1$, then $f_{\beta'}(\ell^k) < 0$ for some $\beta' \in B$ and for infinitely many k;

b) if $\sum_{b\in B} \ell^0(b) < 1$ and $B^0 \equiv \{b \in B \; : \ell^0(b) = 0\} \neq \emptyset$, then $f_{\beta''}(\ell^k) > 0$ for some $\beta'' \in B^0$ and for infinitely many k.

Conditions (a) is easy to verify since, by construction, the function f is such that, for every $k \geq 1$, either there is a $\beta' \in B$ such that $f_{\beta'}(\ell^k) < 0$, or $f_b(\ell^k) = 0$ for all $b \in B$ (which is what we are looking for). Conditions (b) is similarly verified if $B^0 = B$: for every $k \geq 1$, either there is a $\beta'' \in B$ such that $f_{\beta''}(\ell^k) > 0$ or $f_b(\ell^k) = 0$ for all $b \in B$. So consider case (b) with $B^0 \neq B$.

We have to show that for some $\beta'' \in B^0$ and for infinitely many k,

$$f_{\beta''}(\ell^k) = |B|\sum_{\alpha_{-i}} \left[\sum_{j\neq i} \ell^k(\alpha_j) u(x(\beta'', \alpha_{-i}, \ell^k), \alpha_j) p(\alpha_{-i})\right]$$

$$- \sum_{b\in B}\sum_{\alpha_{-i}} \left[\sum_{j\neq i} \ell^k(\alpha_j) u(x(b, \alpha_{-i}, \ell^k), \alpha_j) p(\alpha_{-i})\right] > 0.$$

Although $\ell^0 \notin \mathring{\Lambda}_B$, $x(\alpha, \ell^0)$ is defined whenever α is such that $\ell^0(\alpha_j) > 0$ for some j. Let $A^{\ell^0}_{-i} = \{\alpha_{-i} \in A_{-i} \; : \ell^0(\alpha_j) > 0 \text{ for some } j\}$, by continuity we have :

$$\lim_{k\to\infty} f_{\beta''}(\ell^k) = |B| \sum_{\alpha_{-i}\in A^{\ell^0}_{-i}} \left[\sum_{j\neq i} \ell^0(\alpha_j) u(x(\beta'', \alpha_{-i}, \ell^0), \alpha_j) p(\alpha_{-i})\right]$$

$$- \sum_{b\in B}\sum_{\alpha_{-i}\in A^{\ell^0}_{-i}} \left[\sum_{j\neq i} \ell^0(\alpha_j) u(x(b, \alpha_{-i}, \ell^0), \alpha_j) p(\alpha_{-i})\right] .$$

Letting also $\beta'' \in \operatorname{Arg}\max_{b\in B^0} \sum_{\alpha_{-i}\in A^{\ell^0}_{-i}} \sum_{j\neq i} \ell^0(\alpha_j) u(x(b, \alpha_{-i}, \ell^0), \alpha_j)$ it is sufficient to show that

$$\sum_{b\in B} \left\{ \sum_{\alpha_{-i}\in A^{\ell^0}_{-i}} \sum_{j\neq i} \ell^0(\alpha_j) u(x(\beta'', \alpha_{-i}, \ell^0), \alpha_j) - \sum_{\alpha_{-i}\in A^{\ell^0}_{-i}} \sum_{j\neq i} \ell^0(\alpha_j) u(x(b, \alpha_{-i}, \ell^0), \alpha_j) \right\} > 0.$$

By **H1'** and by the definition of β'', if $b \in B$ is such that $\ell^0(b) = 0$, then the term in brackets in

[8]To prove the theorem, it is sufficient to show that f has the same boundary behavior (and hence the same degree) as the function ρ defined by $\rho_\beta(\ell) = 1 - |B|\ell(\beta) - \sum_b \ell(b)$ the degree of which is nonzero.

the above expression is non negative. If $b \in B$ is such that $\ell^0(b) > 0$ (and there is at least one such b), then for every $\alpha_{-i} \in A^{\ell^0}_{-i}$,

$$\sum_{j \neq i} \ell^0(\alpha_j) u(x(\beta'', \alpha_{-i}, \ell^0), \alpha_j) - \sum_{j \neq i} \ell^0(\alpha_j) u(x(b, \alpha_{-i}, \ell^0), \alpha_j) > 0.$$

The result follows.

Conclusion

This paper shows that to obtain a Pareto-optimal BIC mechanism without transfers and without relying on either a "dictator" or an "arbitrator", one could restrict to symmetric mechanisms, designed for symmetric multiperson decision problems under incomplete information. Our notion of a symmetric multiperson situation has however been limited to the case where all individuals believe that any other is independently drawn from a given distribution, namely to the so-called "independent case".

A more interesting world arises by assuming that all individual beliefs are consistently derived from a symmetric joint probability distribution Π (also called "exchangeable") over the set of possible listings of types $A = B^N$, i.e. such that for any permutation σ on N and every $\alpha \in A$, $\Pi(\alpha) = \Pi(\alpha_\sigma)$. In this more general world the beliefs of an individual are no longer independent of his own type, an assumption which, if taken into account, would allow to enlarge the set of possible interpretations of the results. To take an example if we think of X as the set of possible allocations of some given bundle of resources and if the private ownership on this bundle is part of the characteristics, it seems reasonable to consider that the level of the initial endowment of an individual affects the probability that this individual assigns to the others initial endowments. However, extending the present results to these enlarged situations is an open problem. All our arguments have actually been based on lemma 1 which as such, do not extend to the case where the individuals beliefs are dependent on their own types. Even though there are results obtained in the transferable utility case without independence, in order to extend the property to the non-transferable case without independence one would have to show that monetary transfers are continuous in the scaling factor λ and are symmetric in the symmetric case.

A second set of observations is about "individual rationality". All our arguments are based on the mechanism of theorem 6 in d'Aspremont, Gérard-Varet (1979a) for which, in the transferable utility case, the BIC (or λ-BIC) property is obtained using Pareto-optimality without individual rationality (in the classical sense of ensuring a certain minimal utility level usually interpreted as the utility level the individual would get without participation). In the results obtained here for the non-transferable utility case, a very weak form of individual rationality is ensured by the assumption **H1'**, since, combining a zero-normalization and a non-levelness condition, it ensures a positive weight to every individual of any type and, hence, a positive utility level. It is very weak though, because the assumption implies that, for any type configuration, any Pareto-optimal payoff is individually rational. It is more interesting to notice that both for the non-transferable utility case and for the transferable utility case, usual individual rationality constraints are replaced by individual incentive constraints. In particular in lemma 2 the incentive constraints play a role which is analogous to the role played by the individual incentive constraints in Shapley's (1969) argument.

Classical individual rationality is based on some given reference point, the status-quo point. Here we have not introduced such a point and assumption **H1'** in effect eliminates the possibility of such a point. One might support this general approach on ethical grounds. In that perspective, it might seem justified (following Harsanyi's approach) to impose, in addition, symmetry, in the sense that the relevant features of an individual be completely characterized by his type and that the solution of the problem be only a function of individual characteristics. Indeed this kind of anonymity avoids putting "excessive" weight on some individual in relation to his name (or number) only. And this is certainly a desirable property from a normative point of view. However, it does not exclude that some type receives "excessive" weight. But this is more difficult to assess. Although it would seem justified to give different weights to different individual characteristics, one would have to specify which characteristics receive what weight. Then nothing ensures that the weights distributions which are ethically admissible are also incentive compatible.

References

Aumann, R.J. (1985), "An Axiomatization of the Non-transferable Utility Value", *Econometrica*, 53, 599–612.

Aumann, R.J. and M. Kurz (1977), "Power and Taxes", *Econometrica*, 45, 1137–1161.

Champsaur, P. (1989), "Information, Incentives, and General Equilibrium", in *Contributions to Operations Research and Economics* (Eds. B. Cornet and H. Tulkens). Cambridge : MIT Press, 23–44.

d'Aspremont, C., Crémer, J. and L.-A. Gérard-Varet (1990), "Incentives and the Existence of Pareto-optimal Revelation Mechanisms", *Journal of Economic Theory*, 51, 233–254.

d'Aspremont, C. and L.-A. Gérard-Varet (1979a), "Incentives and Incomplete Information", *Journal of Public Economics*, 11, 25–45.

d'Aspremont, C. and L.-A. Gérard-Varet (1979b), "On Bayesian Incentive Compatible Mechanisms", chapter 15 in *Aggregation and Revelation of Preferences* (Ed. J.J. Laffont). Amsterdam : North-Holland, 269–288.

d'Aspremont, C. and L.-A. Gérard-Varet (1990), "Utilitarian Fundamentalism and Limited Information", in *International Comparisons of Well-Being* (Eds. J. Roemer and J. Elster). Cambridge : Cambridge University Press.

Green, J. and J.J. Laffont (1979), "*Incentives in Public Decision Making*", Amsterdam : North-Holland.

Groves, T. and J.O. Ledyard (1987), "Incentive Compatibility since 1972", in *Information, Incentive and Economic Mechanisms. Essays in Honor of L. Hurwicz* (Eds. T. Groves, R. Radner and S. Reiter). Oxford : Blackwell, 48–111.

Hammond, P.J. (1979), "Straightforward Individual Incentive Compatibility in Large Economies", *Review of Economic Studies*, 46, 263–282.

Harsanyi, J.C. (1955), "Cardinal Welfare, Individualistic Ethics and Interpersonal Comparisons of Utility", *Journal of Public Economics*, 63, 309–321.

Harsanyi, J.C. (1967-68), "Games of Incomplete Information Played by 'Bayesian' Players", Part I-III, *Management Science*, 14, 159–182, 320–334, 486–502.

Harsanyi, J.C. (1977), "*Rational Behavior and Bargaining Equilibrium in Games and Social Situations*". Cambridge : Cambridge University Press.

Johnson, S., Pratt, J.W. and R.J. Zeckhauser (1990), "Efficiency Despite Mutually Payoff-Relevant Private Information : The Finite Case", *Econometrica*, 58, 873–900.

Myerson, R.B. (1985), "Bayesian Equilibrium and Incentive-Compatibility : an Introduction", chapter 8 in *Social Goals and Social Organization.* (Eds. L. Hurwicz, D. Schmeidler and H. Sonnenschein). Cambridge : Cambridge University Press, 229–259.

Schulz, N. (1985), "Existence of Equilibria Based on Continuity and Boundary Behaviour", *Economic Letters*, 19, 101–103.

Shapley, L.S. (1969), "Utility Comparison and the Theory of Games", in *La décision. Aggrégation et dynamique des ordres de préférence.* Paris : Editions du CNRS, 251–263.

III. Basic Questions of Non-Cooperative Game Theory

Equilibrium Selection by Unilateral Deviation Stability*

by

Werner Güth
University of Frankfurt
D–6000 Frankfurt, Germany

Contents

Abstract: From its very beginning, the theory of equilibrium selection has been based on constructive solution procedures as well as on axiomatic characterizations of solutions. HARSANYI and SELTEN have suggested a set of convincing axioms which allows to select a unique solution in all 2 × 2–bimatrix games with two strict equilibria. This paper tries to generalize this approach. Our concept of unilateral deviation stability gives priority to risks of unilateral deviations, i.e. deviations from an intended behavior by single players. We investigate the general properties of unilateral deviation stability and consider some crucial examples.

---

*I would like to thank John C. Harsanyi for encouraging and guiding my work in game theory and for his and Anne Harsanyi's friendship.

I. Introduction

Since the development of equilibrium selection theory is described in GÜTH and KALKOFEN (1989), we will only briefly discuss how our approach is embedded in the tradition of equilibrium selection theory. NASH (1950, 1953) can be seen as the founder of equilibrium selection theory since he first explicitly proposed to solve every (cooperative) game by selecting one of its equilibria as the solution. He, furthermore, proposed an ingenious solution for all unanimity bargaining games which provided an excellent point of orientation for later and more general concepts (see HARSANYI and SELTEN, 1988, as well as GÜTH and KALKOFEN, 1989, and also Section IV.a of this paper). To justify his solution NASH provided a constructive solution procedure als well as an axiomatic characterization of its result (see GÜTH and KALKOFEN, 1989, for details).

Although the later and more general concepts have been primarily based on constructive solution procedures, much effort has been spent to explore also their general properties. The major recent result along these lines seems to be the axiomatic solution for all 2 × 2–bimatrix games with two strict equilibria (HARSANYI and SELTEN, 1988). It is the main purpose of this paper to generalize the axiomatic approach of HARSANYI and SELTEN to all noncooperative games.

In a strictly noncooperative game all players have to determine their decisions independently. If coordination of strategic moves is possible, coordination activities would have to be captured as strategic possibilities by the rules of the game, i.e. the decision whether to coordinate or not would have to be chosen independently by each player. Thus the event that two or more players simultaneously deviate from an intended behavior is highly unlikely in a strictly noncooperative game. Our concept of unilateral deviation stability therefore gives priority to unilateral deviations, i.e. it primarily reflects the risks of deviations by single players form the potential solutions.

By building on the HARSANYI and SELTEN–axioms for 2 × 2–bimatrix games with two strict equilibria, our concept can be shown to satisfy some intuitively convincing general properties although its basic idea is a simple constructive procedure for solving a game. Due to its priority for risks of unilateral deviations, it does not reflect all strategic aspects. Even if this would be viewed as a major deficiency, the concept of unilateral deviation stability may serve as a unique point of reference for judging and comparing other theories of equilibrium selection. The main advantage of unilateral deviation stability is its straightforward constructive idea which seems to be a promising starting point for a purely axiomatic approach of equilibrium selection.

In the following we first discuss the axiomatic approach of HARSANYI and SELTEN (1988) for 2 × 2–bimatrix games with two strict equilibria. These axioms can be used to define the relative strength of one solution candidate on which unilateral deviation stability is based in the comparison games with just two active players. We then introduce ways to aggregate relative strength over the possible comparison games. With the notion of total relative strength we then define maximally stable solution candidates. After elaborating the general properties of unilateral deviation stability, we consider some crucial examples of games with multiple equilibria, namely unanimity bargaining games and market entry games, in order to test and illustrate the selection behavior of unilateral deviation stability. It will be shown that only one of the proposed aggregation methods selects the cooperative NASH–solution in unanimity (bargaining) games. We therefore define unilateral deviation stability by the maximally stable solution candidate according to this aggregation method.

II. The Harsanyi and Selten–axioms for 2 × 2–bimatrix games with two strict equilibria

With no loss of generality every 2 × 2–bimatrix game with two strict equilibria can be represented as follows:

1 \ 2	s_2^1	s_2^2
s_1^1	a, e	d, f
s_1^2	b, h	c, g

$a > b$, $c > d$, $e > f$, and $g > h$

Table II.1: The general model of a 2 × 2–bimatrix game with two strict equilibria $s^1 = (s_1^1, s_2^1)$ and $s^2 = (s_1^2, s_2^2)$

Player 1(2)'s payoff appears in the upper left (lower right) corner of the strategy vector, respectively matrix component $s = (s_1, s_2)$. A strategy vector is an <u>equilibrium point</u> if no player gains by unilaterally deviating from it. If every such unilateral deviation implies a loss for the deviating player, the equilibrium point is said to be <u>strict</u>. Due to our parameter restrictions the game of Table II.1 has two strict equilibria, namely $s^1 = (s_1^1, s_2^1)$ and $s^2 = (s_1^2, s_2^2)$. In addition to s^1 and s^2 the game has a completely mixed and therefore non–strict equilibrium $q = (q_1, q_2)$ with

$$\frac{q_1(s_1^1)}{q_1(s_1^2)} = \frac{g - h}{e - f} \tag{II.1}$$

and

$$\frac{q_2(s_2^1)}{q_2(s_2^2)} = \frac{a - b}{c - d} \tag{II.2}$$

where $q_j(s_j)$ is the probability for choosing the pure strategy s_j. q will be selected as the solution only if any distinction between s^1 and s^2 would be completely arbitrary. This already indicates that the selection of the solution is based on a hierarchy of selection principles in the sense that non-arbitrariness is more important than strictness.

In their pioneering book on equilibrium selection theory, HARSANYI and SELTEN (1988, Section 3.9) propose a very convincing set of axioms to select among the two strict equilibria s^1 and s^2, except for the degenerate cases when only an arbitrary selection among s^1 and s^2 would be possible and when q becomes the solution of the game. We say that two games G and G′ in normal form are <u>isomorphic</u> if G′ can be derived from G by renaming players and/or strategies as well as by positive affine transformations of the players' payoff functions. Let f be such an isomorphism mapping G into G′ and let f(s) be the image of strategy vector s of G in the game G′ according to this isomorphism:

Axiom II.1: "Independence of isomorphic transformations"
If G and G′ are isomorphic, i.e. there exists an isomorphism f such that $G' = f(G)$, and if s is the solution of G, then the solution of G′ is f(s), i.e. the image of s according to the isomorphism f relating G and G′.

Since names of players and strategies are arbitrary and since cardinal utility functions are only uniquely defined up to positive affine transformations, every non-arbitrary selection procedure will have to satisfy Axiom II.1. This, of course, implies the choice of q in case of the pure coordination game with

$$a = c,\ e = g,\ b = d, \text{ and } f = h, \tag{II.3}$$

what can be shown as follows: Assume that s^1 would be the solution of game G in Table II.1 with the additional assumption (II.3). Interchanging the names of strategies of both players yields the same game although this isomorphism f determines a different solution in the game f (G), namely the lower right corner of f (G) whereas the solution of G is the upper left corner. If (II.3) holds, the only possibility to satisfy Axiom II.1 is to select q as the solution.

For a strategy vector s the <u>stability set</u> R (s) is the set of all mixed strategy vectors q to which s is a best reply, i.e. every component s_i of s is a best reply to the behavior $q_{-i} = (q_1,\ldots,q_{i-1},q_{i+1},\ldots,q_n)$ of all other players according to q. Two n-person games $G = (S_1,\ldots,S_n;H)$ and $G' = (S_1,\ldots,S_n;H')$ with the same pure strategy sets $S_1,\ldots,S_n$ but different payoff functions $H(s) = (H_1(s),\ldots,H_n(s))$ and $H'(s) = (H_1'(s),\ldots,H_n'(s))$, respectively, have the same <u>best reply structure</u> if for all $s \in S = S_1 \times \ldots \times S_n$ the stability sets R(s) in G and R′(s) in G′ coincide. A solution is called <u>best reply invariant</u> if it selects the same solution in games which have the same best reply structure:

Axiom II.2: "Best reply invariance"

If for $G = (S_1...,S_n;H)$ and $G' = (S_1,...,S_n;H')$ one has $R(s) = R'(s)$ for all strategy vectors s and if s^* is the solution of G, then the solution of G′ is also the strategy vector s^*.

Since an isomorphic transformation clearly preserves the best reply structure, best reply invariance and independence of isomorphic transformations are mutually consistent.

Let $G = (S_1,...,S_n;H)$ be a game in normal form with multiple strict equilibria $s^1,...,s^m$ in which neither s^2, nor $s^3,...$, nor s^m is the solution, i.e. the solution is either s^1 or some non–strict equilibrium. The game $G' = (S_1,...,S_n;H')$ with $H'(s) = H(s)$ except for

$$H_i'(s^1) > H_i(s^1) \quad \text{for some player i with } s_i^k \neq s_i^j \ (j \neq k;\ j,k=1,...,m) \tag{II.3}$$

is said to result from G by <u>strengthening</u> the solution candidate s^1. A solution concept is <u>monotonic</u> if strengthening s^1 makes s^1 the solution of G′.

Axiom II.3: "Monotonicity"

If $G = (S_1,...,S_n;H)$ and $G'(S_1,...,S_n;H')$ are two games in normal form such that G′ results from G by strengthening the strict equilibrium s^* of G and if no strict equilibrium other than s^* is the solution of G, then s^* is the solution of G′.

Strengthening a candidate is not an isomorphic transformation, nor does it preserve the best reply structure. Monotonicity is therefore consistent with best reply invariance and independence of isomorphic transformations.

For the class of 2 × 2–bimatrix games G with two strict equilibria there exists only one solution function φ which selects a unique solution $\varphi(G)$ for every such game and which satisfies the three porperties, Axioms II.1, II.2, and II.3. In the following we will determine this solution for the general representation of such a game G as given by Table II.1.

The game in Table II.2 is derived from Table II.1 by subtracting for each pure strategy of the other player the own non–equilibrium payoff. So, player 1′s payoffs in the first (second) column of Table II.2 result from substracting the non–equilibrium payoff b (d). Similarly, player 2's payoff entries in the first (second) row of Table II.2 are determined by subtracting the non–equlibrium payoff f (h). According to equations (II.1) and (II.2) the mixed strategy equilibrium $q = (q_1,q_2)$ depends only on the differences of the equilibrium and the non–equilibrium payoff for the two pure strategies of the coplayer, i.e. equations (II.1) and (II.2) determine also the mixed strategy equilibrium for the game of Table II.2. Thus the game of Table II.2 has the same stability sets for all pure strategy vectors as the one of Table II.1. Due to best reply invariance the solution of Table II.1 can be determined by solving the unanimity bargaining game of Table II.2:

1 \ 2	s_2^1	s_2^2
s_1^1	a–b ; e–f	0 ; 0
s_1^2	0 ; 0	c–d ; g–h

Table II.2: A best reply–invariant version of the game in Table II.1.

1 \ 2	s_2^1	s_2^1
s_1^1	$\frac{a-b}{c-d}$; 1	0 ; 0
s_1^2	0 ; 0	1 ; $\frac{g-h}{e-f}$

Table II.3: An isomorphic version of the game in Table II.2.

Due to $c > d$ and $e > f$ the game in Table II.3 results from the one in Table II.2 by dividing player 1's (2's) payoff by the positive number $c - d$ (respectively $e-f$), i.e. by a positive affine transformation of payoffs. Because of independence of isomorphic transformations and best reply invariance, we can solve the game of Figure II.1 by solving the one of Table II.3.

Now for

$$\frac{a-b}{c-d} = \frac{g-h}{e-f} \tag{II.4}$$

invariance with respect to isomorphic transformations implies the mixed strategy solution $q = (q_1,q_2)$, described in (II.1) and (II.2), since the game is completely symmetric. Since strengthening s^1 (or s^2) implies $\frac{a-b}{c-d} > \frac{g-h}{e-f}$ (or the reverse inequality), monotonicity determines the solution $s^1 = (s_1^1,s_2^1)$ for

$$\frac{a-b}{c-d} > \frac{g-h}{e-f} \tag{II.5}$$

and $s^2 = (s_1^2,s_2^2)$ for

$$\frac{a-b}{c-d} < \frac{g-h}{e-f}. \qquad \text{(II.6)}$$

This proves

Theorem II.1: (HARSANYI and SELTEN, 1988, p. 87):
There is only one solution for the game of Table II.1 satisfying axioms II.1, II.2, and II.3, namely the one which selects s^1 in the parameter region (II.5), s^2 in the parameter region (II.6), and $q = (q_1,q_2)$. defined by (II.1) and (II.2), in case of (II.4).

Since the solution according to their general theory of equilibrium theory (HARSANYI and SELTEN, 1988) can differ from the one implied by Theorem II.1, the solution, implied by axioms II.1, II.2, and II.3, will be called the axiomatic Harsanyi and Selten–solution.

The game in Table II.2, and therefore also in Table II.3, is a unanimity game for which the bargaining solution, proposed by NASH (1950, 1953), coincides with the one described by the above Theorem except for the degenerate case (II.4). Since the NASH–bargaining solution is widely accepted, this strongly supports the solution implied by the three axioms (II.1), (II.2), and (II.3).

In our view, independence of isomorphic transformations and best reply invariance are indispensable requirements of every concept of individually rational decision behavior. Furthermore, monotonicity is very convincing and an intuitively desirable rationality requirement. HARSANYI and SELTEN (1988, Section 3.8) themselves criticise montonicity. But their argument is based on correlated behavior (see HARSANYI and SELTEN, 1988, p. 85). In our view, intuition about individually rational decisions should not depend on strategic aspects which only become relevant if players correlate their decisions which in a strictly noncooperative game is impossible (see also GÜTH, 1985, and GÜTH and KALKOFEN, 1989). Unlike the general theory of equilibrium selection proposed by HARSANYI and SELTEN (1988), which does not satisfy the monotonicity requirement, our concept of unilateral deviation stability is based on the intuitively desirable monotonicity axiom (the ESBORA–concept of GÜTH and KALKOFEN (1989) also satisfies a monotonicity property). In the following we will describe a straightforward and – in our view – convincing approach to generalize the axiomatic HARSANYI and SELTEN–solution for 2×2–bimatrix games with two strict equilibria to all finite games.

III. The concept of unilateral deviation stability

Let $G = (S_1,...,S_n;H)$ be an n–person normal form game with finite strategy sets $S_i (i = 1,...,n)$ and payoff function H assigning a payoff vector

$$H(s) = (H_1(s),...,H_n(s)) \tag{III.1}$$

to every strategy vector $s = (s_1,...,s_n)$ in G. As before we only consider strict equilibrium points as initial solution candidates and want to define a solution function φ which is defined over the set $\mathscr{G}$ of all games G for which the set of strict equilibria is non-empty. Non-strict solution candidates will be considered in Section V.

Since the strategy sets are finite, the set of strict equilibria for all $G \in \mathscr{G}$ is finite and non-empty. If a game G in $\mathscr{G}$ has a unique strict equilibrium point, this is the only solution candidate and therefore the solution of G. Let $\mathscr{G}^*$ denote the subset of $\mathscr{G}$ such that all games G in $\mathscr{G}^*$ have more than one strict equilibrium point. We determine a unique solution for all games G in $\mathscr{G}^*$ by generalizing the axiomatic approach of HARSANYI and SELTEN for 2×2-bimatrix games with two strict equilibria. For any game $G \in \mathscr{G}^*$ let C(G) denote the set of strict equilibria. We say that C(G) is the candidate set of G.

Our concept is rather straightforward and intuitively convincing. But, of course, it primarily reflects only strategic risks related to unilateral deviations from a candidate. In a strictly noncooperative game, deviations by more than one player are very unlikely and should therefore be of no primary concern. Even if one does not want to rely on such an extreme noncooperative point of view, it seems worthwhile to elaborate its consequences in order to compare them to those of less extreme, but in our view also less consistent concepts (e.g. HARSANYI and SELTEN, 1988, and GÜTH and KALKOFEN, 1989).

We will first define comparison games for general comparisons of two solution candidates to which the axiomatic HARSANYI and SELTEN-solution is directly applicable. In Section b) we propose three ways of aggregating the relative strength of one solution candidate over all possible comparison games. We do not argue strongly for one versus the other aggregation method since this can be more easily done with the help of the results which they imply for the examples considered in Section IV. Our method to compare relative strength is described in Section c).

III.a) The notion of relative strength

For G in $\mathscr{G}^*$ let $\tilde{s} = (\tilde{s}_1,...,\tilde{s}_n)$ and $\hat{s} = (\hat{s}_1,...,\hat{s}_n)$ be two different strict equilibria. For a strategy vector $s = (s_1,...,s_n)$ the n-1-tuple of strategies $(s_1,...,s_{i-1}, s_{i+1},...,s_n)$ is denoted by s_{-i}. The players in

$$M(\tilde{s},\hat{s}) = \{i = 1,...,n : \tilde{s}_i \neq \hat{s}_i\} \tag{III.2}$$

are called the active players for the comparison of $\tilde{s}$ and $\hat{s}$. If $\tilde{s}$ and $\hat{s}$ are the only solution candidates, only players in $M(\tilde{s},\hat{s})$ are troubled by the strategic uncertainty whether to behave according to $\tilde{s}$ or $\hat{s}$. For two different players i and j in $M(\tilde{s},\hat{s})$ we define the game $G_{ij}(\tilde{s},\hat{s}) = (S_i',S_j';H')$ for comparing $\tilde{s}$ and $\hat{s}$ by

$$S_k' = \{\tilde{s}_k,\hat{s}_k\} \quad \text{for } k = i,j \tag{III.3}$$

and

$$H'(s') = (H_i'(s'),H_j'(s')) \tag{III.4}$$

for all $s' = (s_i',s_j') \in S_i' \times S_j'$ with

$$H_k'(s_i',s_j') = \begin{cases} H_k(\tilde{s}) & \text{if } s_i' = \tilde{s}_i,\ s_j' = \tilde{s}_j \\ H_k(\tilde{s}_k,\hat{s}_{-k}) & \text{if } s_k' = \tilde{s}_k,\ s_l' = \hat{s}_l \\ H_k(\hat{s}_k,\tilde{s}_{-k}) & \text{if } s_k' = \hat{s}_k,\ s_l' = \tilde{s}_l \\ H_k(\hat{s}) & \text{if } s_i' = \hat{s}_i,\ s_j' = \hat{s}_j \end{cases} \tag{III.5}$$

for both players $k = i,j$ where l always denotes the other player $l \in \{i,j\}$. $G_{ij}(\tilde{s},\hat{s})$ is called the comparison game of $\tilde{s}$ and $\hat{s}$ with players i and j.

The basic idea of the comparison game $G_{ij}(\tilde{s},\hat{s})$ is that both players assume that all other players know the solution and behave accordingly. This explains why player k thinks that all other players in $M(\tilde{s},\hat{s})$ will choose the same strategy as player l. A similar argument is used to defend the definition of bicentric priors which is a central element of the HARSANYI and SELTEN–theory of equilibrium selection (HARSANYI and SELTEN, 1988).

One might object against the definition of the game $G_{ij}(\tilde{s},\hat{s})$ that it relies on correlated behavior since player k thinks that all other players behave according to the same solution candidate. But this is not our interpretation of the game $G_{ij}(\tilde{s},\hat{s})$. In the definition (III.4) all strategy vectors are either elements of C(G) or differ from elements of C(G) by at most one unilateral deviation. Thus the payoff function H_k' essentially measures how player k suffers if he is the only one who deviates from a commonly expected behavior.

Since $\tilde{s}$ and $\hat{s}$ are strict equilibria in G, the two pure strategy combinations $\tilde{s}' = (\tilde{s}_i,\tilde{s}_j)$ and $\hat{s}' = (\hat{s}_i,\hat{s}_j)$ are strict equilibria in the comparison game $G_{ij}(\tilde{s},\hat{s})$ of $\tilde{s}$ and $\hat{s}$ with players i and j. Thus the comparison game $G_{ij}(\tilde{s},\hat{s})$ is a 2 × 2–bimatrix game with two strict equilibria.

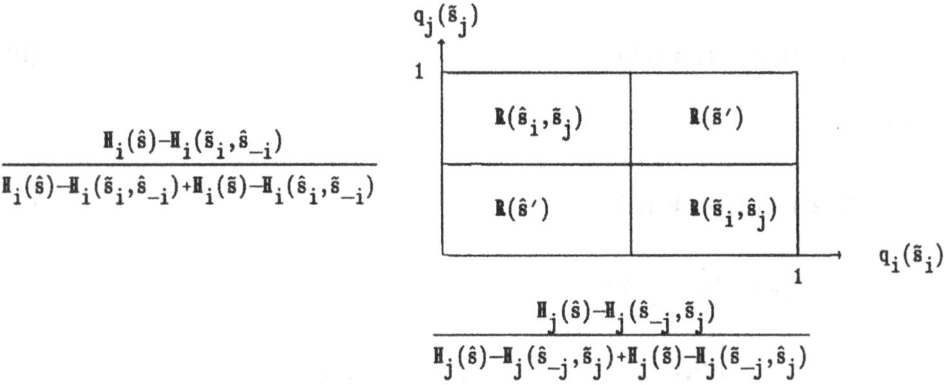

Figure III.1: The stability regions of all four pure strategy combinations in $G_{ij}(\tilde{s},\hat{s})$

For the game $G_{ij}(\tilde{s},\hat{s})$ the stability regions of all four pure strategy vectors $s' = (s_i',s_j')$ are graphically visualized in Figure III.1. The point where all four stability regions meet is the mixed strategy equilibrium point corresponding to equations (II.1) and (II.2). The stability region $R(\tilde{s}')$ is greater than the stability region $R(\hat{s}')$ if condition

$$\frac{H_j(\tilde{s}) - H_j(\tilde{s}_{-j},\hat{s}_j)}{H_j(\hat{s}) - H_j(\hat{s}_{-j},\tilde{s}_j)} > \frac{H_i(\hat{s}) - H_i(\tilde{s}_i,\hat{s}_{-i})}{H_i(\tilde{s}) - H_i(\hat{s}_i,\tilde{s}_{-i})} \tag{III.6}$$

is satisfied. The converse inequality

$$\frac{H_j(\tilde{s}) - H_j(\tilde{s}_{-j},\hat{s}_j)}{H_j(\hat{s}) - H_j(\hat{s}_{-j},\tilde{s}_j)} < \frac{H_i(\hat{s}) - H_i(\tilde{s}_i,\hat{s}_{-i})}{H_i(\tilde{s}) - H_i(\hat{s}_i,\tilde{s}_{-i})} \tag{III.7}$$

is the condition that $R(\hat{s}')$ is larger than $R(\tilde{s}')$. Thus the two conditions (II.5) and (II.6), determining according to axioms II.1, II.2, and II.3 whether $\tilde{s}'$ or $\hat{s}'$ is the solution of $G_{ij}(\tilde{s},\hat{s})$, are nothing else than the conditions (III.6) and (III.7) comparing the size of the stability regions $R(\tilde{s}')$ and $R(\hat{s}')$. This justifies the following definition of the relative strength of $\tilde{s}$ against $\hat{s}$ in the game $G_{ij}(\tilde{s},\hat{s})$:

Definition III.1: The relative strength $R_{ij}(\tilde{s},\hat{s})$ of the strict equilibrium $\tilde{s}$ against $\hat{s}$ in the comparison game $G_{ij}(\tilde{s},\hat{s})$ is the relation of the size of $R(\tilde{s})$and the size of $R(\hat{s})$, i.e.

$$R_{ij}(\tilde{s},\hat{s}) = \frac{(H_j(\tilde{s}) - H_j(\tilde{s}_{-j},\hat{s}_j))\ (H_i(\tilde{s}) - H_i(\hat{s}_i,\tilde{s}_{-i}))}{(H_j(\hat{s}) - H_j(\hat{s}_{-j},\tilde{s}_j))\ (H_i(\hat{s}) - H_i(\tilde{s}_i,\hat{s}_{-i}))} \tag{III.8}$$

With the help of this definition Theorem II.1 can be restated by saying that, according to axioms II.1, II.2, and II.3, the unique solution of the bimatrix game G in Table II.1 is s^1 if $R_{12}(s^1,s^2)$ is greater than 1, the mixed strategy equilibrium q if $R_{12}(s^1,s^2)$ is equal to 1, and s^2 if $R_{12}(s^1,s^2)$ is smaller than 1. In the 2 × 2–bimatrix games $G_{ij}(\tilde{s},\hat{s})$ with two strict equilibria $\tilde{s}'$ and $\hat{s}'$, the relative strength $R_{ij}(\tilde{s},\hat{s})$ therefore measures the relative opposition against $\hat{s}$ in view of the three rationality principles, independence of isomorphic transformation, best reply invariance, and monotonicity. In the following we define the total relative strength $R(\tilde{s},\hat{s})$ in the original game G by aggregating the values $R_{ij}(\tilde{s},\hat{s})$ over all comparison games $G_{ij}(\tilde{s},\hat{s})$ with $i,j \in M(\tilde{s},\hat{s})$.

III.b) Aggregation of relative strength

A natural way of aggregating the values $R_{ij}(\tilde{s},\hat{s})$ for all pairs i and j of active players in $M(\tilde{s},\hat{s})$ is to sum up all values $R_{ij}(\tilde{s},\hat{s})$, i.e.

$$R_+(\tilde{s},\hat{s}) = \sum_{\substack{i,j\in M(\tilde{s},\hat{s})\\ i>j}} R_{ij}(\tilde{s},\hat{s}). \tag{III.b.1}$$

Another obvious possibility would be to multiply the values $R_{ij}(\tilde{s},\hat{s})$, i.e.

$$R_*(\tilde{s},\hat{s}) = \prod_{\substack{i,j\in M(\tilde{s},\hat{s})\\ i>j}} R_{ij}(\tilde{s},\hat{s}). \tag{III.b.2}$$

Since the $R_{ij}(\tilde{s},\hat{s})$ – values depend only on particular aspects of the best reply structure, namely the ones concerning unilateral deviations from solution candidates, both aggregation methods are best reply invariant, i.e. for two best reply invariant games G and G′ and any two strict equilibria $\tilde{s}$, $\hat{s} \in C(G) \cap C(G')$ the total relative strength of $\tilde{s}$ against $\hat{s}$ is the same for both games regardless whether we use the aggregation $R_+(\cdot,\cdot)$ or $R_*(\cdot,\cdot)$. Strengthening $\tilde{s}$ by increasing $H_k(\tilde{s})$ of an active player $k \in M(\tilde{s},\hat{s})$ clearly increases $R_+(\tilde{s},\hat{s})$ and $R_*(\tilde{s},\hat{s})$ and decreases $R_+(\hat{s},\tilde{s})$ and $R_*(\hat{s},\tilde{s})$. Selecting $\tilde{s}$ out of $C(G) = \{\tilde{s},\hat{s}\}$ as the solution of G in case of $R_+(\tilde{s},\hat{s}) > R_+(\hat{s},\tilde{s})$ or $R_*(\tilde{s},\hat{s}) > R_*(\hat{s},\tilde{s})$ is therefore in line with the HARSANYI and SELTEN–axioms II.1, II.2, and II.3 where, of course, monotonicity has to be restricted to active players (in the class of games considered in Section II, both players are active). Both aggregation methods thus generalize the axiomatic HARSANYI and SELTEN–solution to n–person games G with just two strict equilibria as solution candidates.

In addition to the two aggregation methods described above, we also want to consider a weighted aggregation method

$$R_w(\tilde{s},\hat{s}) = \sum_{\substack{i,j \in M(\tilde{s},\hat{s}) \\ i > j}} P_{ij}(\tilde{s},\hat{s})\, R_{ij}(\tilde{s},\hat{s}) \tag{III.b.3}$$

The weights should take into account whether other players are bothered by unilateral deviations of player i or j. For every pair $i,j \in M(\tilde{s},\hat{s})$ and any two solution candidates $\tilde{s},\hat{s} \in C(G)$ the weight $P_{ij}(\tilde{s},\hat{s})$ can be defined as follows:

$$P_{ij}(\tilde{s},\hat{s}) = 1 + \frac{1}{2} \sum_{\substack{k \in M(\tilde{s},\hat{s}) \\ k \neq i,j}} \left[\frac{H_k(\tilde{s}) - H_k(\hat{s}_i,\tilde{s}_{-i}) + H_k(\tilde{s}) - H_k(\hat{s}_j,\tilde{s}_{-j})}{H_k(\hat{s}) - H_k(\tilde{s}_k,\hat{s}_{-k})}\right] \tag{III.b.4}$$

The payoff difference $H_k(\tilde{s}) - H_k(\hat{s}_l,\tilde{s}_{-l})$ expresses how player k is affected by a unilateral deviation of player $l = i,j$ from $\tilde{s}$. This payoff difference is divided by the positive payoff difference $H_k(\hat{s}) - H_k(\tilde{s}_k,\hat{s}_{-k})$ since the weights should not depend on positive affine transformations of players' utilities. As the payoff difference $H_k(\tilde{s}) - H_k(\hat{s}_l,\tilde{s}_{-l})$ for $k \neq l$ can be negative, also the weights $P_{ij}(\tilde{s},\hat{s})$ can be negative. If i and j are the only active players, i.e. $M(\tilde{s},\hat{s}) = \{i,j\}$, the weight $P_{ij}(\tilde{s},\hat{s})$ is equal to 1. Thus selecting $\tilde{s}$ out of $C(G) = \{\tilde{s},\hat{s}\}$ if $R_w(\tilde{s},\hat{s}) > R_w(\hat{s},\tilde{s})$ and $\hat{s}$ in case of the reverse inequality also generalizes the axiomatic HARSANYI and SELTEN–solution for 2×2–bimatrix games with two strict equilibria to n–person games G with just two strict solution candidates.

By the very definition of the weights $P_{ij}(\tilde{s},\hat{s})$ in (III.b.4), the weight $P_{ij}(\tilde{s},\hat{s})$ increases if $\tilde{s}$ is strengthened by increasing the payoff $H_k(\tilde{s})$ for some active player $k \in M(\tilde{s},\hat{s})$ with $k \neq i,j$. The aggregation method $R_w(\cdot,\cdot)$ therefore satisfies the monotonicity axiom II.3. Clearly, the weights

$P_{ij}(\tilde{s},\hat{s})$ – and selecting $\tilde{s}$ out of $C(G) = \{\tilde{s},\hat{s})$ as the solution, if $R_w(\tilde{s},\hat{s}) > R_w(\hat{s},\tilde{s})$ – do not depend on isomorphic transformations of the game G. Unfortunately, this criterion for selecting the solution does not satisfy best reply invariance. To see this, consider the payoff difference for a given mixed strategy combination

$$q_{-k} = (q_1,\dots,q_{k-1}, q_{k+1},\dots,q_n) \tag{III.b.5}$$

resulting from two different strategies $\tilde{s}_k$ and $\hat{s}_k$ of player k:

$$H_k(\tilde{s}_k,q_{-k})-H_k(\hat{s}_k,q_{-k}) = \sum_{s_{-k}} \prod_{j\neq k} q_j(s_j) \left[H_k(\tilde{s}_k,s_{-k}) - H_k(\hat{s}_k,s_{-k})\right] \tag{III.b.6}$$

According to (III.b.6) this payoff difference depends on the payoff function $H_k(\cdot)$ only via the payoff effects $H_k(\tilde{s}_k,s_{-k}) - H_k(\hat{s}_k,s_{-k})$ caused by the different choice $\tilde{s}_k$ or $\hat{s}_k$ by player k himself. Thus whether $\tilde{s}_k$ or $\hat{s}_k$ is a best reply to q_{-k} depends only on the payoff effects due to different choices of player k. As a consequence the best reply structure of a given game G depends purely on payoff differences of the form $H_k(\tilde{s}_k,s_{-k}) - H_k(\hat{s}_k,s_{-k})$ whereas the weights, defined by (III.b.4), depend on payoff differences of the form $H_k(\tilde{s}_l,s_{-l}) - H_k(\hat{s}_l,s_{-l})$ with $l \neq k$. The aggregation method $R_w(\cdot,\cdot)$ satisfies therefore best reply invariance only for the set of pairs (G,C(G)) for which any comparison of two candidates $\tilde{s}$ and $\hat{s}$ in C(G) involves only two active players. Selecting $\tilde{s} \in C(G) = \{\tilde{s},\hat{s}\}$ if $R_w(\tilde{s},\hat{s}) > R_w(\hat{s},\tilde{s})$ is thus a selection method which satisfies axioms II.1 and II.2 in general but axiom II.3 essentially only for the set of 2×2–bimatrix games with two strict equilibria. Nevertheless, this selection criterion also generalizes the axiomatic HARSANYI and SELTEN–solution since it coincides with this solution for the class of 2×2–bimatrix games with two strict equilibria.

Proposition III.1: If $G = (S_1,\dots,S_n;H) \in \mathcal{G}^*$ is an n–person game with a candidate set C(G) containing only two different strict equilibria $\tilde{s}$ and $\hat{s}$, all three selection criteria

(i) $\tilde{s}$ is the solution if $R_+(\tilde{s},\hat{s}) > R_+(\hat{s},\tilde{s})$

(ii) $\tilde{s}$ is the solution if $R_*(\tilde{s},\hat{s}) > R_*(\hat{s},\tilde{s})$

(iii) $\tilde{s}$ is the solution if $R_w(\tilde{s},\hat{s}) > R_w(\hat{s},\tilde{s})$

generalize the axiomatic HARSANYI and SELTEN–solution for

2 × 2–bimatrix games with two strict equilibria, i.e. they all coincide with the axiomatic HARSANYI and SELTEN–solution for this class of games.

An arbitrary aggregation method is denoted by $R(\cdot,\cdot)$. $R(\tilde{s},\hat{s})$ is named the total relative strength of $\tilde{s}$ against $\hat{s}$. The notion of total relative strength allows to solve all games G with $C(G) = \{\tilde{s},\hat{s}\}$ and $R(\tilde{s},\hat{s}) \neq R(\hat{s},\tilde{s})$. In the following it will be described how to select a solution in games G with candidate sets C(G) containing more than just two elements. The problem how to treat degenerate games with $R(\tilde{s},\hat{s}) = R(\hat{s},\tilde{s})$ for at least two solution candidates $\tilde{s}$ and $\hat{s}$ will be discussed in Section V.

III.c) Comparing relative strength

Let $G \in \mathcal{G}^*$ be a game with a candidate set C(G) containing $m(\geq 2)$ elements which all are strict equilibria of G. To compare the various candidates according to their relative strength we use a familiar concept in game theory, namely that of lexicographic minimization (SCHMEIDLER,1969, and GÜTH and KALKOFEN, 1989). For every $s^j \in C(G)$ the vector

$$R(s^j) = \left[R(s^{j_1},s^j),\ldots,R(s^{j_m},s^j)\right] \qquad \text{(III.c.1)}$$

contains all values $R(s,s^j)$ with $s \in C(G)$ and $s \neq s^j$ as components where the ordering is such that the components of $R(s^j)$ are non–increasing, i.e.

$$R(s^{j_k},s^j) \geq R(s^{j_l},s^j) \quad \text{for } k < l. \qquad \text{(III.c.2)}$$

Thus the first component of $R(s^j)$ measures the maximal relative strength of any other solution candidate against s^j. Lexicographic minimization means therefore nothing else than to select the solution candidate $s^* \in C(G)$ for whom the maximal relative strength against it is minimal, i.e.

$$R(s^*) \leq_L R(s) \quad \text{for all } s \in C(G), \qquad \text{(III.c.3)}$$

where the notation '$\leq_L$' means that $R(s^*)$ is lexicographically not greater than R(s). An equilibrium point $s^* \in C(G)$ satisfying condition (III.c.3) is called maximally stable.

For every pair (G,C(G)) we say that game G with the non–empty candidate set C(G) is solvable if there exists a unique candidate $s^* \in C(G)$ with

$$R(s^*) \leq_L R(s) \quad \text{for all } s \in C(G), \tag{III.c.4}$$

i.e. if there exists a unique maximally stable solution candidate in C(G).

It is, of course, obvious that not all games G with non–empty candidate sets C(G) are solvable. As an example consider the game G of Table II.1 with

$$\frac{g-h}{e-f} = \frac{a-b}{c-d} \tag{III.c.5}$$

and the candidate set

$$C(G) = \left\{s^1 = (s_1^1, s_2^1),\ s^2 = (s_1^2, s_2^2)\right\}. \tag{III.c.6}$$

Condition (III.c.5) implies

$$R_{12}(s^1, s^2) = R_{12}(s^2, s^1) = 1$$

and, due to $P_{12}(s^1, s^2) = 1$, also

$$R(s^1, s^2) = R(s^2, s^1) = 1$$

regardless whether $R(\cdot,\cdot)$ is given by $R_+(\cdot,\cdot)$, $R_*(\cdot,\cdot)$, or $R_w(\cdot,\cdot)$. Because of the symmetry of the game, both candidates are equally strong so that the game G with candidate $C(G) = \{s^1, s^2\}$ is not solvable.

We will first concentrate on solvable games by analysing some crucial examples. Thereafter we will discuss how to solve games with candidate sets for which no maximally stable solution candidate exists and how to determine the candidate set for all games.

IV. Examples

The most direct way to justify a solution concept is to explore the basic axioms from which it can be derived. In Section II we therefore have stated the HARSANYI and SELTEN–axioms and argued that they are intuitively convincing. The basic ideas of unilateral deviation stability, discussed in Section III, were all designed to satisfy the HARSANYI and SELTEN–axioms at least for the class of games considered by HARSANYI and SELTEN (1988). A fully satisfactory justification along these lines would, of course, require an axiomatic characterization of unilateral deviation stabiliy.

Another way of demonstrating that a solution concept is reasonable is to apply it to special games or classes of games for which one has quite a good idea what should be the solution (both selection theories, the one of HARSANYI and SELTEN, 1988, and the one of GÜTH and KALKOFEN, 1989, were applied to such crucial examples). Limitations of space do not allow us to apply all three versions of unilateral deviation stability to all the examples of games with multiple strict equilibria discussed by HARSANYI and SELTEN (1988) and GÜTH and KALKOFEN (1989). We therefore restrict ourselves to the more important ones, namely unanimity games and market entry games as studied by SELTEN and GÜTH (1982).

Due to the three possibilities of aggregating relative strength, namely $R_+(\cdot,\cdot)$, $R_*(\cdot,\cdot)$, and $R_w(\cdot,\cdot)$, we have described three solution concepts for solvable games. Since all three versions consider only risks of unilateral deviations from solution candidates, we speak of three versions of unilateral deviation stability. In the following we will try to narrow down the spectrum of aggregation methods by testing them for both classes of games. This will enable us to decide which aggregation method is most reasonable and thereby to define a unique concept of unilateral deviation stability.

IV.a) Unanimity games

Unanimity bargaining gives veto–power to every player. Thus all strategic risks are fully captured by the risks of unilateral deviations. Let

$$I = \{i_1,\ldots,i_K\} \text{ with } K \in \mathbb{N}, K \geq 2, \tag{IV.a.1}$$

be an index set. I is the pure strategy set of all players $l = 1,\ldots,n$ (≥ 2). For every strategy vector

$$s = (s_1,\ldots,s_n) \in I^n \tag{IV.a.2}$$

the payoff of player $l = 1,\ldots,n$ is defined as follows:

$$H_l(s) = \begin{cases} u_l^k \text{ if } s_j = i_k \in I \text{ for } j = 1,\ldots,n \\ 0 \text{ otherwise,} \end{cases} \tag{IV.a.3}$$

where we assume

$$u_l^k > 0 \text{ for all } k = 1,...,K \text{ and } l = 1,...,n. \tag{IV.a.4}$$

Thus all players achieve a positive payoff only if they unanimously agree on choosing the same index i_k representing a specific payoff distribution

$$u^k = (u_1^k,...,u_n^k). \tag{IV.a.5}$$

Formally, the unanimity game G is fully described by

$$G = (I, U = (u^k)_{k=1,...,K}), \tag{VI.a.6}$$

i.e. by the for all players same strategy set I as well as by the list U of payoff distributions for the K possible agreements.

Here we do not consider all unanimity (bargaining) games G but restrict our attention to the set $\mathcal{G}^a$ of games G which satisfy the following non–degeneracy assumption:

$$\prod_{l=1}^{n} u_l^k \neq \prod_{l=1}^{n} u_l^{k'} \text{ for all } k, k' = 1,...,K \text{ with } k \neq k' \tag{IV.a.7}$$

For each game G in $\mathcal{G}^a$ the so–called <u>cooperative NASH–solution</u> (NASH, 1950 and 1953) is uniquely defined and given by the pure strategy equilibrium point $s^{k^*} = (i_{k^*},...,i_{k^*})$ satisfying

$$\prod_{l=1}^{n} u_l^{k^*} > \prod_{l=1}^{n} u_l^k \text{ for all } k = 1,...,K \text{ with } k \neq k^*. \tag{IV.a.8}$$

It will be demonstrated that only one of the three aggregation methods yields a concept which selects s^{k^*} as the solution of games $G \in \mathcal{G}^a$ when the candidate set C(G) is given by the set of all strict equilibria, namely

$$C(G) = \{s^k = (i_k,...,i_k) : k = 1,...,K\}. \tag{IV.a.9}$$

When unilateral deviation stability selects the cooperative NASH–solution in unanimity games, we will say that it satisfies the <u>NASH–property</u> (see also HARSANYI and SELTEN, 1988, as well as GÜTH and KALKOFEN, 1989, who also prove the NASH–property for their selection theories).

In the unanimity game $G \in \mathcal{G}^a$ all players are active when comparing two different candidates s^k and $s^{k'}$ in C(G). Since all non-equilibrium payoffs for pure strategy combinations are 0, definition (III.8) implies

$$R_{ij}(s^k,s^{k'}) = \frac{u_j^k\, u_i^k}{u_j^{k'}\, u_i^{k'}} \tag{IV.a.10}$$

Thus the aggregation method (III.b.1) yields

$$R_+(s^k,s^{k'}) = \sum_{\substack{i,j=1\\ i>j}}^{n} \frac{u_i^k\, u_j^k}{u_i^{k'}\, u_j^{k'}} \tag{IV.a.11}$$

To demonstrate that $R_+(\cdot,\cdot)$ does not satisfy the NASH–property we consider the game with $n = 3$ and just two candidates s^1 and s^2 with $u^1 = (1,1,1)$ and $u^2 = (\frac{1}{10},3,3)$. Clearly, s^1 is the cooperative NASH–solution since $u_1^1\, u_2^1\, u_3^1 = 1 > u_1^2\, u_2^2\, u_3^2 = \frac{9}{10}$. But since the definition (IV.a.11) implies

$$R_+(s^1,s^2) = \frac{1}{0.3} + \frac{1}{0.3} + \frac{1}{9} = \frac{61}{9}$$

and

$$R_+(s^2,s^1) = \frac{3}{10} + \frac{3}{10} + 9 = \frac{96}{10}$$

and therefore $R_+(s^2,s^1) > R_+(s^1,s^2)$, the vector $R_+(s^1)$ according to (III.c.1) is lexicographically larger than $R_+(s^2)$. Thus unilateral deviation stability relying on the aggregation method $R_+(\cdot,\cdot)$ does not select s^1, i.e. it does not satisfy the NASH–property.

For the aggregation method $R_*(\cdot,\cdot)$, defined in (III.b.2), one obtains

$$R_*(s^k,s^{k'}) = \frac{\prod_{i=1}^{n} (u_i^k)^{n-1}}{\prod_{i=1}^{n} (u_i^{k'})^{n-1}} \tag{IV.a.12}$$

so that

$$R_*(s^{k^*},s^k) > R_*(s^k,s^{k^*}) \text{ for all } k = 1,\dots,K,\ k \neq k^*. \tag{IV.a.13}$$

Unilateral deviation stability relying on the aggregation methode $R_*(\cdot,\cdot)$ therefore satisfies the NASH–property.

For the weighted aggregation method $R_w(\cdot,\cdot)$, as described by (III.b.3) and (III.b.4), the weights $P_{ij}(s^k,s^{k'})$ are given by

$$P_{ij}(s^k,s^{k'}) = 1 + \sum_{\substack{l=1 \\ l\neq i,j}} \frac{u_l^k}{u_l^{k'}}. \tag{IV.a.14}$$

Thus one obtains

$$R_w(s^k,s^{k'}) = \sum_{\substack{i,j=1 \\ i>j}}^{n} \left[1 + \sum_{\substack{l=1 \\ l\neq i,j}}^{n} \frac{u_l^k}{u_l^{k'}}\right] \frac{u_i^k}{u_i^{k'}} \frac{u_j^k}{u_j^{k'}} \tag{IV.a.15}$$

With the help of the same example used for $R_+(\cdot,\cdot)$ one can demonstrate that unilateral deviation stability based on the aggregation method $R_w(\cdot,\cdot)$ does not satisfy the NASH–property, either. Our results for unanimity games are summarized by

Proposition IV.1: Of the three versions of unilateral deviation stability described in Section III.b), only the one based on the aggregation method $R_*(\cdot,\cdot)$, defined in (III.b.2), satisfies the NASH–property.

The fact that neither $R_+(\cdot,\cdot)$ nor $R_w(\cdot,\cdot)$ is consistent with the NASH–porperty is the decisive reason why we favor the aggregation approach $R_*(\cdot,\cdot)$. This does not mean that unilateral deviation stability relying on $R_*(\cdot,\cdot)$ is the only way of defining a concept of unilateral deviation stability satisfying the NASH–property. If one, for instance, would use weights equal to 1 if $M(\tilde{s},\hat{s})\backslash\{i,j\}$ is empty and

$$P_{ij}(\tilde{s},\hat{s}) = \prod_{\substack{l\in M(\tilde{s},\hat{s}) \\ l\neq i,j}} \frac{H_l(\tilde{s}) - H_l(\hat{s}_i,\tilde{s}_{-i}) + H_l(\tilde{s}) - H_l(\hat{s}_j,\tilde{s}_{-j})}{2(H_l(\hat{s}) - H_l(\tilde{s}_l,\hat{s}_{-l}))} \tag{IV.a.16}$$

otherwise instead of the weights defined in (III.b.4), one would obtain

$$R_{ij}(s^k,s^{k'})\,P_{ij}(s^k,s^{k'}) = \frac{\prod_{l=1}^{n} u_l^k}{\prod_{l=1}^{n} u_l^{k'}} \tag{IV.a.17}$$

and therefore

$$R_w(s^k,s^{k'}) = (n-1)!\,\frac{\prod_{l=1}^{n} u_l^k}{\prod_{l=1}^{n} u_l^{k'}}, \tag{IV.a.18}$$

i.e. a weighted aggregation method satisfying the NASH–property.

Here we have not considered such other aggregation methods since at least those, we are aware of, suffer from other, even more serious deficiencies than the one of lacking the NASH–property. The weight defined by (IV.a.16) is, for instance, negative if just one player $l \neq i,j$ suffers from both, the deviation by player i from $\tilde{s}$ as well as the one of player j. But if this is true for exactly two players $l \neq i,j$, the weight suddenly becomes positive. One certainly does not want such an undesirable property. We therefore favor unilateral deviation stability based on the aggregation method $R_*(\cdot,\cdot)$. It naturally extends the axiomatic HARSANYI and SELTEN–solution, it satisfies the NASH–property, and other aggregation methods, also satisfying the NASH–property, seem to be less reasonable.

Another way of deriving the aggregation method $R_*(\cdot,\cdot)$ is, of course, to require some obvious axioms for aggregating individual values. Let $x = (x_1,\ldots,x_m)$ with $m \geq 2$ be a vector of individual values $x_i \in \mathbb{R}$ which should be aggregated to a total value $f(x) \in \mathbb{R}$. Clearly, independence of isomorphic transformations implies that f has to be <u>anonymous</u>, i.e.

$$f(\Pi\, x) = f(x)$$

for every permutation $\Pi:\{1,\ldots,m\} \longrightarrow \{1,\ldots,m\}$ where we use the notation

$$\Pi\, x = (x_{\Pi(1)},\ldots,x_{\Pi(m)}).$$

Furthermore, monotonicity requires that f(x) must be an increasing function of all its components x_i $(i=1,\ldots,m)$. By requiring

Axiom IV.1: "Multi–Linearity"

$$f(x_1,x_2,\ldots,x_m) = x_1\, f(1,x_2,\ldots,x_m)$$

we can derive

$$f(x) = f(1,\ldots,1) \prod_{i=1}^{m} x_i \text{ with } f(1,\ldots,1) > 0.$$

Normalizing according to

Axiom IV.2: "Normalization of Neutrality"

$f(1,...,1) = 1$

we get

$$f(x) = \prod_{i=1}^{n} x_i$$

and therefore the aggretation method $R_*(\cdot,\cdot)$ according to which the total relative strength is nothing else than the product of all individual values $R_{ij}(\cdot,\cdot)$. Clearly, axioms II.1, II.2, II.3 as well as (IV.1) and (IV.2) are mutually consistent. Thus all these axioms together imply $R_*(\cdot,\cdot)$ as the only admissible aggregation method.

Except for games G with candidate sets C(G) which are not solvable we now are able to define the concept of unilateral stability: For every pair (G,C(G)) which is solvable according to the aggregation method $R_*(\cdot,\cdot)$, the solution of G with candidate set C(G) is the only maximally stable solution candidate according to the aggregation method $R_*(\cdot,\cdot)$.

IV.b) (Market) Entry games

In (market) entry games every player has to decide whether to enter, e.g. a market, or not. The profitability of being 'in' or 'out' depends on the number of those being in or out, as well as on the individual costs of entry. Market entry is only one of many possible applications of (market) entry games. SELTEN and GÜTH (1982) define (market) entry games as follows: Each player $i = 1,...,n$ can decide between $\delta_i = 1$ and $\delta_i = 0$. The number of players entering the market is

$$m(\delta) = \sum_{i=1}^{n} \delta_i \tag{IV.b.1}$$

for each vector $\delta = (\delta_1,...,\delta_n)$. Payoffs H_i depend on δ_i and $m(\delta)$ according to

$$H_i(\delta) = \begin{cases} R(m(\delta)) - C_i & \text{for } \delta_i = 1 \\ S(m(\delta)) & \text{for } \delta_i = 0 \end{cases} \tag{IV.b.2}$$

where

$$A(m) := R(m) - S(m-1) \tag{IV.b.3}$$

is assumed to be non-increasing in m, i.e.

$$A(1) \geq A(2) \geq ... \geq A(n). \tag{IV.b.4}$$

A(m) is called the <u>incentive for market entry</u> if already m − 1 other players have entered the market; C_i (≥ 0) is player i's <u>cost of market entry</u>. To avoid games with dominated strategies we assume that

$$A(1) > C_i > A(n) \quad \text{for } i = 1,\dots,n, \tag{IV.b.5}$$

i.e. every player would like to be the only player on the market, e.g. the monopolist if players are potential sellers, and every player prefers no market entry if all the others enter the market. A non–degeneracy assumption is that the entry costs of all players are different. By an appropriate renumbering of players we can therefore assume that

$$C_1 < C_2 < \dots < C_n. \tag{IV.b.6}$$

Further non–degeneracy assumptions are

$$A(m) \neq C_i \quad \text{for } i,m = 1,\dots,n \tag{IV.b.7}$$

and

$$C_i + C_j \neq C_k + C_l \quad \text{for all } i,j,k,l = 1,\dots,n \text{ with } (i,j) \neq (k,l). \tag{IV.b.8}$$

A (market) entry game $G = (\mathcal{R}, \mathcal{S}, \mathcal{C})$ is completely described by the vectors

$$\mathcal{R} = (R(1),\dots,R(n)), \tag{IV.b.9}$$

$$\mathcal{S} = (S(0),\dots,S(n-1)), \tag{IV.b.10}$$

and

$$\mathcal{C} = (C_1,\dots,C_n) \tag{IV.b.11}$$

satisfying all these assumptions (for a justification see SELTEN and GÜTH, 1982). The class of such games $G = (\mathcal{R}, \mathcal{S}, \mathcal{C})$ is denoted by $\mathcal{G}^b$.

Let m^* with $0 < m^* < n$ be that unique number for which

$$C(m^*) < A(m^*) \text{ and } C(m^*) > A(m^*+1). \tag{IV.b.12}$$

SELTEN and GÜTH (1982) have shown that the strategy vector

$$\delta^* = (1,\dots,1, 0,\dots,0) \quad \text{with } m(\delta^*) = m^* \tag{IV.b.13}$$

is a strict equilibrium point of $G \in \mathcal{G}^b$ and that all other strict equilibria of G must imply the same number m^* of market entrants as δ^*, i.e. if $\delta = (\delta_1,...,\delta_n)$ is a strict equilibrium of G, then $m(\delta) = m^*$. Beside $m(\delta) = m^*$ a pure strategy vector δ must satisfy $C_i < A(m^*)$ for all players i with $\delta_i = 1$ and $C_i > A(m^*+1)$ for all players i with $\delta_i = 0$ to be a strict equilibrium point. Since every strict equilibrium implies the same number of market entrants, the equilibrium point δ^*, according to which the m^* players with lowest entry costs enter the market, is the most obvious and natural solution of any game G in $\mathcal{G}^b$. We want to show that unilateral deviation stability, as defined by the aggregation method $R_*(\cdot,\cdot)$, selects δ^* and thereby passes the crucial test of selecting the 'right solution' in (market) entry games.

For two different strict equilibria δ and δ' of $G \in \mathcal{G}^b$ the set of active players is given by

$$M(\delta,\delta') = \{i = 1,...,n : \delta_i \neq \delta_i'\}. \tag{IV.b.14}$$

For two different active players $i,j \in M(\delta,\delta')$ we have to distinguish two cases, namely the one where $\delta_i = \delta_j$ and the one with $\delta_i \neq \delta_j$. For $\delta_i = \delta_j = 0$ one obtains

$$R_{ij}(\delta,\delta') = \frac{(C_j - A(m^*+1))\ (C_i - A(m^*+1))}{(A(m^*) - C_j)\ (A(m^*) - C_i)} \tag{IV.b.15}$$

where all factors on the right hand side of equation (IV.b.15) are positive since δ and δ' are strict equilibria of G. If $\delta_i = 1 = \delta_j'$, the corresponding result is

$$R_{ij}(\delta,\delta') = \frac{(C_j - A(m^*+1))\ (A(m^*) - C_i)}{(A(m^*) - C_j)\ (C_i - A(m^*+1)} \tag{IV.b.16}$$

where again all factors appearing on the right hand side are positive. For all other strategy constellations of players i and j the value $R_{ij}(\delta,\delta')$ can be derived from these two formulae by either exchanging δ and δ' or i and j.

We only consider the aggregation method $R_*(\cdot,\cdot)$ since it was the only one of the three methods satisfying the NASH–property. According to (III.b.2) one obtains

$$R_*(\delta,\delta') \quad = \prod_{\substack{i,j \in M(\delta,\delta') \\ \delta_i=\delta_j=0}} \frac{(C_j - A(m^*+1))\ (C_i - A(m^*+1))}{(A(m^*) - C_j)\ \ (A(m^*) - C_i)}$$

$$\times \prod_{\substack{i,j \in M(\delta,\delta') \\ \delta_i=\delta_j=1}} \frac{(A(m^*) - C_j)\ \ (A(m^*) - C_i)}{(C_j - A(m^*+1))\ (C_i - A(m^*+1)}$$

$$\times \prod_{\substack{i,j \in M(\delta,\delta') \\ \delta_i=\delta'_j=1}} \frac{(C_j - A(m^*+1))\ (A(m^*) - C_i)}{(A(m^*) - C_j)\ \ (C_i - A(m^*+1))}$$

$$\times \prod_{\substack{i,j \in M(\delta,\delta') \\ \delta_i=\delta'_j=0}} \frac{(A(m^*) - C_j)\ \ (C_i - A(m^*+1))}{(C_j - A(m^*+1))\ (A(m^*) - C_i)}$$

Now it is easy to see that

$$\frac{A(m^*) - C_i}{C_i - A(m^*+1)} > \frac{A(m^*) - C_j}{C_j - A(m^*+1)} \tag{IV.b.18}$$

is equivalent to $C_j > C_i$ due to $A(m^*) > C_i, C_j > A(m^*+1)$ for all players $i,j \in M(\delta^*,\delta)$. Thus if a pair $i,j \in M(\delta^*,\delta)$ with $\delta^*_i = \delta^*_j = 1$ exists, the corresponding factor in (IV.b.17) for $R_*(\delta^*,\delta)$ is larger than the corresponding factor for $R_*(\delta,\delta^*)$. For the same reason also the factor in (IV.b.17) corresponding to pairs $i,j \in M(\delta^*,\delta)$ with $\delta^*_i = \delta^*_j = 0$ of $R_*(\delta^*,\delta)$ is larger than the corresponding factor of $R_*(\delta,\delta^*)$ if such a pair exists. Similarly, one shows also for the remaining strategy constellations $\delta^*_i = 1 = \delta_j$ and $\delta^*_i = 0 = \delta_j$ that the corresponding component of $R_*(\delta^*,\delta)$ would be greater than the one of $R_*(\delta,\delta^*)$. Since $\delta^* \neq \delta$, at least one pair of players $i,j \in M(\delta^*,\delta)$ exists with $\delta^*_i = 1 = \delta_j$ and therefore $C_i < C_j$ due to $i \leq m^* < j$. This shows that $R_*(\delta^*,\delta) > R_*(\delta,\delta^*)$ for all strict equilibria δ of $G \in \mathcal{G}^b$ with $\delta \neq \delta^*$. Consequently, $R_*(\delta^*)$ is lexicographically smaller than $R_*(\delta)$ for all $\delta \in C(G)$ with $\delta \neq \delta^*$.

Proposition IV.b.1: In (market) entry games unilateral deviation stability, as defined by the aggregation rule $R_*(\cdot,\cdot)$, selects the intuitively reasonable solution candidate δ^*, described in (IV.b.13).

According to Proposition IV.b.1, our concept of unilateral deviation stability selects the obvious solution δ^* in (market) entry games and thereby passes a crucial test of acceptability. Of course, many concepts will select δ^* so that selecting δ^* is only a necessary, but by far not a sufficient condition for a general solution concept. More satisfying justifications will require to apply the theory of unilateral deviation stability to more examples or to explore more thoroughly the general properties of unilateral deviation stability.

V. Non–strict solution candidates

Strict equilibria are natural solution candidates since they are strictly self–enforcing behavioral expectations in the sense that every unilateral deviation would hurt the deviating player. Non–strict solution candidates have to be considered if either any decision between strict equilibria would be completely arbitrary, e.g. in case of a = g, e = c, h = b, and f = d in Table II.1, or if no strict equilibrium point exists. In the following we will briefly describe how to define a candidate set which initially or later may include non–strict equilibria and how to make sure that unilateral deviation stability is well–defined even when comparing non–strict equilibria. In doing so we will rely on ideas originally proposed by HARSANYI and SELTEN (1988) in their pioneering approach to develop a general theory of equilibrium selection.

If a game has no strict equilibria or if its solution should not be a strict equilibrium, as in the case of symmetry mentioned above, one has to look for a natural extension of the notion of a strict equilibrium. Fortunately, HARSANYI and SELTEN (1988) have generalized the concept of strict equilibrium points by their definition of primitive formations. If $G = (S_1,...,S_n;H)$ is a normal form game, a <u>formation</u> F of G is the normal form game $F = (F_1,...,F_n;H|_{\times_{i=1}^{n} F_i})$ with $\emptyset \neq F_i \subseteq S_i$ for $i = 1,...,n$ and $H|_{\times_{i=1}^{n} F_i}$ being the restriction of $H = (H_1,...,H_n)$ to strategy vectors s in F if F is closed in G with respect to the best reply correspondence of G. The latter condition means that for all players i = 1,...,n all pure best replies to all mixed strategy vectors

$$q_{-i} = (q_1,...,q_{i-1}, q_{i+1},...,q_n)$$

with

$$q_j(s_j) = 0 \text{ for all } s_j \notin F_j \ (j \neq i)$$

are contained in F_i. A formation F of G is called <u>primitive</u> if it contains no proper formation, i.e.

if no F′ with $F_i' \subsetneq F_i$ for all i and $F_i' \neq F_i$ for at least one player i is a formation of G. If $s^* = (s_1^*,...,s_n^*)$ is a strict equilibrium point of G, then

$$(\{s_1^*\},...,\{s_n^*\}; H|_{\{s^*\}})$$

is a primitive formation of G. The concept of primitive formations therefore generalizes the notion of strict equilibrium points.

As a finite normal form game, every formation has at least one equilibrium (NASH, 1951). But although it never has been encountered in applying equilibrium selection theory so far, a primitive formation may contain multiple equilibria (for an example see GÜTH and KALKOFEN, 1989). Here we do not want to engage into a discussion how to select one of many equilibria of a primitive formation as the formation solution (see HARSANYI and SELTEN, 1988, as well as GÜTH and KALKOFEN, 1989, for ways to do so) but will simply assume that every primitive formation F contains only one equilibrium which we call the solution of F.

By the very definition of a formation, every intersection of two formations is also a formation. This shows that every game G has a unique list

$$(F^l)_{l=1,...,m} \tag{V.1}$$

of primitive formations whose solutions are denoted by $s^l (l = 1,...,m)$. Regardless whether G has strict equilibria or not, we want to define its candidate set C(G) by

$$C(G) = \{s^1,...,s^m\}. \tag{V.2}$$

Since m is at least 1, the set C(G) is never empty. The comparison of two different solution candidates $s^l \in F^l$ and $s^k \in F^k$ with $F^l \cap F^k = \emptyset$ is well-defined since the stability regions of all four pure strategy vectors in the games $G_{ij}(s^l,s^k)$ with players $i,j \in M(s^l,s^k)$ are generic, i.e. have positive Lebesque-measure. One therefore can apply the concept of unilateral deviation stability to every normal form game G whose primitive formations contain only one equilibrium.

The second problem mentioned above is that one may not be able to select a solution out of C(G) because any such decision would be completely arbitrary. The special games G with a = g, e = c, h = b, and f = d in Table II.1, for instance, have exactly two primitive formations, namely the ones defined by its two strict equilibria s^1 and s^2. Thus we obtain $C(G) = \{s^1,s^2\}$ but every decision for s^1 or s^2 as the solution of G would be completely arbitrary.

Our proposal to deal with such cases (but see also HARSANYI and SELTEN, 1988, as well as GÜTH and KALKOFEN, 1989) is to substitute the initial candidate set C(G) by a new candidate set C′(G) if at least two solution candidates are maximally stable. Let $C^*(G)$ denote the set of maximally stable solution candidates in C(G). If $C^*(G)$ contains at least two elements, we define a second candidate set C′(G) of G where we consider a formation F of G as 'primitive' if F contains no other proper formation than those primitive formations F^l with $s^l \in C^*(G)$, i.e. when defining C′(G), we neglect those (primitive) formations with solutions in $C^*(G)$. The new candidate set C′(G) is the set of solutions of these new 'primitive' formations.

Since some new 'primitive' formations contain more than one of the old primitive formations, it will be rather typical than exceptional that at least some of the new 'primitive' formations have multiple equilibria. On the other hand this problem is alleviated since one naturally will want to exclude all equilibria in $C^*(G)$ as the solution of a new 'primitive' formation. In the game of Table II.1 with a = g, e = c, h = b, and f = d one has

$$C(G) = \{s^1, s^2\} = C^*(G)$$

and therefore the game G itself as the only new 'primitive' formation when defining C′(G). Since s^1 and s^2 are not qualified for becoming the solution of the new 'primitive' formation F = G, the mixed strategy equilibrium point q, defined in (II.1) and (II.2), is the only remaining solution candidate, i.e.

$$C'(G) = \{q\}.$$

We therefore obtain the intuitively convincing solution of the symmetric games G satisfying a = g, e = c, h = b, and f = d.

VI. Conclusions

Equilibrium selection theory tries to resolve strategic uncertainty completely by selecting a unique equilibrium point. Although its program has already been outlined by NASH (1953), it took 35 years before we experienced the first general attempt to accomplish this task, namely the general theory of equilibrium selection by HARSANYI and SELTEN (1988).

There is no doubt that equilibrium selection theory is still in a premature state and that still too few scholars are working in this field. In our view, this is mainly due to the large entry costs. The general theory of HARSANYI and SELTEN as well as the related ESBORA–concept of GÜTH and KALKOFEN are composed of many different constructive procedures and based on quite a number of conceptual ideas. This makes it difficult to understand the theory and even more cumbersome to apply it.

The main motivation of this paper has been to develop a general theory of equilibrium selection which is simple and can be easily applied and which satisfies some natural solution requirements, namely the ones proposed by HARSANYI and SELTEN for solving 2 × 2–bimatrix games with two strict equilibria. We hope that our concept will lower the barriers of entry so that more scholars will engage in developing and applying theories of equilibrium selection. Our preliminary results show that unilateral deviation stability has quite a number of desirable properties although it does not consider the risks of deviations by more than one player.

Of course, not all problems have been settled to make unilateral deviation stability a fully general theory of equilibrium selection. The problem how to select a solution for a primitive formation has already been mentioned. Even more important is to guarantee that the solution is uniformly perfect (see HARSANYI and SELTEN, 1988), i.e. can be approached by equilibria of uniformly perturbed games. The latter requirement can be easily fulfilled by applying unilateral deviation stability not directly to the game itself but to its uniformly perturbed games and by considering the limit, if perturbances vanish, as the solution of the unperturbed game. Other problems are the requirements of cell and truncation consistency and the necessity to reduce a game before perturbing it. Here such problems could not be discussed. But often enough one can simply refer to the familiar ways of taking care of such problems as proposed by HARSANYI and SELTEN.

Unlike the previous general theories of equilibrium selection (HARSANYI and SELTEN, 1988, and GÜTH and KALKOFEN, 1989) we have neglected payoff dominance since it is completely unrelated to the strategic risks considered here. Actually an earlier version of the general theory of HARSANYI and SELTEN (see the description in GÜTH, 1978) also neglected payoff dominance as does the rather recent ad hoc–concept of CARLSSON and VAN DAMME (1989) for 2 × 2–bimatrix games.

References

Hans CARLSSON – Eric VAN DAMME: Global payoff uncertainty and risk dominance, CentER Discussion Paper No. 8933, Tilburg University 1989.

Werner GÜTH: Zur Theorie kollektiver Lohnverhandlungen, Nomos–Verlag, Baden–Baden 1978.

Werner GÜTH: A remark on the Harsanyi–Selten theory of equilibrium selection, International Journal of Game Theory 14, 31–39.

Werner GÜTH – Brigitte KALKOFEN:
Unique solutions for strategic games,
Lecture Notes in Economics and Mathematical Systems, Vol. 328,
Springer–Verlag, Berlin 1989.

John C. HARSANYI – Reinhard SELTEN:
A general theory of equilibrium selection in games,
The MIT Press, Cambridge/Mass. 1988.

John F. NASH: The bargaining problem,
Econometrica 18 (1950), 361–382.

John F. NASH: Non–cooperative games,
Annals of Mathematics 45 (1951), 286–295.

John F. NASH: Two–person cooperative games,
Econometrica 21 (1953), 128–140.

David SCHMEIDLER: The nucleolus of a characteristic function game, SIAM Journal of Applied Mathematics 17, 1163–1170.

Reinhard SELTEN – Werner GÜTH:
Equilibrium point selection in a class of market entry games,
in: Games, economic dynamics, time series analysis (edrs. M. Deistler, E. Fürst, and G. Schwödiauer), Physica–Verlag, Wien 1982, 101–116.

Determinacy and Rational Choice

Russell Hardin
Departments of Political Science and Philosophy, University of Chicago
5828 South University Avenue, Chicago, IL 60637, USA

Suppose we have a fully determinate rational choice theory that will tell each of us what to do in a particular interactive choice context and suppose each of us knows the positions of all the others. From our theory I can calculate not only what I ought to do but also what each other agent in the interaction ought rationally to do. Given knowledge of what each other agent should do, I can check whether I could in fact do better by not following the theory. If that theory yields determinate solutions, in the sense that it tells each of us explicitly what to do, these solutions must be in equilibrium for rational players who have full knowledge of the payoff structure, who have this theory, and who assume that their co-players are rational and have this theory.

If a purported solution were not in equilibrium, then, by definition, at least one of the players would have an incentive to choose a different strategy in order to receive a better payoff. But a coherent theory cannot be used against itself to tell me to do other than what it commends. Hence, each of us must find that what the theory tells us to do is, once all others are also following the theory, an equilibrium outcome.

John Harsanyi has long argued for determinacy in game theoretic rational choice theory (Harsanyi 1956, 1977) and, not surprisingly, he has also generally argued for equilibrium outcomes as rational, as in his recent work with Reinhard Selten (Harsanyi and Selten 1988). Against this general position, I wish to argue that primitive notions of rationality do not entail determinacy when followed by all agents to an interaction. The imposition of determinacy as a part of the notion of rationality is an extra-rational move that is not compelling in many contexts. In particular, it is not compelling in contexts, such as iterated Prisoner's Dilemma, in which our theory must give contingent instructions on how to play into the future.

First, focus on the underlying intuition that stands firm through all the problems we may face: A rational person prefers more value to less or, absent value assignments, has transitive preferences. For present purposes, I will assume cardinal utility in keeping with the overwhelming practice in game theoretic discussions, so that we may dispense with reference to transitivity in a strictly ordinal theory of preference. In cardinal contexts,preference for x over y merely means the value of x is greater than the value of y. Many critics of this notion of rationality make the simplistically one-step, materialist mistake of supposing the principle implies one should prefer more to less of anything valued at all. But, obviously, I may not prefer more dinner to less tonight because I may not prefer the consequences of still more once I have consumed enough. Indeed, I am unlikely even to have appetite for more beyond some level of consumption. Cardinal preference for more is strictly preference for more value or utility and is therefore trivially transitive.

If my preferences are transitive in this cardinal sense, it follows that, if I am choosing from n possibilities, there must be one or more possibilities that I prefer to all others, or that I hold indifferent to all, so that in this case preferring more to less value is equivalent to maximizing. Our more general problem is how to choose under varied constraints when we have to compromise with the principle of maximization. An obvious a priori case in which there can be no compellingly unique compromise is that in which I am offered a choice from an infinite set of possibilities none of which has the greatest value (e.g., $1000, $2000, $3000, ...). But this, unfortunately, is not a practical problem.

Maximization is commonly treated as the basic meaning of rationality or, at least, as part of the meaning. In particular contexts we have no difficulty giving clear meaning to the notion of maximizing in individual choice over various alternatives. For example, given a choice between two outcomes, one of which I prefer to the other, I know what it means to maximize. If my choices or actions determine results only probabilistically, the notion of maximization is already in question. Most choice theorists seem to find the generalization to maximum expected utility an obvious extension of the primitive notion of rationality, although some people do not find that move obvious or even right. The central problem of rationality in economics is that we cannot so readily or agreeably generalize our notion of maximization to many other contexts. Most especially, we cannot generalize it to cover contexts in which choices of different agents interact in producing joint outcomes for the agents. The theory of games was invented to model such choice problems.

The difficulty in generalizing our notion of maximization to cover all games is that each agent is to maximize his or her own utility function over all the outcomes while different agents may have quite different utility functions over those outcomes. It is not always impossible to maximize over two functions at once, but it is not possible to do so in general or to specify a general rule for doing so. A trivial case in which genuine multiple maximization is possible is when our maxima are all in the same outcome even though our functions are not otherwise identical, as in certain coordination games. Despite special cases such as these, however, it is still true that there is mathematically no solution to the general problem of multiple maximization. Hence, much of game theory, and of choice theory and economics more generally, is directed at finding compromises to multiple maximization that seem to be what one would deduce from the principle of individual maximization. As compelling as some such compromises might be in certain contexts, they are still not as compelling as maximization over a single function that has a greatest value at some point or points.

The first success of game theory was Neumann's saddlepoint theorem for two-person constant sum games. In such a game, the two players have inversely related utility functions over the outcomes. Clearly, it is impossible to maximize both functions at a single outcome. But, because the game is constant sum, these two functions are not independent, so that there might be a plausible compromise to simultaneous maximization, maximization subject to a constraint. In the two-person constant sum game both players face the same constraint, defined by the inverse of each player's own utility function over the outcomes. Neumann's maximin strategy seems compelling to virtually every game theorist who has considered the problem. One may say it is a compelling compromise that seems consistent with what we would expect from maximizing players, given that maximization is mathematically impossible for both at once. But the description of that solution — maximin — is not a good prescription for a solution to the general case in which the functions over which we are simultaneously trying to maximize may be independent so that we do not face analogous constraints.

There are other contexts in which multiple maximization — that is, maximization over more than one function at once — seems to have compelling implications. In a market of many sellers and many buyers, whose utility functions over all possible outcomes of trade cannot be simultaneously maximized, we may nevertheless expect a very restricted range of possible outcomes to result. Again, although the argument may be a bit convoluted, this result follows from what seems similarly to be a compelling compromise implicit in or consistent with the notion of individual maximization. All players in the market face the constraint that large numbers with similar interests typically imply the failure of collective action to collude on prices for sale or purchase.

Equilibrium

As it happens, all of the results above — those that follow from literal maximization and those that follow from some compromise more or less deducible from it — are equilibria in their particular games or interactions. For many contemporary economists that is not surprising, because they suppose equilibrium is inherently bound up in rationality. But the bonds go in only one direction. Equilibrium is defined in rational terms; rationality is not defined in terms of equilibrium. An equilibrium is an outcome from which no one would have incentive to deviate, by choosing a different strategy, even if it were possible to second-guess all the others in the interaction after knowing what choices they have made.

That the bonds do not go the other way can be seen from the facts that (1) there can be multiple equilibria with differing payoffs to various players, and (2) there can be equilibria that are disastrous for all concerned and that are Pareto inferior to other possible outcomes. In both cases, it may be true that some subgroup has an interest in switching collectively out of the strategies that put them in the equilibrium. This move is often prohibited in individualistically defined equilibrium theories. Yet, much of the social interaction we wish to explain is just such subgroup cooperation for the subgroup's interest. In their solution theory, Harsanyi and Selten resolve many of the cases under (2) by invoking their principle of "collective rationality," according to which, "in the absence of special reasons to the contrary, rational players will choose an equilibrium point yielding all of them higher payoffs, rather than one yielding them lower payoffs" (Harsanyi and Selten 1988, p. 356).

Equilibrium is defined in terms of individual rationality. If we are in equilibrium, then it would not be rational for any of us individually to choose a different strategy. Frank Hahn has suggested an alternative notion of general economic equilibrium that makes the bonds go both ways. We may restate his notion to fit game theoretic choice problems: Agents in an interactive choice are in equilibrium if their choices generate messages that do not cause agents to change the theories they hold or the policies they pursue. Hahn says the central point of this conception of equilibrium is that one abandon one's theory when it is "sufficiently and systematically falsified" (Hahn 1984, p. 59). At first reading, this may sound like an empiricist view. I wish to read it, rather, as a theoretical view.

Robert Aumann (1985, pp. 28-9) says that a game theoretic "solution concept should be judged more by what it does than by what it is; more by its success in establishing relationships and providing insights into the workings of the social processes to which it is applied than by consideration of a priori plausibility based on its definition alone." Selten (1985, p. 81) replies that: "Descriptive theories need to be compared with reality, whereas normative theories cannot be tested empirically. The justification of normative theories of rationality must be sought in compelling intuitive arguments." Similarly, Richard Jeffrey (1983, p. 166) notes that the Bayesian theory of preference is ours — the theorists' — but not necessarily the agent's. Such a theory is, in this respect, not an analog of, say, the neurological theory of vision, which is also not the agent's theory. Rational choice theory is not descriptive, as the neurological theory is, but normative. This means, incidentally, that the normative theory of rational choice is very hard — indeed, it is beyond most people including theorists.

As I understand them, Selten's and Jeffrey's view on this issue is far more compelling than Aumann's. In what follows, I will argue entirely from intuitive tests of arguments. In particular, in keeping with Hahn's notion of equilibrium, I will test Harsanyi's determinacy principle for game theoretic solution theory. There are, in any case, virtually no data on real-life, many-times but finitely iterated Prisoner's Dilemmas.

Iterated Prisoner's Dilemma

Few problems in game theory have generated as much discussion as the iterated Prisoner's Dilemma, especially when it is iterated a fixed number of times. Very early it was proposed that, if it is rational to defect in a single-play Prisoner's Dilemma, then it is also rational to defect in every play in a fixed iteration of plays of the game. The argument, from so-called backwards induction, is that it must be rational to defect on the last play because there are no longer any incentives from future plays to bring into consideration. But then that last play cannot provide incentives in prior plays and both players must know this to be true, so that it must be rational to defect on the next to last play. And so on back to the first play. (Formal demonstrations of this transparent result abound [e.g., Friedman 1986, pp. 95-6].) There have been many practical and theoretical objections to that conclusion and there have been many and varied proposals for seemingly more sensible resolutions (Luce and Raiffa 1957, pp. 97-102; Radner 1980; Hardin 1982, pp. 138-54; Sorenson 1988, pp. 344-61).

Determinacy yields an argument for the rationality of continuous defection that is much simpler than the traditional backwards induction. From the symmetry of the players, a determinate theory must prescribe the same strategy to each player. We may suppose it permits mixed strategies, so long as they are well defined. If it prescribes cooperation on the final, nth ($n >> 1$) play of the series, either player can do better by violating the theory. Similarly, suppose that, on the final play, it prescribes a probabilistic mix of strategies, $pC + (1 - p)D$, where C is cooperation, D is defection, and $1 \geq p > 0$. Then either player can do better by defecting on that play. If it prescribes defection on all plays after some number q ($n > q \geq 1$) with cooperation or a mix of cooperation and defection on play $q - 1$, then either player can do better by defecting on play $q - 1$. Hence, the theory cannot prescribe either cooperation or a mix of cooperation and defection on the final play and it cannot prescribe either of these on any play that precedes a final series of one or more defections. It must therefore prescribe defection on every play. Hence, either we should defect on every play in a fixed iteration Prisoner's Dilemma or we can have no acceptable determinate theory.

Against the first half of this conclusion, many have noted that the players in a Prisoner's Dilemma that is iterated many times could do better than the mutual always-defect outcome. That is just the core meaning of rationality: to choose more rather than less. Already Luce and Raiffa (1957, p. 100) supposed that rational players would cooperate for the first 90 or so plays of a 100-play Prisoner's Dilemma. Unfortunately, it is not clear what it would mean to maximize in this context. Nor may anything like the Luce and Raiffa solution be in equilibrium under a given equilibrium concept. Of course, the always-defect strategy also does not maximize. We could say that a particular strategy would maximize only against some particular strategy followed by the other player. If that other player's strategy were determinate, then I could know how to maximize. If we constrain rational choice to yield only the always-defect strategy, then it follows that each can maximize with respect to that strategy by playing it.

To deal with such problems as the apparent irrationality of always-defecting in a finitely iterated Prisoner's Dilemma, game theorists have proposed many equilibrium concepts to single out better results than the sometimes dismal Nash equilibrium. These concepts include proper, perfect, subgame perfect, sequential, and persistent equilibria. These are not equivalent and some may fairly be called arcane. All of the debate here is over what counts as a good rational choice theory.

Given that we — rational choosers — have difficulty finding and agreeing on such a theory, we should hedge our own rational choices by considering the possibility that the theory we hold at the moment will eventually seem wrong. If my theory says I should defect continuously in iterated Prisoner's Dilemma, I should hold open the possibility that you will convince me of the superiority of another theory that recommends otherwise. Hence, when we begin playing our many-times iterated Prisoner's Dilemma, if you cooperate on the first or some early move, I should reconsider my theory as Hahn suggests. Of course, because of the symmetry of our reasoning under any correct theory, I should also consider what would be the effect of my own cooperative choice on the first play (Hardin 1982, p. 149; for related but more complex views, see Bicchieri 1988). If you are, say, twice as modest as the average academic game theorist, you will likely reconsider the wisdom of continual defection once I have cooperated and thereby wrecked your expectations. Here we do not invoke the trembling hand, which might mistakenly choose the wrong strategy. Rather, we invoke a trembling intellect that, rightly, has reason to doubt its own theory. In this case, by violating my own theory, I can expect to improve my own long-term payoff. But that just means my theory is self-contradictory. If a theory recommends its own violation, it is not in equilibrium under the Hahn notion of equilibrium. The theory not only trembles, it collapses.

Where does this leave us? Evidently, we are stuck with indeterminacy in game theoretic solution theory and in rational choice theories more generally. Our theory cannot tell players in an iterated Prisoner's Dilemma fixed to run, say, 100 plays that they should cooperate for most of the plays and then start defecting at the 98th play even if the other is still cooperating on the 97th play. It can only say that efforts at cooperation may be individually rational until far along, perhaps until nearly the final play of the series. If the series is short — say, two or three plays — the theory might commend defection all the way if it commends defection in the single-play game. (And, of course, if the payoff for cooperating while the other defects is very large relative to the difference between the both-cooperate and the both-defect payoffs, it may not be rational to cooperate at all.)

In single-play Prisoner's Dilemma, there is an additional consideration that makes defection seem compelling: Defection dominates cooperation, which is the only other pure strategy. (It therefore also dominates every mixed strategy.) In iterated Prisoner's Dilemma, always-defect does not dominate all other pure strategies. It may be said to chain-dominate them, as in the backwards induction argument. For example, Tit-for-tat all the way is dominated by Tit-for-tat with defection on the last play (call this T). The latter is dominated by Tit-for-tat with defection on the last two plays (call this T - 1). And so forth. But chain dominance is not a compelling principle for choice. It leaves open the possibility that, for example, T - k is inferior to T. In pairwise choices over various strategies, always-defect might lose to many other strategies. In any case, there are many strategies which it does not dominate. To force its selection by imposing determinacy or equilibrium conditions on final outcomes goes well beyond the primitive notion that it is rational to choose more rather than less value or to have transitive preferences.

The sure thing or dominance principle sounds misleadingly like a transitivity requirement over strategy choices (not outcomes). Strategy A dominates strategy B, we eliminate B. Now among the strategies in the newly reduced set, C is the strategy our theory selects. But this does not entail transitivity of dominance or even of strategy rankings. If B were still in the set, it might be chosen over C because it beats C for some strategy choices of other players or because it would elicit more beneficial strategy choices from others. Under our theory without the elimination of dominated strategies, we prefer C to A to B to C.

If A, B, and C were states of affairs or values, we would want our preferences over them to be transitive. Does it make sense to have intransitive preferences over strategies? It would be wrong to have intransitive preferences over strategies if the strategies determine outcomes directly, for example, if all other players' strategy choices are already known. But this is because, under these conditions, intransitive rankings of strategies entail intransitive preferences over states of affairs.

More generally, however, it is odd even to speak of preferences over strategies. What we normally speak about is preferences over alternative states or over different values. Then we may rank strategies according to the payoffs we expect them to bring us. My choice of any strategy will bring me a return that is a function of what strategy you choose. To determine what I expect my return to be in an iterated interaction, I must estimate what you will likely do in response to my strategy choice. Then I may say in the ordinary way that I prefer the likely state of affairs that follows from my choosing one strategy to the likely state of affairs that follows from my choosing the always-defect strategy. In a many-times iterated Prisoner's Dilemma with reasonably favorable ratios of the payoffs, one suspects that almost everyone would choose to play cooperatively to some extent to engender cooperation.

A surprising number of recent equilibrium concepts seem to have been motivated at least partly by the compelling intuitive sense that the always-defect strategy in finitely iterated Prisoner's Dilemma is not rational. To suppose it is rational is to suppose that very good game theorists would play always-defect for hundreds, thousands, or more plays if only the trial is to terminate after a set number of plays. That is an absurd supposition. This is only an intuitive claim for cooperation, but it is a firmer intuition than any intuition that rational solution theory must be determinate or must produce standard equilibrium outcomes in ongoing contexts. It appears to be an intuition shared by many, perhaps most, contemporary game theorists, who seem to be motivated by the evident failure of the standard equilibrium argument of backward induction.

Concluding Remarks

Perhaps the demand for determinacy comes from earlier simplifying assumptions of complete and perfect information. If we all have complete information about what all others are going to do, and they have complete information about us, the outcome of our interactive choices must be determinate, at least as from now. But that means it must be an equilibrium. If it were not, at least one of us would have incentive to act in some way other than that in which complete and perfect information says he or she will act. One might suppose that this is paradoxical and therefore conclude that the assumption of complete information on future choices is an incoherent notion. (This is one response to some ways in which Newcomb's problem is stated [Hardin 1986].) Alternatively, one might conclude that the notion is not incoherent, that there are no actual choices to be made just because all actions are determinate. This move raises the usual conundrums of determinism in philosophy. Not the least troubling of its implications is that choice theory cannot be prescriptive because there are no agents to whom it could prescribe anything.

Whatever motivates that demand, however, requiring determinacy in rational choice theory is less compelling than the urge to prefer more to less value in our interactions. To require determinacy sets up the cute claim that it is rational to act irrationally in iterated Prisoner's Dilemma, because it is beneficial to act cooperatively despite the rational dictum always to defect. As Harsanyi notes, for game theory we must go beyond the rationality postulates of individual decision theory (Harsanyi, 1977, p. 10), or we cannot resolve our problems. Iterated Prisoner's Dilemma seems to provide a major litmus test for the general applicability of various extensions of the notion of rationality. Determinacy fails that test.

References

Aumann, Robert J. 1985. "What Is Game Theory Trying to Accomplish?" Pp. 28-76 in Kenneth J. Arrow and Seppo Honkapohja, eds. Frontiers of Economics. Oxford: Basil Blackwell.

Bicchieri, Cristina. 1988. "Self-Refuting Theories of Strategic Interaction: A Paradox of Common Knowledge." Erkenntnis 11: 1-17.

Friedman, James W. 1986. Game Theory with Applications to Economics. New York: Oxford University Press.

Hahn, Frank. 1984. "On the Notion of Equilibrium in Economics." Pp. 43-71 in Frank Hahn, ed., Equilibrium and Macroeconomics. Cambridge, Mass.: MIT Press.

Hardin, Russell. 1982. Collective Action. Baltimore, Md.: Johns Hopkins University Press.

-----. 1986. "Pragmatic Intuitions and Rational Choice." Pp. 27-36 in A. Diekmann and P. Mitter, eds. Paradoxical Effects of Social Behavior: Essays in Honor of Anatol Rapoport. Heidelberg: Physica-Verlag.

Harsanyi, John C. 1977. Rational Behavior and Bargaining Equilibrium in Games and Social Situations. Cambridge: Cambridge University Press.

-----. 1956. "Approaches to the Bargaining Problem Before and After the Theory of Games: A Critical Discussion of Zeuthen's, Hicks', and Nash's Theories." Econometrica 24: 144-57.

Harsanyi, John C., and Reinhard Selten. 1988. A General Theory of Equilibrium Selection in Games. Cambridge, Mass.: MIT Press.

Jeffrey, Richard C. 1983. The Logic of Decision. Chicago: University of Chicago Press, 2nd edition.

Luce, R. Duncan, and Howard Raiffa. 1957. Games and Decisions. New York: Wiley.

Radner, Roy. 1980. "Collusive Behavior in Noncooperative Epsilon-Equilibria of Oligopolies with Long but Finite Lives." Journal of Economic Theory 22: 136-54.

Selten, Reinhard. 1985. "Comment." Pp. 77-85 in Kenneth J. Arrow and Seppo Honkapohja, eds. Frontiers of Economics. Oxford: Basil Blackwell.

Sorensen, Roy A. 1988. Blindspots. Oxford: Oxford University Press.

Inductive Reasoning in Repeated Games

Robert Sugden
School of Economic and Social Studies, University of East Anglia
Norwich NR4 7TJ, United Kingdom

It is generally agreed that rational players in a one-off Prisoner's Dilemma will defect. It is true that both players would do better if they both cooperated; but each is still acting rationally when he or she defects. Most decision theorists have no trouble in accepting this conclusion; and I shall not be challenging it. But what if two rational individuals play a finite sequence of n Prisoner's Dilemmas? There is a well-known argument of backward induction or backward recursion which leads to the conclusion that rational players will defect in every round - no matter how large n may be. This argument was presented by Duncan Luce and Howard Raiffa (1957, pp.97-102) over thirty years ago. Ever since then, the formal validity of this conclusion has been widely accepted. John Harsanyi and Reinhard Selten (1988, p.345), for example, say that they regard 'backward-induction rationality' as 'an essential aspect of game-theoretic rationality when dealing with sequential games'.

This conclusion, however, has often been seen as paradoxical. Intuitively, it seems that if n is sufficiently large, a player would be well-advised to follow a strategy such as tit-for-tat, which might signal to his opponent a conditional willingness to cooperate. Luce and Raiffa, while endorsing the backward recursion argument as formally correct, remark that they do not think it 'reasonable' to single out this solution to the game. By this they mean that they do not predict that, even after full consideration of the theoretical argument for defection, intelligent players would in fact defect in every round. Similarly, Selten (1978, pp.138) says that the backward recursion argument is 'game theoretically correct' but the advice to play tit-for-tat in all but the very last rounds of the game 'seems to be the better guide to practical behavior'.

One common response to this paradox has been to contrast the unbounded rationality that game theorists attribute to individuals in their theories with the necessarily bounded rationality of real human beings. Thus, it is said, the backward recursion argument is valid for perfectly rational individuals, while being inappropriate if the object is to advise ordinary human beings how to play games against one another. Since we are ordinary human beings ourselves, it is not surprising that we have an intuitive sense that the backward recursion argument is wrong, and indeed it gives <u>us</u> bad advice; but it would give good advice to the unboundedly rational agents to whom it is addressed. Selten (1978, especially p.147) himself seems to take this line, offering a sketch of a theory of 'limited rationality' which would explain why people might play tit-for-tat. The theory presented by David Kreps and Robert Wilson (1982) and by Kreps, Paul Milgrom, John Roberts and Wilson (1982) may be interpreted as belonging to the same line of thought, in that it introduces a small element of uncertainty about the rationality of the players.

One might respond instead by questioning the validity of the backward recursion argument itself. Perhaps the conception of rationality which underlies the argument is not so much super-human as incoherent. This is essentially what has been argued in different ways by Philip Reny (1986), Ken Binmore (1987), Cristina Bicchieri (1989), Philip Pettit and myself (1989), Kaushik Basu (1990) and Giacomo Bonanno (1991). The present paper belongs to the same tradition. However, it has a slightly different emphasis. It focuses on the role of inductive reasoning in repeated games, with 'induction' being used in the philosophical and scientific sense of 'making general inferences from particular instances'. (The term 'backward induction' appeals to a different sense of 'induction'; to avoid confusion, I use the term 'backward recursion' in place of 'backward induction'.)

The backward recursion argument is best-known in relation to two particular games: the finitely-repeated Prisoner's Dilemma, and Selten's (1978) finitely-repeated Chain Store Game. (The Chain Store Game involves two players, the <u>monopolist</u> and the <u>entrant</u>. First the entrant chooses whether to enter the monopolist's market or not; if the decision is to enter, the monopolist chooses between a <u>cooperative</u> and an <u>aggressive</u> response. Given the fact of entry, it pays the monopolist to be cooperative; but if he could make a credible threat to be aggressive, this would deter entry and give him a still better payoff. If this game is repeated a finite number of times, a backward recursion argument shows that the entrant will always choose to enter and that the monopolist will always make a cooperative response.) For my purposes, however, it is convenient to analyse the implications of backward-recursive reasoning for a different game.

The Hybrid Game

Consider the game shown in Figure 1. This game is a cross between the Prisoner's Dilemma and the Assurance Game (Amartya Sen, 1967); I shall call it the <u>Hybrid Game</u>. To dramatize the game, suppose that both players are self-interested and risk-neutral, so that utility can be measured in money units.

Each player is offered the choice between 'cooperating' and 'defecting'. A player who cooperates must pay £1 to the bank; a player who defects pays nothing. The bank then pays £100 to A if and only if B cooperates; it also pays £100 to B if and only if both players cooperate. Thus the payoffs for A are as in the Prisoner's Dilemma: he is better off it both cooperate than if both defect, but defection is a dominant strategy. The payoffs for B, in contrast, are as in the Assurance Game. B, like A, is better off if both cooperate than if both defect, but defection is not a dominant strategy for her: if A cooperates, her best reply is to cooperate too.

Figure 1: The Hybrid Game

		Player B	
		cooperate	defect
Player A	cooperate	99, 99	-1, 0
	defect	100, -1	0, 0

First suppose that this game is played only once. If A is rational, he will play his dominant strategy, which is to defect. If B believes A to be rational, she will expect him to defect. Her best reply to this strategy is to defect too. Thus if each player is rational and believes the other to be rational, both players will defect. In this respect, the Hybrid Game is like the Prisoner's Dilemma.

There is, however, a difference between the two games. In the Hybrid Game, it would be in B's interest to cooperate if she expected A to cooperate. Thus if there were some mechanism by which A could publicly and irrevocably commit himself to the strategy of cooperation, it would be in his interest to make such a commitment. There would be no need for negotiation between the players: A could be confident that his own unilateral commitment would be sufficient to induce B to cooperate too. In the Prisoner's Dilemma, in contrast, the ability to make binding commitments is not sufficient to allow a cooperative solution. There also has to be some mechanism by which each player can make his own commitment to cooperate conditional on the other's making a similar commitment. Intuitively, then, we might expect the problem of reaching the cooperative outcome to be less intractable for a repeated Hybrid Game than for a repeated Prisoner's Dilemma.

I shall analyse the game which consists of exactly one hundred rounds of the Hybrid Game. After each round, each player knows which move (i.e. 'cooperate' or 'defect') was chosen by his or her opponent in that round. Each player's payoff for the whole game is the sum of his or her payoffs in the hundred rounds. We may now use an argument of backward recursion to show that, if the rationality of the players is a matter of common knowledge, each player will defect in every round; each will end up with a payoff of zero.

This conclusion seems even more surprising than the analogous conclusion for the finitely-repeated Prisoner's Dilemma. Suppose you are playing this game as A, facing an opponent whom you believe to have excellent powers of reasoning. (To avoid begging the question, I do not say that your opponent is rational in the sense in which this word is used in game theory.) Consider the following advice that someone might give to you: 'Irrespective of what B does, cooperate in each of Rounds 1,..., 99 and defect in Round 100. The very worst that can happen to you is that B never cooperates, in which case your payoff is -99. But you can be fairly confident that this strategy will induce B to start playing cooperative moves fairly early in the game - say, no later than Round 30 - and that she will continue to cooperate until quite late in the game - say, at least up to Round 80. So you can be fairly confident of getting a payoff of at least 5000. And even if B cooperates only once in the whole game, your payoff will still be positive'. We need not ask whether this is the best advice that you could be given. It is sufficient to ask whether you could rationally believe that the expected utility of following this advice - of following what I shall call the <u>unconditionally cooperative strategy</u> - would be greater than zero. On the face of it, this seems an entirely reasonable belief.

There is a paradox here, similar to the Chain Store paradox and to the paradox of the finitely-repeated Prisoner's Dilemma. We can prove that players whose rationality is common knowledge will defect in every round. But at the same time, it seems that a player who does not defect in every round might reasonably expect to do better than a player who does. How can these two conclusions be reconciled?

<u>The Argument for Defection: Common Knowledge of Rationality</u>

As a first step towards answering this question, I shall set out the backward recursion argument, as carefully as I can. Let me begin by offering a definition of the concept of <u>common knowledge of rationality</u> (CKR). I shall say that there is CKR if all three of the following conditions are satisfied:

K1. The mathematical description of the game is common knowledge.

K2. Each player is rational in the sense of expected utility theory, and this is common knowledge.

K3. Any logical or mathematical theorem that can be proved about the game is common knowledge.

For my present purposes, the 'mathematical description of the game' is the game tree, with the utility payoffs for both players specified. A player is 'rational in the sense of expected utility theory' if he assigns subjective probabilities to all events and then maximizes expected utility. (The choice of any particular strategy by the players' opponent is to be understood as an event.) These subjective probabilities must be consistent with what the player knows to be true; thus, for example, each player must attach a subjective probability of zero to the event that his opponent is irrational. A proposition is common knowledge if each player knows it to be true, each knows the other knows it to be true, and so on indefinitely (David Lewis, 1969).

In game theory it is conventional to assume K1 - K3. These conditions are equivalent to the assumptions from which Douglas Bernheim (1984) and David Pearce (1984) derive the solution concept of rationalizability. Some theorists argue that the concept of common knowledge of rationality includes more than K1 - K3. Robert Aumann (1987), for example, argues that we are entitled to make the 'common prior assumption' - to assume that if two players receive precisely the same information, they each attach the same subjective probability to any given event. For my purposes, there is no need to examine such arguments; it is sufficient to use the weak version of CKR contained in K1 - K3. The conclusions I shall draw will necessarily hold for any stronger version of CKR.

Let us assume that CKR holds and then examine the implications of this assumption for the finitely-repeated Hybrid Game. Consider the last round of the game (Round 100) in any part of the game tree that can be reached, given the truth of CKR. (In other words: consider how Round 100 will be played, given a history of rational play in the previous ninety-nine rounds. Since CKR is true by assumption, we can deduce that when Round 100 comes to be played, there will be a history of ninety-nine rounds of rational play.) Since A is rational, he will defect in Round 100. Since B knows A to be rational, she knows that A will defect in Round 100. And so, since she is rational, she will defect in Round 100 too.

Now consider Round 99 in any part of the game tree that can be reached, given CKR - that is, given a history of ninety-eight rounds of rational play. First suppose that it is consistent with CKR for A to cooperate in this round. In Round 99 itself, the payoffs to A from defecting strictly dominate those from cooperating. Thus it can be rational for A to cooperate in Round 99, only if he believes that by so doing, he can increase his payoff in Round 100.

But we have already proved that whatever move A makes in Round 99, provided it is consistent with CKR, B will defect in Round 100. Thus it cannot be the case that, by making a move in Round 99 that is consistent with CKR, A can increase his payoff in Round 100. Therefore the original supposition is false: A's cooperating in Round 99 is inconsistent with CKR. But CKR is true. So A will defect in Round 99. This is a result that can be proved about the game using premises that B knows to be true, and so by CKR, B knows this result to be true.

Now consider B. Suppose that it is consistent with CKR for B to cooperate in Round 99, given that previous play in the game has been consistent with CKR. B knows that A's move in Round 99 will be consistent with CKR, and that the only such move is defection. Thus B's payoff in Round 99 will be greater if she defects than if she cooperates. So it can be rational for B to cooperate in Round 99, only if she believes that by so doing, she can increase her payoff in Round 100. But she knows that whatever move she makes in Round 99, provided it is consistent with CKR, A will defect in Round 100. Thus it cannot be the case that by making a move in Round 99 that is consistent with CKR, B can increase her payoff in Round 100. Therefore the original supposition is false: B's cooperating in Round 99 is inconsistent with CKR. But CKR is true. So B will defect in Round 99.

This argument can be repeated for Round 98, and so on back to Round 1. Thus CKR implies that <u>both players will defect in every round</u>. Notice, however, that nothing has been said about what would happen if either player were to cooperate in any of Rounds 1,..., 99. What has been proved is that this eventuality will not occur.

This leaves us with the problem of understanding <u>why</u> A defects in every round, rather than playing some other strategy. In particular, we may ask why he does not play the unconditionally cooperative strategy. We know that A is rational in the sense of expected utility theory. So A must believe that the expected utility he derives from playing the strategy he actually does play is no less than the expected utility he would derive from playing the unconditionally cooperative strategy. Thus we must have proved something about A's beliefs about how B would respond, were A to play the unconditionally cooperative strategy.

Now we come to a difficulty. Given that CKR is true, how can A form <u>any</u> rational belief about how B would respond to a move that is inconsistent with CKR? A knows that B is rational, so he must expect her to make a rational response to any move he makes. But if A cooperates in Round 1, B is faced with the occurrence of an event that she <u>knows</u> will not occur. And this is an inconsistent hypothesis. The notion of a rational response to such a hypothesis is incoherent. We cannot ask a theory to tell us what would happen if the theory were false. (This point is common to the arguments of Reny (1986), Binmore (1987), Bicchieri (1989), Pettit and Sugden (1989), Basu (1990) and Bonanno (1991)). So in the context of the game we are considering, CKR implies the truth of an incoherent proposition. The correct conclusion to draw, I suggest, is that CKR is itself an incoherent assumption.

The Argument for Defection: Entrenched Common Belief in Rationality

To this is may be objected that I am interpreting the concept of knowledge too literally. Might we be able to rescue the backward recursion argument by replacing the concept of knowledge with that of subjective certainty? (We might say, for example, that B is subjectively certain that A will defect in Round 1: she attaches a subjective probability of zero to the event 'A cooperates in Round 1'. Then it would be meaningful to ask what beliefs B might rationally form after having observed that A cooperated in Round 1.) Consider the concept of entrenched common belief in rationality (ECBR) as a substitute for CKR. I shall say that ECBR is true if the following conditions hold.

B1. Each player knows the mathematical structure of the game.

B2. Each player is rational in the sense of expected utility theory.

B3. Each player knows the truth of every logical or mathematical theorem that can be proved about the game.

B4. B1 - B3 are matters of common belief at the start of the game; they remain matters of common belief as long as the hypothesis that they are matters of common belief can be maintained without inconsistency; that this is so is itself a matter of common belief.

A proposition is a 'matter of common belief' if each player believes it to be true, each player believes that the other believes it to be true, and so on. Conditions B1 - B3 express the same concept of rationality as do Conditions K1 - K3, and require that players are rational in this sense. The difference is that B1 - B3 say nothing about the players' rationality being common knowledge. Instead, B4 makes their rationality a matter of common belief; this common belief is 'entrenched' in the sense that it will be maintained as long as it is not contradicted by anything else that the players know to be true. (The idea of entrenchment has some similarities with Elon Kohlberg and Jean-Francois Merten's (1986) concept of forward induction: if a player assigns zero probability to an event which then occurs, and if the occurrence of that event is consistent with the existence of a common belief in rationality, then when the player revises his beliefs to account for the occurrence of the event, he must continue to believe in there being a common belief in rationality). I present the concept of ECBR as a suggested reconstruction of what game theorists have intended (or should have intended) when saying that the rationality of the players is 'common knowledge'.

I have already proved that CKR implies that in the finitely-repeated Hybrid Game, both players will defect in every round. This proof can easily be adapted to show that ECBR implies the same conclusion.

Suppose that B1 - B4 are true. Then both players are rational; and at the start of the game, this is a matter of common belief. Thus each player's move in Round 1 will be consistent with his or her rationality and with there being a common belief in rationality. So the common belief in rationality is not contradicted by either player's obervation of the moves made in Round 1; because of B4, this common belief will remain in place at the start of Round 2. We may now repeat the argument: each player's move in Round 2 will be consistent with his or her rationality and with there being a common belief in rationality, and so this common belief will remain in place in Round 3. And so on: the common belief in rationality will remain intact throughout the game. (The common belief can be destroyed only if one or other player makes a move that is inconsistent with ECBR; since both players are rational, neither will choose to make such a move, and since each believes the other to be rational, neither will believe that the other will choose to do so.)

We can now adapt the original proof that the players will defect in every round, substituting 'ECBR' for 'CKR' and 'believe' for 'know'. Now, however, there is no incoherence in asking how B would respond, were A to cooperate in (say) Round 1. We - the theorists who have defined the game - know that A will defect in Round 1. But B does not <u>know</u> this; she merely <u>believes</u> certain propositions which imply that A will defect. Were A to cooperate in Round 1, B would revise her beliefs. There would no longer be a common belief in rationality, but this would not prevent B from forming a new set of consistent beliefs and maximizing expected utility in the light of these.

Let us take stock of what we have proved. On the assumption of ECBR, we have proved that each player will follow a strategy that leads him or her to defect in each round, provided the other player has defected in all previous rounds. We have also proved that each player will initially believe that the other will follow such a strategy, and will continue to hold this belief as long as the other player's moves do not contradict it. Thus each player will go into the game expecting that both players will defect in every round. And so relative to his or her beliefs, each player's expected utility at the start of the game must be zero.

Consider player A. His chosen strategy, which generates an expected utility of zero, must be optimal in relation to his beliefs. (This is implied by the assumption that A is rational.) So there can be no other strategy open to A which would yield a higher expected utility. Now consider an alternative strategy that A could play: the unconditionally cooperative strategy of cooperating in Rounds 1, ..., 99 and defecting in 100, irrespective of how B responds. Our proof implies that A's beliefs are such that the expected utility of this strategy is no greater than zero. Any beliefs which imply that the unconditionally cooperative strategy has a positive expected utility are inconsistent with ECBR.

Now consider how B would respond to A's unconditionally cooperative strategy. More particularly, consider any Round i. Let p(i) stand for B's judgement of the probability that A will cooperate in Round i, conditional on A's having cooperated in each of Rounds 1, ..., i - 1 and on B's having defected in each of these rounds. If p(i) > 0.01, the expected payoff for B in Round i will be greater if she cooperates than if she defects. Thus it will be rational for her to cooperate, provided she believes that her cooperation in Round i will not make A less likely to cooperate in later rounds. This latter belief seems entirely reasonable, and so from now on I shall assume that B holds it. Now suppose we could show that for at least one round i, p(i) > 0.01. This would imply that if A followed the unconditionally cooperative strategy, and if B had not cooperated before Round i, it would be rational for her to cooperate in Round i. Thus if B were rational, she would cooperate in at least one round (either in Round i or in some preceding round). And if she were to cooperate in at least one round, A's payoff for the whole game would be greater than zero. So if A believes that p(i) > 0.01 for at least one round i, and if he believes that B will act rationally in this round, then the expected utility to him of the unconditionally cooperative strategy is greater than zero. But ECBR implies that the expected utility of the unconditionally cooperative strategy is _not_ greater than zero. Thus ECBR implies that A does not hold such beliefs. We may still, however, ask _why_ A does not hold such beliefs. I shall consider this question later in the paper.

The Argument for Defection: Common Knowledge of Incomplete Rationality

I have argued that CKR is an incoherent assumption, and have suggested ECBR as a more satisfactory way of expressing what game theorists have intended. An alternative way of dealing with the problem of incoherence is proposed by Selten (1975, especially pp.35-36). Selten's idea is that we should '[look] at complete rationality as a limiting case of incomplete rationality'; we should then assume there to be common knowledge that the players are _almost_ completely rational.

This approach requires us to formulate a theory of how players may occasionally act non-rationally. The truth of this theory is assumed to be common knowledge between the players. Selten proposes a theory in which each player is subject to occasional lapses of rationality or trembles when 'rationality breaks down' and 'some unspecified psychological mechanism' takes over. Such lapses last only as long as it takes the player to make one move - in the context of the Hybrid Game, they last for the duration of one round. They occur with very low probability, and the probabilities of lapses in different rounds are independent of one another. For a player suffering a lapse of rationality, every move that could possibly be made has a non-zero probability. This idea can be modelled by using the concept of a perturbed game. A perturbed game is one in which every possible move has some non-zero minimum probability. These minimum probabilities are treated as constraints on the actions of the players. (Thus, for example, we might assume that there is a 0.002 probability that player A will suffer a tremble when making his first move, and that if he does, he is equally likely to cooperate or to defect. This would be modelled by setting a minimum probability of 0.001 for each move.) The players are then assumed to act rationally, subject to these constraints. Formally, then, there is no departure from rationality; the possibility of trembles is modelled as part of the structure of the game. Selten proposes that we think of a game in which there is complete rationality as the limit of a sequence of perturbed games as all minimum probabilities converge to zero.

We may now define the concept of common knowledge of incomplete rationality (CKIR) by the following conditions:

I1. The mathematical structure of the game, including the minimum probabilities implied by some given theory of trembles, is common knowledge.

I2. Subject to the constraints set by the minimum probabilities, each player is rational in the sense of expected utility theory; and this is common knowledge.

I3. Every logical or mathematical theorem that can be proved about the game is common knowledge.

If we assume CKIR and take all minimum probabilities to be vanishingly small -- this amounts to taking Selten's limiting case - then the backward recursion argument goes through very simply. Because of the role of the minimum probabilities, every part of the game tree can be reached with some positive probability: every conceivable course of play is consistent with CKIR. In Round 100, irrespective of the history of play, it is rational for A to defect. B knows that A is rational, and so knows that he is almost certain to defect. (There is a vanishingly small probability that he will suffer a tremble.) Thus it is rational for her to defect too. In Round 99, irrespective of the history of play, each player knows that the other is rational. Each knows that, irrespective of the moves that are made in Round 99, the other is almost certain to defect in Round 100. Thus it is rational for A to defect in Round 99. B knows this, and so knows that A is almost certain to defect. So it is rational for her to defect too. And so on, back to Round 1. The implication is that each player will choose to defect in every round, irrespective of the history of play; cooperative moves will be made only as a result of trembles.

We may now ask what will happen if A plays the unconditionally cooperative strategy. Clearly, B will defect in every round (unless she trembles): we have already proved that B will defect in every round, irrespective of the history of play. Suppose that i - 1 rounds have been played. In each of these rounds, A has cooperated and B has defected. Given what B has been assumed to know, she can explain these observations in only one way: she must infer that A has trembled in i - 1 successive rounds. But she knows that the probabilities of trembles in different rounds are independent of one another, and so her observations do not affect her judgement of the probability that A will tremble in Round i. And she knows that if A does not tremble in Round i, he will defect. So p(i), B's judgement of the probability that A will defect in Round i, must be equal to the minimum probability of defection in that round, as given by the theory of trembles. And that probability, we have assumed, is vanishingly small. Thus CKIR implies that p(i) is vanishingly small for all i. A knows that this is the case, and so if he plays the strategy of unconditional cooperation, his expected utility is close to -99. In this sense, we can explain why A does not choose this strategy. Nevertheless, as I shall now try to show, there remains a sense in which B's beliefs - and hence A's beliefs about B's beliefs - are unreasonable.

Inductive Reasoning

A firm is testing a new machine for performing some operation. On any one trial of the machine, it may either succeed or fail. A series of 100 trials is carried out on successive days. The characteristics of each trial are as near to identical as it is possible to arrange, but of course, the machine has been subjected to more use before later trials than before earlier ones. Suppose that i - 1 of the trials have been carried out, and that the machine has failed on every one. What subjective probability p(i) would it be reasonable to assign to the event that the machine will fail on the ith trial?

There is no uniquely correct answer to a question of this kind. The standard Bayesian response is to say that the answer depends on one's prior probabilities. But I shall suggest that, on the basis of familiar principles of inductive reasoning, some answers can be dismissed as unreasonable. Consider, for example, the judgement that for all i = 1, ..., 100, p(i) = 0.01. Before the trials begin, it might be entirely reasonable to subscribe to a theory which has exactly this implication. (The theory might be that the failures are caused by some random process: on every trial, there is an independent probability of failure to equal to 0.01. According to this theory, the probability of failure on the ith trial is independent of previous observations.) But would it be reasonable to continue to accept this theory after observing failures on each of the first 20, or 30, or 40 trials? Surely not. Given these observations, we are entitled to say that the original theory should be rejected.

Suppose we have observed thirty successive failures on the first thirty trials. It seems undeniable that this represents a pattern in our observations, which might be described as 'the machine's failing on every trial' or perhaps 'a very strong tendency for the machine to fail on any trial'. This is more than an ex post rationalization of a set of random observations. To see this, suppose we had only the first fifteen of these thirty observations. How would we describe the pattern in these observations? Surely, in the same terms. The second fifteen observations are then consistent with a pattern that we could have identified by looking only at the first fifteen. This test (which corresponds with the idea of 'post-sample prediction' in econometrics) gives us reason to believe that the pattern we have found is not spurious.

On the basis of our observations, and without invoking any theory of the workings of the machine itself, we may reasonably propose the hypothesis that the probability of failure on each trial is greater than the 0.01 implied by the original theory. This hypothesis has been generated inductively: it is a hypothesis for organizing the data that our observations have given us. Suppose we form this hypothesis after making the first fifteen observations. This becomes our alternative hypothesis: the null hypothesis is that the original theory is correct. We then observe fifteen more successive failures. If the null hypothesis is true, the probability of observing fifteen successive failures is 1 in 10^{30}. It would be wholly unreasonable to maintain the null hypothesis in the face of this evidence. More generally, it would be unreasonable to hold any set of beliefs which implied $p(i) \leq 0.01$ for all i = 1, ..., 100.

Notice that I am not claiming that the mere fact of thirty successive observations of failure makes it unreasonable to believe that the thirty-first trial has a high probability of success. An understanding of the workings of the machine might give us some reason to expect a high probability of failure on early trials and a low probability of failure on later ones. If so, then from the outset we might conceivably have accepted a hypothesis such as the following: p(i) = 0.99 for i = 1, ..., 30; p(i) = 0.01 for i = 31, ..., 100. The evidence of the first thirty trials gives us no strong reason to reject this hypothesis. And so we should be entitled to maintain the hypothesis, with its implication that the probability of failure on the thirty-first trial is only 0.01. But we are entitled to do so, only because we have a reason for expecting the probability of failure to change after thirty trials - a reason that is independent of the fact that the thirty-first trial happens to be the next one. So it is not necessarily unreasonable to attach a low probability to p(i) for some i, even where i is relatively large. But it is unreasonable to attach a low proability to every p(i).

Suppose we reinterpret the example so that it refers to the reasoning processes of a human being. Let us substitute for the machine, a reputable mathematician who is confronted with a series of a hundred arithmetic problems. Each 'trial' is now a new problem given to the mathematician. The problems are similar in structure and complexity. (Suppose each problem requires a five-digit number to be multiplied by a two-digit one.) An incorrect answer counts as a 'failure'. The mathematician will be paid £100 for each correct answer. The probability p(i) is defined as before: it is the probability of failure on the ith trial, conditional on our having observed failures on each of the first i - 1 trials. Would it be reasonable to believe p(i) ≤ 0.01 for all i = 1, ..., 100? The answer, I think, must be 'No' - for exactly the same reasons as before. At the outset, it might be entirely reasonable to accept a hypothesis which implies an independent probability of failure on each trial, less than or equal to 0.01. (Even the best mathematicians make occasional mistakes.) But it would be unreasonable to maintain this hypothesis after observing failures on each of the first say, thirty trials.

Inductive Reasoning in the Hybrid Game

Now consider the implications of this kind of reasoning for the finitely-repeated Hybrid Game. Let us leave open the question of what the players believe about one another's rationality. (The object of this exercise is to ask whether the beliefs that are implied by ECBR or CKIR are reasonable. So we should not include ECBR or CKIR among the premises from which we deduce propositions about reasonable and unreasonable beliefs.) Suppose that i - 1 rounds of the Hybrid Game have been played. In each of these rounds, A has cooperated and B has defected. Let p(i) be B's judgement of the probability that, given this history of play, A will cooperate in Round i. Would it be reasonable for B to believe $p(i) \leq 0.01$ for all i = 1, ..., 100? The same arguments apply to this case as to the two previous ones: such beliefs would be unreasonable.

It may be tempting to think that inductive reasoning is inappropriate when forming beliefs about the actions of a rational agent. But this, I suggest, would be a mistake. Suppose that B has observed A to cooperate in Rounds 1, ..., 30. There is a pattern in these observations, and it is reasonable to expect that there exists some explanation for it. If A is <u>not</u> a rational agent, then the explanation will be non-rational. This case is clearly analogous with that of the machine, and inductive reasoning is legitimate. If, in constrast, A <u>is</u> a rational agent, then presumably the explanation will have some reference to his reasons for action: A must have had good reason for cooperating in Rounds 1, ..., 30. Unless there is something special about Round 31, we should expect the same reasons to apply in this round, and so we should expect A to have good reason to cooperate in Round 31. Thus we are entitled to use inductive reasoning, irrespective of whether or not A is a rational agent.

As an illustration of this general point, consider how B might respond to the observation that A has cooperated in Rounds 1, ..., 30. She might think: 'I feel sure A is rational. So he must be cooperating with the object of inducing me to cooperate. But as a rational agent, he will not cooperate right up to Round 100. Perhaps he will start defecting in Round 31'. These thoughts are perfectly reasonable as they stand, but we must ask what probability it would be reasonable for B to assign to the 'perhaps'. It would be unreasonable to maintain a hypothesis that implies a low value of p(i) for all i = 1, ..., 30: such a hypothesis fails to account for the observed pattern of A's play in these rounds, and so must be rejected. Either there is something special about Round 31, or p(31) cannot reasonably be judged to be close to zero.

What if B feels sure, not only that A is rational, but also that A feels sure that she is rational, and so on indefinitely? The answer is simple: that is a hypothesis that has already been refuted by the evidence of Rounds 1, ...,30. Why has A cooperated in Round 30? Either because he is irrational, or because he believed that by so doing, he could induce B to cooperate in some later round, say Round 31. But why would we expect B to cooperate in Round 31? Either because B is irrational, or because B expects A to cooperate in Round 31. But then why would B expect A to cooperate in Round 31? This method of reasoning cannot be continued indefinitely, since the game ends at Round 100. A's cooperation in Round 30 cannot be explained without rejecting the hypothesis that there is a common belief in rationality: this fact is consistent only with a limited 'depth' of beliefs by the players about one another's rationality.

Perhaps, it might be suggested, B subscribes to the hypothesis that there is a finite limit to the depth of A's beliefs. In itself, this hypothesis is quite defensible - it may, indeed, be more defensible than the hypothesis that beliefs extend to unlimited depths (Binmore, 1987; Bicchieri, 1988; Michael Bacharach, 1990). But only a very special hypothesis about the exact limits of the depth of beliefs would both be consistent with the observation that A has cooperated in Rounds 1, ..., 30 and imply that, with high probability, A will defect in Round 31. Thus we arrive at the same conclusion as before: either there is something special about Round 31, or p(31) cannot reasonably be judged to be close to zero. Neither case provides a justification for setting p(i) close to zero for all i = 1, ..., 100.

A different possibility is that B believes A to be subject to lapses of rationality, and attributes cooperative moves to such lapses. This makes the case of the Hybrid Game completely analogous with the cases of the machine and of the mathematician. At the start of the game, it might be reasonable for B to accept the hypothesis that in every round there is an independent probability, of 0.01 or less, that A will cooperate. This hypothesis might be implied by a theory of lapses something like Selten's. But it would be unreasonable to maintain this hypothesis after observing A to cooperate on Rounds 1, ..., 30.

It is sometimes argued that Selten's model of incomplete rationality is implausible. Binmore (1987, pp.182-184), for example, argues that a satisfactory theory of mistakes would 'attribute the primary source of mistakes to imperfections within the reasoning process itself'. Such a theory, he says, would imply the possibility of systematic errors, and thus that mistakes made by the same player in different rounds of a game might be correlated. He concludes that any worthwhile theory of games must incorporate a theory of bounded rationality, and that the validity of any particular theory of bounded rationality is an empirical matter. 'Abstract mathematical reasoning alone' (such as, presumably, lies behind Selten's theory of trembles) is inadequate (Binmore, 1988, p.48).

Binmore may well be right in all this; but the point I am making is rather different. Binmore's argument is about the nature of non-rational behaviour. In contrast, I am asking whether A might rationally choose the strategy of unconditional cooperation. The possibility I am considering is that A might deliberately choose to put a pattern into his play - the pattern of cooperating in every round - in the belief that B will recognize this pattern and expect it to continue. We have no reason to expect that a deliberately-chosen pattern of play will be capable of being explained by a theory of mistakes, however well-grounded that theory may be - unless we beg the question by presupposing that the deliberately-chosen pattern is an irrational one. Indeed, a rational player might know what the best-available theory of mistakes implies, and might deliberately put into his play a pattern that the theory cannot explain.

For the same reason, the present argument must be distinguished from that of Kreps and Wilson (1982). Kreps and Wilson analyse the finitely-repeated Chain Store Game. They change the game slightly, assuming that there is a small probability that the monopolist's payoffs are different from those of the standard game: in this case, the monopolist benefits from following the aggressive strategy in any round in which the entrant chooses to enter. Kreps and Wilson describe this as a model of 'incomplete information', but at one point they suggest an alternative interpretation in which there is a small probability that the monopolist irrationally acts if his payoffs were different from what they actually are (p.256). This amounts to a model of incomplete rationality in which, in Binmore's terms, errors are systematic. A small probability of this particular form of irrational play is enough to generate a 'sequential equilibrium' in which entry will not occur until the final rounds of the game.

This kind of argument, however, seems <u>ad hoc</u>. Kreps and Wilson do not deny this. Their readers, they say, may suspect that by cleverly choosing the nature of the uncertainty or irrationality that is assumed, 'one can get out of a game-theoretic analysis whatever one wishes'. They reply: 'we certainly share these suspicions. If this is so, then the game-theoretic analysis of this type of game comes down eventually to how one picks the initial incomplete information' (p.276).

My argument, in contrast, does not require any particular prior hypothesis about the nature of irrational play. If A chooses the strategy of unconditional cooperation, he is putting a pattern into his play. It is sufficient for my argument that B can be expected to recognise this pattern. It does not matter whether she explains the pattern as the result of rational choice or as the result of some kind of systematic mistake - or, indeed, whether she can explain it at all. Nor does it matter whether, at the start of the game, she foresaw the possibility of observing such a pattern: it is enough that she recognises it when she sees it.

So whatever else we assume about B's beliefs, familiar principles of inductive reasoning imply that it would be unreasonable for her to believe that $p(i) \leq 0.01$ for all $i = 1, ..., 100$. Thus if A is sufficiently confident of B's rationality, he has good reason to believe that he if chooses the unconditionally cooperative strategy, B will cooperate at least once. And this implies that A has good reason to believe that the expected utility of the unconditionally cooperative strategy is greater than zero.

Resolving the Paradox

Recall what we have proved so far. If we assume either common knowledge of rationality, or entrenched common belief in rationality, or common knowledge of incomplete rationality, we can prove that A will defect in every round of the finitely-repeated Hybrid Game, and that B will respond by defecting in every round too. Thus A's expected utility from playing the game is zero. This implies that A must believe that the expected utility of playing the unconditionally cooperative strategy is no greater than zero. If we have begun by assuming CKR, this implication is incoherent, and so we must conclude that CKR is itself an incoherent assumption. If we have begun by assuming ECBR or CKIR, we can deduce that, for each $i = 1, \ldots, 100$, A does not believe that B believes $p(i) > 0.01$. But we have also shown that, according to principles of inductive reasoning, it would be unreasonable for B to believe $p(i) \leq 0.01$ for all i.

It seems, then, that there is a fundamental conflict between the principles of inductive reasoning on the one hand, and ECBR and CKIR on the other. My own view is that we should treat the validity of inductive reasoning as a fixed point, since it provides the foundation for so much of what we regard as knowledge. It is doubtful whether the human race could have survived at all if it had not been endowed with the capacity for such reasoning. Any acceptable conception of rationality ought to include the principles of both deductive and inductive reasoning. In contrast, ECBR and CKIR are no more than technical assumptions within game theory. My provisional conclusion, then, is that in the context of the finitely-repeated Hybrid Game, ECBR and CKIR are false. This implies that we may not assume that either ECBR or CKIR are true for all games.

Some readers will perhaps be more tempted by the conclusion that the principles of inductive reasoning, at least in the form I have used them, are not universally valid. So let me offer an independent reason for scepticism about ECBR and CKIR.

Underlying both ECBR and CKIR is a theory of rational choice in which individuals assing utility indices to all consequences, assign subjective probabilities to all events, and then maximize expected utility. Given this starting-point, we can show that a particular act is rational for some individual only by showing that, in terms of his utility function and his subjective probabilities, that act has at least as great an expected utility as every other feasible act. But when we come to interpret this idea, there is an important distinction between decision-making in games against rational opponents and decision-making in games 'against nature'.

In games against nature, the events to which subjective probabilities are assigned are sets of states of nature, which can be defined independently of the acts between which decisions have to be made. This is an essential feature of Leonard Savage's (1954) axiomatic derivation of expected utility theory. In Savage's theory, the set of all possible states of nature and the st of all possible consequences are defined independently of one another, and any assignment of consequences to states of nature (formally, any mapping from the set of states of nature on to the set of consequences) is an act. Savage's procedure for deriving a person's subjective probabilities works by constructing particular kinds of acts and then looking at that person's preference ranking of those acts. Suppose, for example, we wish to know whether a person judges event E to be more or less probable than event F. We find two consequences, x,y, such that x is strictly preferred to y. We then construct two acts g, h, where g gives x in event E and y otherwise, and h gives x in event F and y otherwise. If the person prefers g to h, then E is defined to be subjectively more probable than F. By using only two such consequences x, y, it is possible to determine the subjective probability of every event; these probabilities can then be used in the analysis of any decision problem. Thus we may think of subjective probabilities as existing prior to the specification of the particular decision problems we wish to analyse. The same idea is implicit in axiom systems such as that of John von Neumann and Oskar Morgenstern (1947), which take probabilities as given.

But this idea cannot necessarily be carried over to games between individuals who hold beliefs about one another's rationality. At first sight, it seems that we may interpret a player's choosing a strategy in a game as an event. Suppose we are interested in A's subjective probability judgements about B's choice of strategy in a particular game. We may construct a matrix in which the rows represent strategies available to A, the columns represent strategies available to B, and the entries represent the consequences for A of the possible combinations of strategies. This looks like an act/event matrix of the kind that defines a decision problem in Savage's theory. But is it one? If it were, we should be able to define the probability of each 'event' (i.e. each of B's strategies) without knowing anything about the 'acts' (i.e. A's strategies and their consequences in each event) that happen to constitute A's decision problem. But if A believes that B believes that he is rational, we may not be able to do this. The problem is that B's decision may depend on what she expects A to do, and this may depend on the nature of the strategies between which A has to choose. If this is the case, then it is not possible to define the probabilities of B's strategies independently of the nature of A's strategies. Savage-style expected utility theory is no longer appropriate.

This is much more than a technical quibble. If we assume ECBR or CKIR, we are assuming that A chooses between strategies so as to maximize expected utility: this is the criterion that determines what it is rational for him to do. In order to carry out this calculation, we need to know the probabilities that A assigns to B's strategies. But if the rationality of the players is a matter of common knowledge or common belief, B's decision may depend on what it is rational for A to do. (In the case of the Hybrid Game, how B will respond to a cooperative move by A may depend on whether she interprets this move as rational or irrational.) In general, then, we need to know the probabilities of B's strategies in order to work out what it is rational for A to do, but we need to know what it is rational for A to do in order to determine these probabilities. We are left with an infinite regress: there may be no way of determining what it is rational for A to do. But neither ECBR or CKIR admits this possibility. Each of these concepts includes the assumption that each player is rational 'in the sense of expected utility theory', with the other player's strategies being treated as events to which subjective probabilities can be assigned. It is a property of expected utility theory that probabilities can always be assigned to events, and that every finite decision problem has a solution. The implication of all this seems to be that ECBR and CKIR, like CKR, may be incoherent if they are interpreted as assumptions that can be made about all games.

Conclusion

The finitely-repeated Hybrid Game, like the finitely-repeated Prisoner's Dilemma and the finitely-repeated Chain Store Game, is a well-defined game which real people could play. It is natural to ask the question: 'What is it rational for each player to do in such a game?' In seeking to make this question more precise, game theorists have asked: 'What would it be rational for each to do, if there were common knowledge of rationality?', or 'What would it be rational for each to do, if there were an entrenched common belief in rationality?', or 'What would it be rational for each to do, if there were common knowledge of imperfect rationality?' The answers to these questions have often seemed counter-intuitive.

I have argued that, at least for the Hybrid Game, these answers are much more than counter-intuitive. If we accept some very basic principles of inductive reasoning, we must conclude that these answers are false: from the premise that the rationality of the players is a matter of common knowledge or common belief, we can derive implications that are self-contradictory. And this implies that the premise itself must be false. To ask what it would be rational for players to do, were the premises true, is to ask a meaningless question.

ACKNOWLEDGEMENTS

This paper was written as part of the Foundations of Rational Choice Theory project at the University of East Anglia, which is supported by the Economic and Social Research Council of the UK (award R 000 23 2269). The ideas contained in the paper developed out of discussions with many people, including Michael Bacharach, Giacomo Bonanno, David Copp, Robin Cubitt, Jean Hampton, Martin Hollis, Judith Mehta, Philip Pettit and Chris Starmer.

REFERENCES

Aumann, Robert J. (1987), 'Correlated equilibrium as an expression of Bayesian ignorance', Econometrica, vol. 55, pp.1-18.

Bacharach, Michael (1990), 'Backward induction and beliefs about oneself', mimeo, Christ Church, Oxford.

Basu, Kaushik (1990), 'On the non-existence of a rationality definition for extensive games', International Journal of Game Theory, vol. 19, pp.33-44.

Bernheim, B. Douglas (1984), 'Rationalizable strategic behavior', Econometrica, vol. 52, pp.1007-1028.

Bicchieri, Cristina (1988), 'Strategic behavior and counterfactuals', Synthese, vol. 76, pp.135-169.

Bicchieri, Cristina (1989), 'Self-refuting theories of strategic interaction: a paradox of common knowledge', Erkenntnis, vol. 30, pp.69-85.

Binmore, Ken (1987), 'Modelling rational players: Part I', Economics and Philosophy, vol. 3, pp.179-214.

Binmore, Ken (1988), 'Modelling rational players: Part II, Economics and Philosophy, vol. 4, pp.9-55.

Bonanno, Giacomo (1991), 'The logic of rational play in games of perfect information', Economics and Philosophy, forthcoming.

Kohlberg, Elon and Jean-Francois Mertens (1986), 'On the strategic stability of equilibria', Econometrica, vol. 54, pp.1003-1037.

Kreps, David M. and Robert Wilson (1982), 'Reputation and imperfect information', Journal of Economic Theory, vol. 27, pp.253-279.

Kreps, David M., Paul Milgrom, John Roberts and Robert Wilson (1982), 'Rational cooperation in the finitely repeated prisoner's dilemma', Journal of Economic Theory, vol. 27, pp.245-252.

Harsanyi, John C. and Reinhard Selten (1988), A General Theory of Equilibrium Selection in Games (Cambridge, Mass.: MIT Press).

Lewis, David K. (1969), Convention: A Philosophical Study (Cambridge, Mass.: Harvard University Press).

Luce, R. Duncan and Howard Raiffa (1957), Games and Decisions (New York: Wiley).

Neumann, John von and Oskar Morgenstern (1947), Theory of Games and Economic Behavior, 2nd edition (Princeton: Princeton University Press).

Pearce, David G. (1984), 'Rationalizable strategic behavior and the problem of perfection', Econometrica, vol. 52, pp.1029-1050.

Pettit, Philip and Robert Sugden (1989), 'The backward induction paradox', Journal of Philosophy, vol. 86, pp.169-182.

Reny, Philip (1986), 'Rationality, common knowledge and the theory of games', Ph.D dissertation, Princeton University.

Savage, Leonard (1954), The Foundations of Statistics, (New York: Wiley).

Selten, Reinhard (1975), 'Reexamination of the perfectness concept for equilibrium in extensive games', International Journal of Game Theory, vol. 4, pp.22-55.

Selten, Reinhard (1978), 'The chain store paradox', Theory and Decision, vol. 9, pp.127-159.

Sen, Amartya (1967), 'Isolation, assurance and the social rate of discount', Quarterly Journal of Economics, vol. 81, pp.112-124.

IV. Game Models

Price-Quantity Oligopoly with Adjustment Costs

James W. Friedman
Department of Economics, University of North Carolina
Chapel Hill, NC 27599-3305

1. Introduction

Economists have long discussed whether output or price is the "correct" decision variable for an oligopolistic firm. The earliest reference to this issue is probably Bertrand (1883); more recent contributors are Alger (1979), Shubik and Levitan (1980), Kreps and Scheinkman (1983), Davidson and Deneckere (1986), Benassy (1986a, 1986b), and Friedman (1988). In Friedman (1988) I analyzed three versions of an n-firm differentiated products oligopoly in which firms choose both prices and output levels, and firms that have not produced enough to meet their demand cause a spillover demand that benefits firms having excess production. Except for Benassy the papers listed above deal with homogeneous product duopoly. Benassy (1986b) is concerned with much the same questions as Friedman (1988) and he uses a similar framework. The present paper extends Benassy (1986b) and Friedman (1988) by allowing firms to adjust their output at some extra cost after obtaining price information. In Friedman (1988), three decision frameworks were examined: a) Game S, a model of simultaneous price and output choice, b) Game Q, a model of sequential choice with output first, and c) Game P, a model of sequential choice with price first. In Benassy (1986b) frameworks (a) and (c) were examined. With attention restricted to pure strategy equilibria, the following results were obtained: An equilibrium never exists in the simultaneous choice model. In the output-first model, equilibrium may exist and, when it does, it coincides with the equilibrium of a straight output model. In the price-first model equilibrium may not exist and, when it does, it coincides with the equilibrium of a straight price model.*

The three frameworks capture different assumptions about the relative flexibility of price and output choice. If, for example, output is chosen prior to price, then price is the

---

* Friedman (1988) incorrectly asserts that equilibrium always exists in Game P. Local conditions are, indeed, satisfied, but existence of equilibrium is not assured. At the price-only equilibrium prices the maximum outputs of the firms could be small enough that a firm has a more profitable price than its price-only equilibrium price. The same elements are at work here as in the simultaneous choice model. This is correctly perceived and analyzed in Benassy (1986b).

more flexible instrument. Obviously flexibility is more appropriately represented in an intertemporal model where the period t price is chosen at a later time that the period t output level; however, a single shot model with various stages of decision does capture the relevant issues and provides guidance for the development of a multiperiod model. A typical scenario with price being the more flexible is where production schedules require considerable advanced planning and commitment while price can be freely varied on short notice. Firms in such a situation can often alter their production plans on short notice at some extra cost, and they will wish to do so if they find the rate of sales turning out to be quite different from what they had expected.

This phenomenon is captured below in single shot two and three stage markets. For Game S, where firms select prices and output levels simultaneously, each firm is permitted to adjust output after the initial selections of price and output are made and announced. For Game Q the firms choose output levels which become public knowledge, then they choose prices which also become public knowledge, and finally, they may adjust output levels. In both the modified Games S and Q the adjustments to output incur an adjustment cost. A similar modification to Game P would be trivial because initial output in Game P is chosen following the selection of prices, making further adjustment useless. When Game S is modified, existence of equilibrium may be restored provided that the adjustment cost is not too great. When Game Q is modified, an array of equilibria become possible, depending on the size of the adjustment cost. The polar cases are zero adjustment cost and infinite adjustment cost. The former renders the original output choice irrelevant and effectively gives us Game P again. The latter makes no change in the original Game Q. For a market in which Games P and Q both have equilibria, as adjustment costs start from zero and rise, the outcome starts at the (unmodified) Game P equilibrium and then moves continuously toward the (unmodified) Game Q equilibrium.

One sometimes hears or reads discussions about firms "choosing" whether price or output will be their strategic variable, with the remaining variable automatically set at the market clearing level. I believe this is an inappropriate approach to the *price versus quantity as strategic variable* question. In general such a choice cannot be costless. A real firm will inevitably select both, so the issue is over whether just one of them has all the strategic clout. The results of Benassy (1986b) and Friedman (1988) suggest that the variable chosen first has the strategic clout when choice is sequential. Given these results, what determines the order of choice? In a one period model a sequential ordering seems artificial; however, it corresponds to temporal flexibility in a multiperiod model. If, for example, period t output must be chosen in period t - 1 while period t price is selected in period t, output is *chosen first* and is *less flexible* than price. Such a multiperiod analysis is carried out in Friedman (1989). The results of the multiperiod model with output chosen a period early are approximately the same as the results of Game Q.

Can firms then choose the degree of flexibility of one variable relative to the other? It is conceivable they could under some circumstances, depending on the nature of the technologies available for producing their output and on the timing of output decisions that the chosen technology imposes. Conceivably a firm may buy greater output flexibility by choosing a higher cost technology. Such a technology choice is not pursued here, although there may be merit in analyzing it. The framework in the present paper follows that of Benassy (1986b) and Friedman (1988) in assuming the firms' technologies to be given.

Section 2 presents the demand and cost structures that are used throughout the paper. Section 3 defines the quantity-only and price-only equilibria. These are benchmark outcomes; for, when equilibrium exists in Game Q, the outcome coincides with that of the quantity-only model, and the outcome in Game P coincides with that of the price-only model. Then Section 4 has an analysis of the final stage game played when outputs have been provisionally selected, prices have been chosen, and the firms can, at some cost, change production plans. The results of this section are needed in Sections 5 and 6. Section 5 contains the analysis of a modified version of Game S that allows subsequent output adjustment. Section 6 is parallel to Section 5 for Game Q, and Section 7 contains concluding remarks.

2. The Demand and Cost Setup

Each of n firms chooses both its selling price and its output level. A firm's actual sales equal its production only if the amount demanded of it is no smaller than the amount it produces. If a firm's production falls short of the amount demanded of it, then rival firms will experience an increase in demand, called *spillover demand*.

The set of firms is denoted $N = \{1, 2, ..., n\}$. For firm i, let s_i be sales, x_i be production, q_i be amount demanded, all three in physical units, and let p_i be price. Vectors of sales, output, etc., are denoted by s, x, q, and p. The n firms form a differentiated products market and their demand functions are developed below. The demand faced by each firm in the absence of any spillover demand (i.e., if $x_i \geq q_i^0$ for all $i \in N$) is

$$q_i^0 = f_i(p) \tag{1}$$

The set of firms is partitioned into those that produce less than q_i^0 (denoted L), more than q_i^0 (denoted M), and exactly q_i^0 (denoted K). For firms in K and L demand is given by eq. (1), so that $q_i = q_i^0$ for $i \in K \cup L$, and sales are given by

$$s_i = \min \{x_i, q_i^0\} \tag{2}$$

If L is empty or M is empty (or both), then sales for all firms are given by eq. (2). Otherwise for firm $i \in M$, spillover demand is related to functions $h_{ij}(p, M)$ which are defined for each set of firms $M \subset N$ ($M \neq N$), each $i \in M$, and each $j \in L$. The function $h_{ij}(p, M)$ specifies the number of additional units demanded of firm i per unit of demand shortfall experienced by firm j. The demand faced by a member of M is

$$q_i = q_i^0 + \Sigma_{j \in L} h_{ij}(p, M) \cdot (q_j^0 - x_j) \tag{3}$$

Sales for firms in M are given by

$$s_i = \min \{x_i, q_i\} \tag{4}$$

For notational simplicity eqs. (1) to (4) are summarized and the sales of firm i are denoted

$$s_i = \delta_i(p, x) \tag{5}$$

Conditions on this demand system are given below as Assumptions 1 and 2. The main economic content of these assumptions are that the firms' products are gross substitutes, each firm has a *choke price* above which it sells nothing, and when firm stocks out (when $x_i < q_i^0$), then firms having excess supply ($x_i > q_i^0$) experience an increase in demand.*

ASSUMPTION 1 The functions $f_i(p)$ and $h_{ij}(p, M)$ are defined, continuous, single valued, and nonnegative for all $p \in R^n_+$, $i \in N$, $M \subset N$ where $i \in M$, and $j \in N \backslash M$. There is a finite, positive price, p_i^0, such that $f_i(p \backslash p_i') = 0$ and $h_{ij}(p \backslash p_i', M) = 0$ for all $i \in N$, $p \in R^n_+$, $p_i' \geq p_i^0$, and M containing i.

---

* Some notational conventions followed in the paper are given below. For sets the notation A\B means *the points in* A *that are not in* B. For vectors, in contrast to sets, the notation $x \backslash y_i$ means the vector x with its ith coordinate set equal to y_i. That is $x \backslash y_i = (x_1, \ldots, x_{i-1}, y_i, x_{i+1}, \ldots, x_n)$. R^n_+ denotes the nonnegative orthant of R^n, and R^n_{++} denotes the positive orthant. Vector inequalities are denoted as follows: For x and y in R^n, $x \leq y$ means $x_i \leq y_i$ for all i, $x < y$ means $x \leq y$ and $x \neq y$, and $x << y$ means $x_i < y_i$ for all i. For partial derivatives the commonly used superscript notation is used. Thus $f_i^j = \partial f_i / \partial p_j$ and $f_i^{jk} = \partial^2 f_i / \partial p_j \partial p_k$, etc.

As firm i's demand will be zero if its price exceeds p_i^0, the price domain can be limited to $\Omega = \{p \in \mathbb{R}_+^n | p \leq (p_1^0,..., p_n^0)\}$. Let the subset of Ω on which $f_i(p)$ is positive be denoted $\Omega_i = \{p \in \Omega | f_i(p) > 0\}$.

ASSUMPTION 2 Each $p_i f_i(p)$ is bounded and concave in p_i on Ω_i, and each $p_j h_{ij}(p, M) f_j(p)$ is bounded, concave, and strictly positive on $\Omega_i \cap \Omega_j$ and is zero elsewhere. The following derivatives exist and satisfy these conditions:
a) $f_i^i(p) < 0$ on Ω_i and $f_i^i(p) = 0$ otherwise,
b) $f_i^j(p) > 0$, $j \neq i$, on $\Omega_i \cap \Omega_j$ and $f_i^j(p) = 0$ otherwise,
c) $f_i^i(p) + \Sigma_{j \neq i} f_i^j(p) < 0$ everywhere on Ω_i.

It is assumed that firm i experiences a strictly increased level of demand if some firm $j \neq i$ cannot meet its total demand and firm i has excess production (i.e., $x_i > q_i^0 > 0$). Condition (a) is the usual downward sloping demand condition stating that a firm's demand falls as its price rises, given $q_i^0 > 0$. Condition (b) requires that the firms' products be gross substitutes. Condition (c) assures that a firm's demand will fall if all firms raise prices together by requiring that *own* effects dominate cross effects in demand.

Assumptions 1 and 2 ensure that inverse no-spillover-demand functions exist. Let $f = (f_1,..., f_n)$ be the demand system and denote the inverse demand system by $F = (F_1,..., F_n)$ where $p_i = F_i(q)$. Let $\mathring{\Theta} = \{q \in \mathbb{R}_{++}^n | q = f(p), p \in \Omega\}$ and let Θ be the closure of $\mathring{\Theta}$.

LEMMA 1 (Friedman 1988). Under Assumptions 1 and 2 the inverse demand system, F, is a single-valued, continuous function from Θ into Ω.

Let the total cost of production for firm i be denoted by $C_i(x_i)$. The profit function of firm i is

$$\pi_i(\cdot) = p_i \delta_i(p, x) - C_i(x_i) \tag{6}$$

Conditions on the cost functions are given below in Assumption 3. The main provisions are that fixed cost is nonnegative, and marginal cost is both strictly positive and nondecreasing.

ASSUMPTION 3 The ith firm's production cost function, C_i, is continuous on $\mathbb{R}_+$, continuously differentiable on $\mathbb{R}_{++}$, and nonnegative and convex in x_i. Marginal cost has a strictly positive lower bound. That is, for all $x_i > 0$, $C_i'(x_i) \geq \zeta > 0$. $i \in N$.

Assumptions 1 to 3 imply that rational producers will never choose output above some finite upper bounds, $\dot{x}_i^0$. Production above this level guarantees the firm will have lower profit than if it produced nothing, irrespective of the prices chosen by the firm or its rivals.

LEMMA 2 (Friedman 1988). Under Assumptions 1 to 3 there is an $x_i^0 < \infty$ such that if $x_i > x_i^0$ then for any $p \in \Omega$, $\pi_i(p) < -C_i(0)$. $i \in N$.

Given Assumption 1 and Lemma 2, it is natural to restrict x_i to be an element of $[0, x_i^0]$ and to restrict p_i to be an element of $[0, p_i^0]$. Let $\Xi = \{x \in R_+^n \mid x \leq (x_1^0,..., x_n^0)\}$.

3. The Price-Only Equilibrium, the Output-Only Equilibrium, and some Previous Results

A straight price model is based on the profit functions

$$\pi_i^p(p) = p_i f_i(p) - C_i(f_i(p)) \tag{7}$$

along with the understanding that firms only choose prices; output is automatically equal to demand. Denote by p^c a noncooperative equilibrium of this game.

A straight output model is based on the profit functions

$$\pi_i^q(q) = q_i F_i(q) - C_i(q_i) \tag{8}$$

along with the understanding that firms only choose output; prices are automatically set to clear the market. Denote by q^c a noncooperative equilibrium of this game.

From Vives (1985) it is known that, for any p^c, there is a q^c such that $p^c \ll F(q^c)$. For the Games S, Q, and P it is known from Benassy (1986b) and Friedman (1988) that output must exactly equal demand for all firms at equilibrium (in the absence of adjustment costs). These facts allow a quick summary of previous results.

For Game S, a firm's strategy is a price and output pair drawn from $\Sigma_i^S = [0, p_i^0] \times [0, x_i^0]$. Another necessary condition of equilibrium in Game S is that prices must equal p^c, the equilibrium price vector of the price-only model. However, with such prices holding, with $x = f(p^c)$, and with spillover demands present in the model, no firm is optimizing. Any firm i, given the outputs and prices of the others, would find its profit to be larger if it charged a slightly higher price and produced enough to meet its demand at that price. To see why, suppose an equilibrium (p^*, x^*) where $x^* = f(p^*)$. Three possibilities must be examined according to whether $\partial\pi_i^p/\partial p_i$ is a) equal to, b) less than, or c) greater than zero. Taking (a) first, recall that $\partial\pi_i^p(p^*)/\partial p_i = 0$ means that without any demand spillovers, the firm is optimizing and, even with spillovers, a price reduction would reduce profit. A price increase, however, would result in increasing above x_j^* the demand faced by the rival firms j. The rival firms would be unable to meet this additional demand, causing the demand of firm i to fall by *less than would occur in the absence of spillovers*. This, in turn, means the profit function of firm i has a kink at the putative equilibrium price p^*. The slope of the profit function is positive at $(p^*\backslash p_i)$ when $p_i < p_i^*$ and converges to zero as p_i rises to p_i^*, but the slope is also positive for $p_i > p_i^*$ and converges to a strictly positive limit as p_i falls to p_i^*. For case (b),

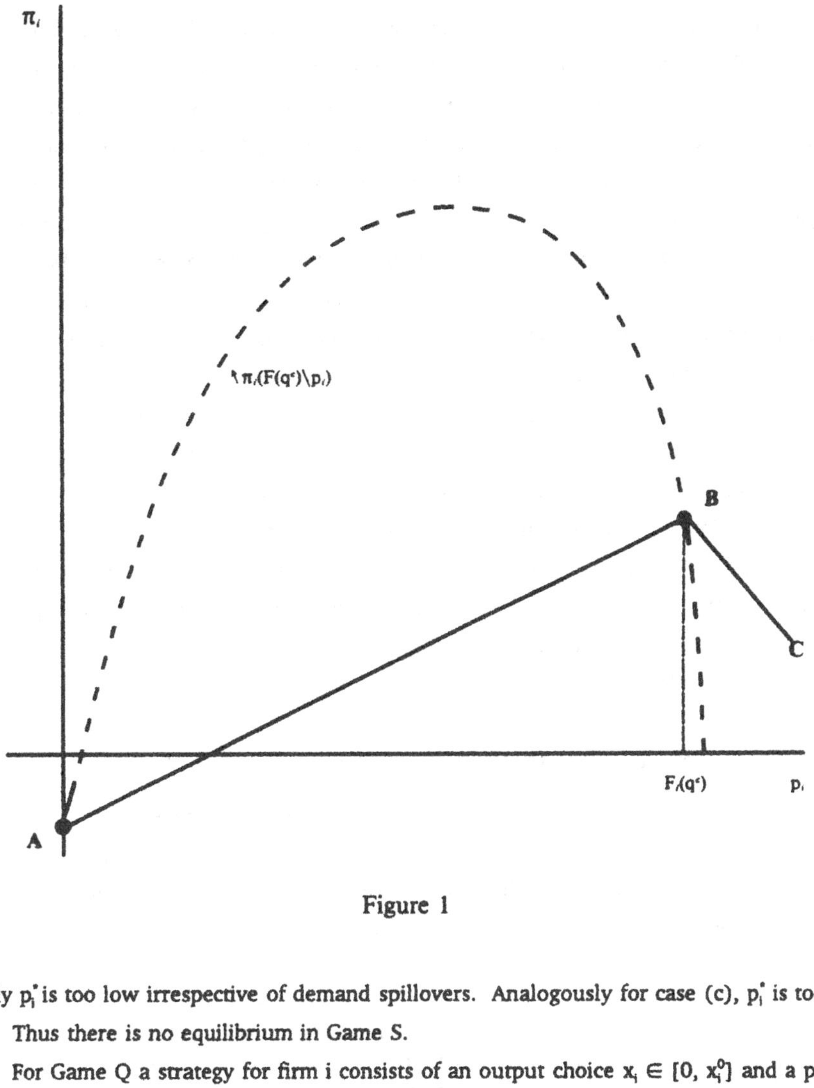

Figure 1

clearly p_i^* is too low irrespective of demand spillovers. Analogously for case (c), p_i^* is too high. Thus there is no equilibrium in Game S.

For Game Q a strategy for firm i consists of an output choice $x_i \in [0, x_i^0]$ and a price choice rule, ω_i that associates a price in $[0, p_i^0]$ with each possible $x \in \Xi$. Again at e-quilibrium $x = f(p)$. The latter and Vives (1985) imply that the equilibrium output choice must be an output-only equilibrium (i.e., some q^c). The price decision rule used at the second stage for firm i simply sets p_i equal to $F_i(q^c)$ unless the firm's total revenue is maximized at a higher price, in which case p_i is chosen to maximize revenue. The situation is depicted in Figure 1. In the figure the broken curve gives profit to firm i given that $q_j = q_j^c$ for $j \neq i$, firm i produces exactly the amount demanded of it, and there are no spillovers. Given both the existence of spillovers and the precommitment to an output of q_i^c, the solid curve gives the relationship between π_i and p_i. The upward-sloping, linear segment reflects the fact that cost is fixed (because production is committed); therefore, maximizing profit means maximizing revenue. The Vives Theorem assures that at equilibrium revenue is maximized by selling total output at $F_i(q^c)$. The remaining portion of the solid curve starts where price is just high enough to cause demand and production to be equal. As price rises from here, spillover effects, along with the inability to reduce output to match reduced demand, cause the profit curve to be different from the broken curve. Existence of a pure strategy equilibrium for Game Q rests upon the slope of the BC portion of the solid curve in Figure 1. If the slope is negative, as shown, then equilibrium exists; however, if the slope near B is positive for at least one firm, then a pure strategy equilibrium does not exist. The mechanism preventing equilibrium is the same as in Game S.

Finally Game P is the situation in which firms post prices, receive orders, and produce either enough to meet their orders or else the amount equating marginal cost to price (whichever is smaller). Equilibrium behavior requires choosing some price vector p^c at the first stage, then, at the second stage, choosing output to clear the market (i.e, $q_i = f_i(p^c)$), except if price would exceed marginal cost at such an output level.*

4. The Final Stage Game, When Output can be Adjusted

In this section equilibrium behavior is examined when firms have made prior choices of p and x, but their output choices, x, are not final. They may adjust them, at some cost. This subgame depicts the situation the firms face at the final stage of both a modified Game S and a modified Game Q. In modified Game S, the players will choose p and x simultaneously,

---

* In general it is possible that a firm will receive more orders than it wishes to fill at its posted price. It would then produce only to the point where price equals marginal cost. As Benassy (1986b) notes, starting from the price-only equilibrium prices, as one firm i raises its price, it will initially induce other firms $j \neq i$ to increase their stage two output; however, each firm j would stop increasing its output when its marginal cost reaches its own price. At this price spillover demands would begin to benefit firm i as it increases its price further. It is possible that the spillover demands will be large enough and start soon enough that firm i will maximize its profit at a price above its price-only equilibrium price.

then their choices become commonly known, and they can simultaneously adjust their output levels to differ from the original x. This adjustment can be expressed as a decision rule for each firm i that associates some final output y_i with any given (p,x). In modified Game Q, each firm i chooses x_i, a decision rule to select p_i as a function of x at stage two, and a decision rule to select y_i as a function of (p, x) at stage three. Thus with modified versions of both games, the firms face a final stage in which they have (p, x) committed from the past and choose final outputs, y, as functions of the observed (p, x).

The total cost firm i has incurred is $C_i(y_i) + \gamma_i(y_i - x_i)$ where $\gamma_i(y_i - x_i)$ is the cost of adjustment. The vector of final output levels is denoted y and the conditions governing γ_i are given below. The principal conditions on the adjustment costs are a) a firm incurs no such cost if $y_i = x_i$ and b) the marginal cost of adjustment is both strictly positive and increasing.

ASSUMPTION 4 The cost of adjusting output for firm i is given by γ_i, a nonnegative, strictly convex, scalar valued function, whose minimum is achieved at $\gamma_i(0) = 0$. For all $z_i \neq 0$, $\gamma_i(z_i)$ is twice continuously differentiable. Right and left limits of $\gamma_i'(z_i)$ as z_i converges to zero exist and have the same absolute value. This common limit value is denoted $\gamma_i'(0)$.

The profit function of firm i at the final stage is thus

$$\pi_i = p_i \delta_i(p, y) - C_i(y_i) - \gamma_i(y_i - x_i) \tag{9}$$

with p and x previously determined. Denote by $\phi_i(y, p, x)$ the best reply for firm i to (y, p, x). Thus $\phi_i(y, p, x)$ is the profit maximizing final output for firm i given that the n firms have already selected the price vector p, the provisional output vector x, and the other firms have selected final output y_j $(j \neq i)$. $\phi_i(y, p, x)$ is characterized below in Lemma 3. The presence of adjustment costs with $\gamma_i'(0) > 0$ imparts stickiness to output adjustment in some circumstances, due to the difference of $2\gamma_i'(0)$ between the marginal cost of increasing output and the marginal cost saving from a decrease in output.

LEMMA 3 The function $\phi_i(y, p, x)$ is single-valued and is characterized as follows:

a) $\phi_i(y, p, x) = q_i$ if $p_i \geq C_i'(q_i) + \gamma_i'(q_i - x_i)$ and $x_i \neq q_i$

b) $\phi_i(y, p, x) = x_i$ if $C_i'(x_i) + \gamma_i'(0) \geq p_i \geq C_i'(x_i) - \gamma_i'(0)$ and $x_i \leq q_i$

c) $\phi_i(y, p, x) = x_i$ if $0 \geq C_i'(x_i) - \gamma_i'(0)$ and $x_i \geq q_i$

d) $p_i = C_i'(\phi_i(y, p, x)) + \gamma_i'(\phi_i(y, p, x) - x_i)$ if $p_i < C_i'(q_i) + \gamma_i'(q_i - x_i)$ and $x_i \neq q_i$

e) $p_i = C_i'(\phi_i(y, p, x)) + \ {}'_i(\phi_i(y, p, x) - x_i)$ if $p_i < C_i'(q_i) + \ {}'_i(0)$ and $x_i = q_i$

f) $0 = C_i'(\phi_i(y, p, x)) + \gamma_i'(\phi_i(y, p, x) - x_i)$ if $C_i'(q_i) + \gamma_i'(q_i - x_i) \leq 0$, $C_i'(x_i) - \gamma_i'(0) \geq 0$, and $x_i > q_i$

PROOF The proof is driven by two facts: Marginal revenue is zero when $x_i > q_i$ and is p_i when $x_i < q_i$. It remain, then, to see what the marginal cost, marginal revenue relationship implies. Marginal cost at x_i is $C_i'(x_i) + \gamma_i'(0)$ for an output increase, but is $C_i'(x_i) - \gamma_i'(0)$ for an output decrease.

Condition (a) applies when the firm can adjust its production either upward or downward (according to whether $x_i < q_i$ or $x_i > q_i$) to equality with demand and have its price be no less than its marginal cost. Condition (b) applies when output is at or below demand and price is between the marginal cost associated with increasing output and the marginal cost associated with decreasing output. Marginal revenue is p_i and it exceeds marginal cost for a production decrease, but falls short for an increase. Condition (c) applies when output is at or above demand and the marginal cost associated with decreasing output is non-positive. Despite production exceeding demand, it does not pay to reduce production because total cost will increase. Condition (d) applies when $x_i \neq q_i$ and the firm's price is below the marginal cost that would prevail if output equalled demand (i.e., if $y_i = q_i$), so that the firm should produce less than demand, selecting output to make price equal to marginal cost. This has two subcases according to whether $x_i > q_i$ or $x_i < q_i$, but in both cases final output is brought below demand. Condition (e) is in the same spirit as condition (d), but is for the borderline situation where $x_i = q_i$ and reduction of output below demand is profitable. Condition (f) applies when provisional output exceeds demand and the marginal cost that would prevail if output equalled demand (i.e., if $y_i = q_i$) is negative. Then reducing output to equal demand is excessive and the firm should only reduce output to where marginal cost is zero. The cases (a) to (f) cover all possible values of (y, p, x) and concavity of the profit function guarantees that $\phi_i(y, p, x))$ is single valued; therefore, the lemma is proved. ■

Equilibrium play in the final stage requires the selection of a final output vector that is a fixed point of $\phi(\cdot, p, x)$ for the given values of (p, x). This fixed point is unique, as shown below.

THEOREM 1 Under Assumptions 1 to 4, $\phi(y, p, x)$ has a unique fixed point, denoted $\psi(p, x)$, for any given (p, x).

PROOF The proof proceeds by the following steps. First, a given (p, x) implies a unique partition of N into sets K, L, and M. The firms in L are the creators of spillover demand and the firms in M are the recipients. Optimal production y_i for a firm in sets K or L depends only on (p, x) and is invariant to choices of y_j by rival firms. This fact determines some values of the fixed point of $\phi(\cdot, p, x)$ and also determines precisely the spillover demands caused by firms in L. Therefore, the total demand for firms in M is determined and, for each firm, the optimal y_i is unique and independent of the optimal y_j of rival firms.

Conditions (a) to (f) from Lemma 3 are used to partition the set of firms, N. The firms in the set L, that would produce less than $f_i(p)$, are those for which i) $C_i'(x_i) + \gamma_i'(0) \geq p_i$ and $x_i < f_i(p)$, those for which ii) $p_i < C_i'(f_i(p)) + \gamma_i'(f_i(p) - x_i)$ and $x_i > f_i(p)$, and those for which iii) $p_i < C_i'(f_i(p)) - \gamma_i'(0)$ and $x_i = f_i(p)$. The firms in (i) do not find it profitable to raise their output from x_i all the way to meet their demand. The firms in (ii) and (iii) have prices low enough that their marginal cost for reducing output below demand exceeds their price.

The firms in the set K, that would produce exactly $f_i(p)$ even if their demand were larger, are those for which i) $x_i < f_i(p)$ and $p_i = C_i'(f_i(p)) + \gamma_i'(f_i(p) - x_i)$ and ii) $x_i = f_i(p)$ and $C_i'(f_i(p)) + \gamma_i'(0) \geq p_i \geq C_i'(f_i(p)) - \gamma_i'(0)$. The firms in (i) find it profitable to raise their output to equal demand; however, price equals marginal cost for them at $y_i = f_i(p)$, so they would not raise output further if their demand were larger. The firms in (ii) keep $y_i = x_i$ because price is both below the [illegible] above the marginal cost for lowering output. Therefore, such a firm would not raise its output if its demand were larger. The remaining firms comprise the set M of firms that would produce more than $f_i(p)$ if their demand were larger.

The optimal production ($\phi_i(y, p, x)$) for firms in L and K depends only on (p, x) and not on y. That is, a firm $i \in K \cup L$ wishes to sell no more than its base demand $f_i(p)$. This fact and the level of the base demand are totally determined by x_i and p; therefore, as the y_j ($j \neq i$) vary, $\phi_i(y, p, x)$ for such a firm does not vary. Thus for $i \in K \cup L$ the best response value of y_i can be denoted by $\psi_i(p, x)$ and note that $\phi_i(y, p, x) \equiv \psi_i(p, x)$ for all y. Using these $\psi_i(p, x)$ in eq. (3) the demand facing firms in M is exactly established given that all firms behave optimally. (I.e., given that firms in L and K set $y_i = \psi_i(p, x)$ and that firms in M set $y_i > f_i(p)$). The value of $\phi_i(y, p, x)$ is uniquely determined for all $i \in M$ once p and x are known and firms $j \in L \cup K$ set $y_j = \psi_j(p, x)$. The unique fixed point of $\phi(y, p, x)$ with respect to y can be denoted $\psi(p, x)$. ■

The optimal final stage behavior, $\psi(p, x)$, is further characterized in Lemma 4 below.

LEMMA 4 The function $\psi_i(p, x)$ has the following properties:

a) if $p_i > C_i'(q_i) + \gamma_i'(q_i - x_i)$ and $x_i > q_i$ then for $j \neq i$

$$\frac{\partial \psi_i}{\partial p_j} > 0, \frac{\partial \psi_i}{\partial p_i} < 0, \frac{\partial \psi_i}{\partial x_i} = 0, \text{ and } \frac{\partial \psi_i}{\partial x_j} = 0$$

b) if $p_i > C_i'(q_i) + \gamma_i'(q_i - x_i)$ and $x_i < q_i$ then for $j \neq i$

$$\frac{\partial \psi_i}{\partial p_j} > 0, \frac{\partial \psi_i}{\partial p_i} < 0, \frac{\partial \psi_i}{\partial x_i} = 0, \text{ and } \frac{\partial \psi_i}{\partial x_j} < 0$$

c) if $C_i'(x_i) + \gamma_i'(0) > p_i$ and $x_i < q_i$ then for $j \neq i$

$$\frac{\partial \psi_i}{\partial p_j} = 0, \frac{\partial \psi_i}{\partial p_i} = 0, \frac{\partial \psi_i}{\partial x_i} = 1, \text{ and } \frac{\partial \psi_i}{\partial x_j} = 0$$

d) if $p_i > C_i'(x_i) - \gamma_i'(0)$ and $x_i > q_i$ then for $j \neq i$

$$\frac{\partial \psi_i}{\partial p_j} = 0, \frac{\partial \psi_i}{\partial p_i} = 0, \frac{\partial \psi_i}{\partial x_i} = 1, \text{ and } \frac{\partial \psi_i}{\partial x_j} = 0$$

e) if $p_i < C_i'(q_i) + \gamma_i'(q_i - x_i)$ and $x_i \neq q_i$ then for $j \neq i$

$$\frac{\partial \psi_i}{\partial p_j} = 0, \frac{\partial \psi_i}{\partial p_i} > 0, \frac{\partial \psi_i}{\partial x_i} > 0, \text{ and } \frac{\partial \psi_i}{\partial x_j} = 0$$

f) if $C_i'(q_i) + \gamma_i'(q_i - x_i) < 0$ and $x_i > q_i$ then for $j \neq i$

$$\frac{\partial \psi_i}{\partial p_j} = 0, \frac{\partial \psi_i}{\partial p_i} = 0, \frac{\partial \psi_i}{\partial x_i} > 0, \text{ and } \frac{\partial \psi_i}{\partial x_j} = 0.$$

PROOF In cases (a) and (b) $\psi_i(p, x) = q_i$, so only variables that affect q_i will affect y_i. Changes in any price will affect q_i as will a change in x_j for a firm j for which $y_j < q_j$. The direction of change on y_i will be the same as the direction of change on q_i. In case (c) $\psi_i(p, x) = x_i$ and no variable other than x_i will affect y_i. Case (d) is like (c). In case (e) $\psi_i(p, x)$ is determined by the relationship $p_i = C_i'(y_i) + \gamma_i'(y_i - x_i)$; therefore, only p_i and x_i can affect the value of y_i. In both instances, the relationship is positive. Finally, in case (f) y_i is determined by $0 = C_i'(y_i) + \gamma_i'(y_i - x_i)$, so only changes in x_i can affect y_i. ■

5. Equilibrium in Game S with Adjustment Costs

Game S modified to include adjustment costs is a two stage game in which firms choose provisional output levels (x) and (p) prices simultaneously at stage 1. This information becomes generally known prior to stage two decisions. The second stage decision is to select final levels of output (y). Thus a strategy for a player i in this game is a triple, $(p_i, x_i, \eta_i(p, x))$ where $p_i \in [0, p_i^0]$, $x_i \in [0, x_i^0]$, and $\eta_i(p, x)$ is any function from $\Omega \times \Xi$ to $[0, x_i^0]$. From the preceding section, $\eta_i = \psi_i$ for $i \in N$ at a subgame perfect equilibrium point.

An interesting, though intuitive, result that carries over from Benassy (1986b) and Friedman (1988) is that a subgame perfect equilibrium requires that $\psi(p, x) = x = f(p)$. That is, best reply behavior for the firms requires that no adjustment to output be made at the second stage and that output for each firm exactly equal its demand. This is proved below.

LEMMA 5 Under Assumptions 1 to 4, at a subgame perfect pure strategy equilibrium, $\psi(p, x) = x = f(p)$.

PROOF The method of proof is to consider an arbitrary (p,x) and check for firm i 1) whether or not $\psi_i(p, x) = x_i = f_i(p)$ and, if not, 2) whether x_i or p_i could be altered to increase the firm's profit. The conditions (a) - (f) in Lemma 3 cover all possibilities. Suppose first that $y_i < q_i$. This can happen under (c) or (e). The derivative of firm i's profit with respect to its price is

$$\frac{\partial\pi_i}{\partial p_i} = \delta_i + p_i\left[\frac{\partial\delta_i}{\partial p_i} + \sum_{j=1}^{n}\frac{\partial\delta_i}{\partial\psi_j}\frac{\partial\psi_j}{\partial p_i}\right] - [C_i' + \gamma_i']\frac{\partial\psi_i}{\partial p_i} \tag{10}$$

Using the fact that $y_i < q_i$ can only occur under cases (c) and (e), so that only derivatives with respect to p_i and x_i can be nonzero, eq. (10) reduces to $\partial\pi_i/\partial p_i = \delta_i > 0$ for case (c) and

$$\frac{\partial\pi_i}{\partial p_i} = \delta_i + p_i\left[\frac{\partial\delta_i}{\partial p_i} + \frac{\partial\delta_i}{\partial\psi_i}\frac{\partial\psi_i}{\partial p_i}\right] - [C_i' + \gamma_i']\frac{\partial\psi_i}{\partial p_i} \tag{11}$$

for case (e). Using the facts that $p_i = C_i' + \gamma_i'$ and $\partial\delta_i/\partial\psi_i = 1$ eq. (11) becomes

$$\frac{\partial\pi_i}{\partial p_i} = \delta_i + p_i\frac{\partial\delta_i}{\partial p_i} \tag{12}$$

which is positive because $\partial\delta_i/\partial p_i = \partial\psi_i/\partial p_i > 0$ and, of course, $\delta_i > 0$. Thus, a firm cannot be behaving optimally if it chooses (p_i, x_i) in a way that leads to selecting $y_i < q_i$. This means, in turn, that cases (c) and (e) cannot arise under equilibrium behavior. Furthermore, as long as $y_i \geq q_i$, changes in x_i will have no effect on y_j ($j \neq i$); therefore, $x_i < q_i$ can never be optimal for firm i.

So, from the preceding paragraph all firms i must, in equilibrium, be choosing (p_i, x_i) so that $y_i \geq q_i$ and $x_i \geq q_i$. Consider now whether $y_i > q_i$ is compatible with equilibrium. The derivative of firm i's profit with respect to its provisional output is

$$\frac{\partial\pi_i}{\partial x_i} = p_i\sum_{j=1}^{n}\frac{\partial\delta_i}{\partial\psi_j}\frac{\partial\psi_i}{\partial x_i} - [C_i' + \gamma_i']\frac{\partial\psi_i}{\partial p_i} + \gamma_i' \tag{13}$$

Cases (d) and (f) are the only times when $y_i > q_i$ can occur. In case (d) firm i can reduce x_i to equality with q_i and no change will take place in any y_j ($j \neq i$) or in the revenue of firm i. Only the cost of firm i will fall, increasing its profits. That is, eq. (13) reduces to $-C_i'$. In case (f) eq. (13) becomes

$$\frac{\partial\pi_i}{\partial x_i} = \frac{-C_i'\gamma_i'' + \gamma_i'C_i''}{C_i'' + \gamma_i''} \tag{14}$$

In case (f) $y_i < x_i$ so that $\gamma_i' < 0$. Both C_i and γ_i are strictly convex, so that eq. (14) is negative. Thus, if $y_i > q_i$ the firm is better off with a lower value of x_i. So equilibrium requires (p, x) such that $\psi(p, x) = q = f(p)$. That being so, and knowing that $x_i < q_i$ cannot be optimal, it is likewise not optimal to select $x_i > q_i$. If the latter holds, the firm can reduce costs with no effect on its own revenue or on the decisions of other firms by choosing $x_i = q_i$. ■

Now turn to the question of existence of equilibrium in Game S with adjustment costs. An obvious candidate for equilibrium is the strategy combination $(p^c, f(p^c), \psi)$. From Benassy (1986b) and Friedman (1988) it is already known that $(p^c, f(p^c))$ is the only possible pure strategy equilibrium point in Game S without adjustment costs.

LEMMA 6 The only possible pure strategy equilibrium for Game S with adjustment cost is $(p^c, f(p^c), \psi)$. Equilibrium occurs at $(p^c, f(p^c), \psi)$ only if $p_i^c > C_i'(f_i(p^c)) + \gamma_i'(0)$ for all $i \in N$.

PROOF That equilibrium can only occur at $(p^c, f(p^c), \psi)$ follows from the observation that equilibrium requires choosing provisional output (x) to equal demand and choosing final output (y) to equal demand. Prices are chosen subject to these conditions, and the only prices that could be optimal are those of the price-only equilibrium.

To see that $p_i^c > C_i'(f_i(p^c)) + \gamma_i'(0)$ for all $i \in N$, suppose that $p_j^c \leq C_j'(f_j(p^c)) + \gamma_j'(0)$ for some firm j. This means that firm j will be a provider of spillover demand to any firm i that chooses a price above p_i^c. More precisely, for any other firm $i \neq j$, $(p_i^c, f_i(p^c), \psi_i)$ is not a best reply to $(p^c, f(p^c), \psi)$; for, given $(p_k^c, f_k(p^c), \psi_k)$ for all firms $k \neq i$, firm i will increase its profit above $p_i^c f_i(p^c) - C_i'(f_i(p^c))$ by selecting a price somewhat above p_i^c and producing exactly enough to meet demand. ■

The content of Lemma 6 is illustrated in Figure 2. The solid curve ABH shows $\pi_i^P(p^c \backslash p_i)$, and the point B indicates an equilibrium of the price-only game. Such an equilibrium price must prevail in Game S, with or without adjustment costs, if there is an equilibrium. Suppose that $p_j^c \leq C_j'(f_j(p^c)) + \gamma_j'(0)$ for all firms j. Then no firm would adjust its output if firm i were to increase its price above p_i^c; therefore, the actual relationship between π_i and p_i would be ABCD. From all other firms $(j \neq i)$ spillovers would flow to firm i as soon as p_i rose above p_i^c.

If, on the other hand, the conditions in Lemma 6 are met by all firms, then, as firm i raises its price above p_i^c while all the other firms follow $(p_k^c, f_k(p^c), \psi_k)$ for $k \neq i$, a critical price is reached for each firm k, p_i^k, such that

$$p_k^c = C_k'(f_k(p^c \backslash p_i^k)) + \gamma_k'(f_k(p^c \backslash p_i^k) - f_k(p^c)) \tag{15}$$

Suppose p_i' is the price at which eq. (15) holds for all firms $k \neq i$. Then the actual relationship between π_i and p_i would be ABFG. In this case firm i would not have an incentive to increase price above p_i^c because its profit on the FG branch of the profit curve never attains the profit associated with B. The smaller a firm's adjustment cost, the larger is the critical value of p_i^k in eq. (15). If the critical values are sufficiently high for all firms, then spillover demands kick in too late to spoil the equilibrium at (p^c, f(p^c), ψ). A final detail: of course the critical values of p_i^k in eq. (15) can differ across firms. If some firms would make no adjustments whatever, so that the first contingent demands kick in at B, and then some other firms satisfy eq. (15) at p_i'', the profit of curve of firm i would be ABCE.

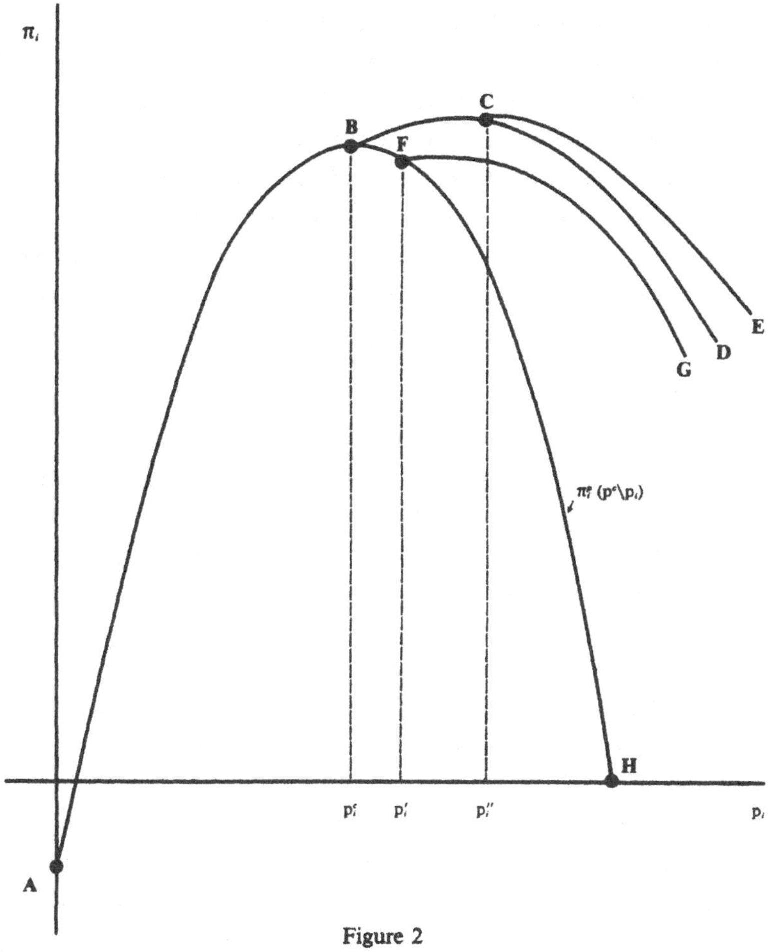

Figure 2

In summary, $\gamma_i(y_i - x_i) = 0$ for all values and all firms is a limiting case in which equilibrium occurs at (p^c, f(p^c), ψ) and another limiting case occurs if $\gamma_i'(0)$ is so large for each firm that no adjustment could ever take place. The latter is the situation in Benassy

(1986b) and Friedman (1988). There are adjustment costs in between at which equilibrium exists and others at which it does not. The principles sketched in discussing Lemma 6 and Figure 2 govern which occurs.

6. Equilibrium in Game Q with Adjustment Costs

In Game Q with adjustment costs, the decision process has three stages. First, x is chosen and made public, then p, and finally y. Thus prices will be chosen as a function of x and then y will be chosen as a function of (p, x). The most fruitful way to think about this game is to note that, after x is chosen, the game is similar to Game P of Benassy (1986b) and Friedman (1988); a game in which price choice is followed by output choice. In the present game, following the choice of x the subgame has cost functions that differ from those in the original Game P. In Game P the cost functions are simply the C_i, but in the subgame the cost functions are $C_i(y_i) + \gamma_i(y_i - x_i)$, with y_i being the output variable and x_i being a parameter. These cost functions need not show nondecreasing marginal cost for $y_i \in [0, x_i]$ and there is a discontinuous jump upward in marginal cost at $y_i = x_i$. Thus after x is chosen, the profit functions are

$$\pi_i = p_i\delta_i(p, y) - C_i(y_i) - \gamma_i(y_i - x_i) \tag{16}$$

The polar cases of this model are zero adjustment costs and infinite adjustment costs. In the former, the choice of x has no relevance because the cost of producing y_i is the same for a firm no matter what the previous x_i choice has been. Therefore, when equilibrium exists in this game it is the same as in Game P without adjustment costs; prices are p^c and output is $f(p^c)$. In the latter case, adjustments are too expensive to be made, so that y = x, and the game is precisely Game Q without adjustment costs. Then, when equilibrium exists, it occurs at prices $F(q^c)$ and output is q^c. At both extremes one of the three stages of decision, either the first or the last, is effectively dropped. In between are equilibria that lie between the two extremes.

A strategy for player i consists of a triple (x_i, θ_i, η_i) where $x_i \in [0, x_i^0]$, θ_i is a function from Ξ to $[0, p_i^0]$, and η_i is a function from $\Xi \times \Omega$ to $[0, x_i^0]$. While I am not in a position to give conditions on the basic model that guarantee existence of an equilibrium, it is possible to characterize some of the conditions that an equilibrium must satisfy. First, $\eta_i = {}_i$, where ψ_i is given above in Lemma 4. At equilibrium, $\theta_i(x)$ must maximize π_i with respect to p_i given x, the θ_j $(j \neq i)$, and the ψ_i $(j \in N)$. Thus $\theta_i(x) \in [0, p_i^0]$ is that p_i which maximizes

$$\pi_i = p_i\delta_i[\theta(x)\backslash p_i, \psi(\theta(x)\backslash p_i)] - C_i[\psi_i(\theta(x)\backslash p_i)] - \gamma_i[\psi_i(\theta(x)\backslash p_i) - x_i] \tag{17}$$

A strategy combination (x^*, θ, ψ) is an equilibrium of the modified Game Q if θ selects the market clearing price vector and ψ calls for no adjustment of output. That is, $\theta(x^*) = F(x^*) \equiv p^*$ and $\psi(\theta(x^*), x^*) = \psi(p^*, x^*) = x^*$. (x^*, θ, ψ) is subgame perfect if, for all $x \in \Xi$, $\theta(x) = F(x)$ and $\psi(\theta(x), x) = x$.

Consider now some characteristics of an equilibrium output vector, x^*, and the associated equilibrium price vector, p^*. At stage two, when prices are chosen, the following conditions must hold

$$\frac{\partial \pi_i}{\partial p_i} = f_i(p^*) + [p_i^* - C_i'(f_i(p^*)) - \gamma_i'(0)]f_i^i(p^*) \geq 0 \tag{18}$$

$$\frac{\partial \pi_i}{\partial p_i} = f_i(p^*) + [p_i^* - C_i'(f_i(p^*)) + \gamma_i'(0)]f_i^i(p^*) \leq 0 \tag{19}$$

Eq. (18) is the condition that the firm cannot lower its price and increase its profit and eq. (19) is the condition that the firm cannot raise its price and increase its profit. Concavity of firm i's profit in p_i assures that both eqs. (18) and (19) fall as p_i increases. Equilibrium requires that eqs. (18) and (19) both be satisfied. The larger are the $\gamma_i'(0)$ the larger is the set of prices that satisfy these conditions. Vives' (1985) theorem assures that eq. (19) is always satisfied with strict inequality at $p = F(q^c)$. On the other hand, if the marginal adjustment cost (i.e., $\gamma_i'(0)$) goes to zero, the two equations become identical and must be met with equality. The latter applies in (unmodified) Game P and requires a price of p^c.

In addition to the price choice conditions, eqs. (18) and (19), at an equilibrium, the firms must also be simultaneously maximizing profits with respect to x. An understanding of the behavior of the equilibrium values, (p^*, x^*), is facilitated by expressing $\partial\pi_i/\partial x_i$ in terms of prices. This is done below supposing that $q_i = x_i$ which means that output choice is considered here when price can be set to clear the market and no further output adjustment is made.

$$\frac{\partial \pi_i}{\partial q_i} = F_i(q) + q_i F_i^i(q) - C_i'(q_i) \tag{20}$$

Eq. (20) must be less than or equal to zero at equilibrium, which can be expressed in terms of prices as

$$f_i(p) + [p_i - C_i'(f_i(p))]\frac{1}{F_i^i} \geq 0 \tag{21}$$

Eq. (21) is satisfied with equality at equilibrium if the adjustment cost is very large. Then the equilibrium price and output vectors are $x^* = q^c$ and $p^* = F(q^c)$. Thus, supposing again that equilibrium always exists, the relationship between (p^*, x^*), $(F(q^c), q^c)$, and $(p^c, f(p^c))$ is parameterized by $\gamma = (\gamma_1'(0),..., \gamma_n'(0))$ in the following way: If γ is large enough that eq. (18) is satisfied for all $i \in N$ at $F(q^c)$, then eq. (21) holds with equality and $(p^*, x^*) = (F(q^c), q^c)$. As γ gets smaller, as soon as

$$f_i(F(q^c)) + [F_i(q^c) - C_i'(q_i^c)) - \gamma_i'(0)]f_i'(F(q^c)) < 0 \tag{22}$$

for one or more firms, eq. (18) becomes the binding constraint for those firms, while eq. (21) is satisfied with strict inequality for those firms. As γ continues to fall, eventually eq. (22) holds for all firms, the equilibrium prices continue to fall for all firms and this process continues until $\gamma = 0$ and equilibrium prices have gone to p^c, with accompanying equilibrium production levels of $f(p^c)$.

7. Concluding Comments

In many markets it is likely that price and output are not equally flexible for a firm, and, where this is true, output is probably the variable that is usually least flexible. Flexibility should be thought of in temporal terms, although the analysis above lacks an explicit temporal structure. Thus, where output is less flexible than price, it must be decided earlier in time than price. Assuming the output decision to be completely locked in is extreme in the sense that a firm can probably alter its production plans as some (extra) cost. For example, a manufacturer of durables may find demand at present prices to be above expectation, and may be able to speed up the rate of production in the very short run by acquiring extra material inputs via air freight rather than truck or rail, or may be willing to resort to higher-than-usual cost suppliers, etc.

The present paper has dealt with models in which subsequent adjustment may be made to a variable. The aim has been only to understand the strategic consequences of later adjustment; thus, being "surprised" by the level of demand is ruled out by our assumptions. The results, however, are the following. If output and price are chosen simultaneously, then, the ability to make a later adjustment of output may restore existence of equilibrium to a situation in which equilibrium would not otherwise exist. If output is chosen after price, then the subsequent adjusting of either one is irrelevant. If output is chosen after price, then the resulting equilibrium is sensitive to the cost of later adjustment of output. Interestingly, zero adjustment cost corresponds to a structure with price chosen first, and, as adjustment costs rise, the equilibrium moves from the price-first equilibrium to that associated with output first.

The sensitivity of equilibrium in the output-first model to the size of adjustment costs raises an interesting question about the preferences of the firms. The output-first equilibrium corresponds to higher prices, lower outputs, and generally higher profits than does the price-first equilibrium. Thus, adding flexibility by way of lowering adjustment costs in the output-first model tends to move the equilibrium toward that of the price-first model which means toward lower prices and lower profits. The firms would collectively be better off if they could not easily make later adjustments. Of course, an *individual firm* will always be better off with more flexibility, given fixed flexibility for its rivals.

References

Alger, Daniel R., 1979, *Markets where Firms Select Both Prices and Quantities: An Essay on the Foundations of Microeconomic Theory*, unpublished doctoral dissertation, Northwestern University.

Benassy, Jean-Pascal, 1986a, "On the Role of Market Size in Imperfect Competition: A Bertrand-Edgeworth-Chamberlin Synthesis."

Benassy, Jean-Pascal, 1986b, "On the Existence of Bertrand-Edgeworth Equilibria with Differentiated Commodities," Chapter 4 in *Contributions to Mathematical Economics, in Honor of Gérard Debreu*, eds. W. Hildenbrand and A. Mas-Colell (North-Holland, Amsterdam).

Bertrand, Joseph, 1883, Review of Cournot (1838), *Journal des Savants*, 499-508. English translation in *Cournot Oligopoly: Characterization and Applications*, pp. 73-81, edited by Andrew Daughety, translated by J. W. Friedman, Cambridge University Press (New York, 1988).

Cheng, Leonard, 1985, "Comparing Bertrand and Cournot Equilibria: A Geometric Approach," *Rand Journal of Economics* 16; 146-152.

Cournot, Augustin, 1838, *Recherches sur les Principes Mathématiques de la Théorie des Richesses* (Hachette, Paris), translated by N. T. Bacon.

Davidson, Carl and Raymond Deneckere, 1986, "Long-run Competition in Capacity, Short-run Competition in Price, and the Cournot Model," *Rand Journal of Economics* 17; 404-415.

Edgeworth, Francis Y., 1925, "The Pure Theory of Monopoly," in *Papers Relating to Political Economy* Vol. I (Macmillan, London), 111-142.

Friedman, James W., 1977, *Oligopoly and the Theory of Games* (North-Holland, Amsterdam).

Friedman, James W., 1986, *Game Theory with Applications to Economics* (Oxford, New York).

Friedman, James W., 1988, "On the Strategic Importance of Prices versus Quantities," *Rand Journal of Economics* 19; 607-622.

Friedman, James W., 1989, "Prices and Quantities in an Infinite Horizon Oligopoly," to appear in *Market Strategy and Structure*, (eds. J. M. Alec Gee and George Norman) Harvester-Wheatsheaf.

Gale, David and Hukukane Nikaido, 1965, "The Jacobian Matrix and the Global Univalence of Mappings," *Mathematische Annalen* 165; 81-93.

Hotelling, Harold, 1929, "Stability in Competition," *Economic Journal* 39; 41-57.

Kreps, David, and Jose Scheinkman, 1983, "Quantity Precommitment and Bertrand Competition Yield Cournot Outcomes," *Bell Journal of Economics* 14; 326-337.

Shubik, Martin and Richard Levitan, 1980, *Market Structure and Behavior* (Harvard, Cambridge).

Singh, Nirvikar and Xavier Vives, 1984, "Price and Quantity Competition in a Differentiated Duopoly," *Rand Journal of Economics* 15; 546-554.

Vives, Xavier, 1985, "On the Efficiency of Bertrand and Cournot Equilibria with Product Differentiation," *Journal of Economic Theory* 36; 166-175.

A Demand Commitment Model of Coalition Bargaining

Reinhard Selten
Institut für Gesellschafts- und Wirtschafts-wissenschaften, wirtschaftstheoretische Ab-teilung I, Adenauerallee 24–42, 5300 Bonn 1

ABSTRACT

A non-cooperative model of coalition bargaining in the context of characteristic functions with side-payments is introduced. Temporary commitments to payoff demands are a salient feature of the model. For every coalition game, the model generates a finite extensive game with perfect information. For two basic cases – unanimity games and 3-person quota games without the grand coalition – the non-cooperative analysis yields results akin to cooperative game theory. The analysis is based on subgame perfectness alone without the use of further selection criteria. The equal payoff division is the only subgame perfect equilibrium outcome in the case of the zero-normalized unanimity game. In the case of the 3-person quota game without the grand coalition, the players receive their quotas in 2-person coalitions, if a subgame perfect equilibrium is played; each of the three 2-person coalitions occurs with probability 1/3.

Introduction

The first one to propose an extensive game model of bargaining in the context of characteristic function games was John C. Harsanyi (1974). It seems to be fitting to offer a contribution in the same area to this volume in his honour. Several other models of characteristic function bargaining can be found in the literature (Selten 1981, Binmore 1985, Gul 1989, Chatterjee, Dutta, Ray, and Sengupta 1990). All these models permit plays of infinite length. From the experimental point of view, this is undesirable. Strictly speaking only finite games can be played in the laboratory.

In his experimental study Maschler (1978) observed the following phenomenon in face to face bargaining on characteristic function games: sometimes a player declares that he commits himself to a payoff demand at which he is willing to enter any coalition. Apparently, such declarations can be made credible even in the absence of enforcability by formal rules.

Temporary commitments to payoff demands as formal moves are a central feature of the model presented here. The only other strategic possibility of a player is to form a coalition whose other members are committed to payoff demands; these other members then receive their payoff demands and the player who forms the coalition receives what is left over from the coalition value.

The game is organized as a succession of "stages". A stage consists of at most two "rounds". In a round a player acts at most once. The order in which the players act is determined by random moves. Commitments to payoff demands are valid for the current round only. If a coalition is formed the round ends and a new stage begins provided that the remaining players still have a profitable coalition they can form; otherwise the game ends. A player cannot commit himself to a payoff demand if he is the last one to act after all other players not yet in coalitions formed. However, he may refuse to form a coalition. In this case a second round follows if the current one is the first one in its stage; if it is already the second one, the game ends in conflict.

The bargaining game generated by the model is a finite extensive game with perfect information. The finiteness of the choice sets results from the restriction of payoff demands to integer multiples of a "bargaining unit": One may think of this restriction as due to a convention which rejects haggling over money amounts smaller than one bargaining unit as an unacceptable waste of time.

The model specifies small costs attached to types of choices taken. It costs one bargaining unit to commit oneself to a payoff demand. It costs less than that to form a coalition. The cost incurred by a player who forms a coalition is an increasing function of the number of members. A player who refuses to form a coalition after all other players still in the game have committed themselves to a payoff demand, faces a cost equal to the cost of forming a coalition of all players still in the game. The costs induce a motivation for short play. A more formal description of the rules and the cost structure will be given in section 2.

To some extent, Binmore's (1985) "market model" is similar to the model presented here. However, Binmore's model permits infinite plays. The author has constructed a finite model of demand commitment bargaining, different from the one presented here, which has been implemented at the Bonn laboratory of experimental economics. Several researchers who have seen this laboratory model have investigated variants of it. Interesting results have been obtained. Elaine Bennett and Eric van Damme (1991) have explored the apex game. Eyal Winter (unpublished) has applied a variant to convex characteristic function games. For subgame perfect equilibria with certain additional properties the Shapley value emerges as the only equilibrium outcome.

It was the aim of this paper to construct a laboratory procedure whose game theoretic analysis yields results akin to cooperative game theory at least in the following two basic cases:

a) the unanimity game
b) the 3–person quota game without the grand coalition.

In zero-normalized unanimity games, one expects equal payoff divisions and in 3-person quota games without the grand coalition, one expects "quota-divisions" which means that a 2-person coalition is formed in which both players receive their quotas. One would like to obtain these natural cooperative solutions as the only possible outcomes of subgame perfect equilibrium play.

As has been explained above, the model presented here involves small costs connected to types of choices. We refer to payoffs without consideration of these costs as "basic payoffs". If we speak of equal payoff divisions we mean equal basic payoffs and in the case of quota divisions we mean 2-person coalitions involving quotas as basic payoffs.

In order to convey a clearer impression of the results of this paper, it is necessary to explain our framework for the description of a coalition situation. The notion of a coalition game used here involves an integer valued characteristic function and a bargaining unit of the form $1/m$ where m is a positive integer. As in the case of the 3-person quota game without the grand coalition not every coalition may be permissible.

We say that an amount is "negotiable" if it is an integer multiple of the bargaining unit. We impose a "divisibility condition" on the surplus of the value of a permissible coalition over the sum of the one-person values of its members: this surplus divided by the number of the members of the coalition must be negotiable. The divisibility condition implies the negotiability of equal shares in zero-normalized unanimity games and of the quotas in 3-person quota games and thereby excludes uninteresting indivisibility effects.

In order to avoid cumbersome border cases of little interest we shall only look at 3-person quota games whose quotas are not smaller than 3 bargaining units. We include this restriction into the definition of a 3-person quota game. For our purposes it is convenient to do this.

The main results on the model presented here concern the consequences of subgame perfect equilibrium play in the two basic cases. In the case of the zero-normalized unanimity game all subgame perfect equilibrium plays result in equal payoff divisions. In the case of the 3-person quota game without the grand coalition all subgame perfect equilibrium plays result in quota divisions. In both cases there are many subgame perfect equilibria, but they disagree with each other at unreached information sets only. The probability distribution over the end points is always the same. We refer to this property as "determinateness". All subgame perfect equilibrium plays end in the first round. The costs of playing a second round are avoided.

In the model of demand commitment bargaining used in the Bonn laboratory up to now players strategically determine the next decision maker, except at the beginning of the round. In subgame perfect equilibria of 3-person quota games without the grand coalition the player who is the first one to act in a round may easily be indifferent as far as the question is concerned whom to choose as the next decision maker. This offers an opportunity for costless punishment of past behavior.

Such supergame effects make it possible to construct subgame perfect equilibria with outcomes far away from quota divisions. An example will be discussed after the derivation of our main results.

The model described here replaces the strategic choice of the next decision maker by a random draw. This simplification avoids the supergame effects mentioned above, but it is not yet sufficient for our results. The specification of the small costs attached to types of choices is also of importance. The structure of these costs has been specifically designed to avoid indifferences, to create incentives for short play, and to achieve exact agreement of subgame perfect equilibrium outcomes with the natural cooperative solutions in the two basic cases.

The cost structure of the model may not be the most natural one. Especially one of the cost rules may be criticized as arbitrary. This rule concerns the situation of a player who as the last uncommitted one among those still in the game must either choose to form a coalition or refuse to do so. The cost of refusal is equal to the cost of forming a coalition of all players still in the game. Admittedly, this rule serves the purpose to obtain the right result in the case of the zero–normalized unanimity game.

This paper mainly tries to support experimental research on the influence of game theoretic properties of coalition bargaining procedures on observed behavior. From this point of view such properties are more important than a strong intuitive appeal of all details of the rules.

Observed behavior in characteristic function experiments is badly explained by classical solution concepts of cooperative game theory (Selten 1987, Uhlich 1989, Uhlich and Rockenbach 1990). This may partly be due to the laboratory procedures used. The procedures involving anonymous interaction via computer terminals have the structure of non–cooperative extensive game models of coalition bargaining. Such models should be analyzed with the help of non–cooperative game theory. Unfortunately, little is known about the non–cooperative properties of some of these laboratory procedures; e.g. that of Kahan and Rapoport (1984). Nevertheless, it seems to be a safe conjecture, that all those used in experiments up to now permit a wide range of subgame perfect equilibrium outcomes in the two basic cases.

A very simple procedure proposed by the author (1981) and used in the Bonn laboratory (Uhlich 1990) has been explored extensively. For this model, which is based on payoff division proposals (not on demand commitments) Elaine Bennett (1991) has proved a "folk theorem" which shows that essentially all individually rational outcomes can be reached by subgame perfect equilibria. Of course, it is possible to reduce the set of all possible equilibrium outcomes by the requirement of stationarity and additional selection criteria (Selten 1981, Moldovanu 1991). However, this does not help if one wants to compare experimental results with predictions derived from the basic rationality assumption of subgame perfect equilibrium alone.

John C. Harsanyi and the author have developed a general theory of equilibrium selection in games (1988). From this work it should be clear that the author is not opposed to equilibrium selection in the framework of rational game theory. However, experimental research has to deal with the bounded rationality of human beings. It is not clear under which conditions human subjects can learn to play subgame perfect equilibria in the laboratory, even if the game is determinate in the sense explained above. The model presented here has the aim to support experimental research on this question in the context of coalition bargaining.

In this paper attention is concentrated on unanimity games and 3-person quota games without the grand coalition. Of course, this is only the beginning of further work on broader classes of coalition games. Some of the features of the model presented here are motivated by more complex cases than those covered here.

1. Coalition games

In this section we introduce the notion of a coalition game used in this paper.

Notation: The number of elements of a finite set M will be denoted by $|M|$. For the sake of notational simplicity we shall often ignore the distinction between a set with only one element and this element. Thus we shall write v(i) instead of v({i}) and $S \cup i$ instead of $S \cup \{i\}$ as well as $S \backslash i$ instead of $S \backslash \{i\}$. We shall also write ij instead of {i,j} and ijk instead of {i,j,k} where this can be done without the danger of misunderstandings.

The notion of a coalition game: In the context of this paper a coalition game

$$G = (N,P,v,g,\gamma) \tag{1}$$

is defined as a quadruple with the following components:

N the player set is a finite set of at least two integers, interpreted as names of players. The non-empty subsets of N are called coalitions. The elements of a coalition C are also called members of C. A one-person coalition is a coalition with only one member. Coalitions with at least two members are called genuine coalitions.

P the set of permissible coalitions is a subset of the set of all coalitions which contains at least all one-person coalitions. The coalitions in P are called permissible.

v the characteristic function is a function which assigns an integer v(C) to every permissible coalition C. The value v(C) of C is interpreted as an amount of money. The surplus s(C) of a permissible coalition is defined as follows:

$$s(C) = v(C) - \sum_{i \in C} v(i) \tag{2}$$

The following <u>profitability condition</u> must be satisfied for every genuine permissible coalition C:

$$s(C) > 0 \tag{3}$$

g the <u>bargaining unit</u> is a number such that 1/g is a positive integer. g is interpreted as an amount of money. The following <u>divisibility condition</u> must be satisfied for every genuine permissible coalition: $s(C)/|C|$ is an integer multiple of g.

γ the <u>cost structure</u> assigns a <u>coalition cost</u> γ_m to every $m = 1,\dots,|N|$, interpreted as the number of members in a coalition. The following condition must be satisfied:

$$0 < \gamma_m < \gamma_{m+1} < g \text{ for } m = 1,\dots,|N|-1 \tag{4}$$

<u>Comments:</u> It is convenient not to define $N = \{1,\dots,n\}$, even if this can be done without loss of generality. In our model a situation may arise after the formation of a coalition which amounts to the application of the model to a new coalition game whose player set consists of the remaining players.

We impose the profitability condition in order to avoid uninteresting border cases. The divisibility condition excludes the possibility that the restriction of payoff demands to integer multiples of the bargaining unit prevents equal payoff divisions and quota divisions as realizable outcomes. This avoids uninteresting indivisibility effects.

<u>Further notational conventions and definitions:</u> Unless we explicitely say something to the contrary, all definitions and notational conventions refer to a fixed but arbitrary coalition game $G = (N,P,v,g,\gamma)$. A number α is called <u>negotiable</u> if it is an integer multiple of g. For every $i \in N$ let P_i be the set of all permissible coalitons with $i \in C$. The set of all genuine permissible coalitions is denoted by P_+. Let S be a coalition (not necessarily a permissible one). We use the symbols $P(S)$, $P_i(S)$, and $P_+(S)$ in order to denote the intersections of P, P_i, and P_+, respectively, with the set of all subsets of S.

<u>Permissible demands:</u> A number x_i is a <u>permissible payoff demand,</u> or shortly a <u>permissible demand</u> for player i, if x_i is negotiable and satisfies the following inequality:

$$0 \leq x_i - v(i) \leq \max_{C \in P_i} s(C) \tag{5}$$

The set of all permissible demands of player i is denoted by X_i. For every coalition M the symbol $X_i(M)$ denotes the set of all negotiable X_i which satisfy (5) with $P_i(M)$ instead of P_i on the right hand side.

Demand systems: A demand system

$$x_S = (x_i)_{i \in S} \tag{6}$$

is a function which assigns a permissible demand for player i to every $i \in S$. Let C be a subset of S. We use the symbol $x_S(C)$ for the sum of the x_i over the members of i:

$$x_S(C) = \sum_{i \in C} x_i \tag{7}$$

Zero-normalization: The zero-normalization of a coalition game $G = (N,P,v,g,\gamma)$ is the coalition game $G' = (N,P,s,g,\gamma)$ with s defined by (2). It is clear that G' has all the properties of a coalition game. A coalition game $G = (N,P,v,g,\gamma)$ is called zero-normalized if we have

$$v(i) = 0 \quad \text{for every } i \in N \tag{8}$$

If this is the case G coincides with its zero-normalization.

After the next section we shall restrict our attention to zero-normalized coalition games. It will always be easy to see, how results can be transferred to coalition games not normalized in this way.

2. The model

In this section the model of coalition bargaining proposed in this paper will be described with the help of the flow chart of figure 1. A dynamic notation is used. Statements of the form $A \to B$ indicate that B assumes the value of A. Arrows at connecting lines show the direction of flow. Rectangles permit only one continuation whereas rhomboids contain questions whose answers YES or NO determine by which branch the game is continued.

For every coalition game $G = (N,P,v,g,\gamma)$ the model generates a finite extensive game $\Gamma(G)$ with perfect information. The structure of this game is described by the flow chart. However, the fact that the game $\Gamma(G)$ has perfect information is not expressed by the flow chart and must be kept in mind.

As has been explained in the introduction, bargaining proceeds as a succession of stages. A stage may have one or two rounds. A stage may end with the formation of a coalition; if no coalition is formed in the second round the game ends. In the following the rectangles and rhomboids of the

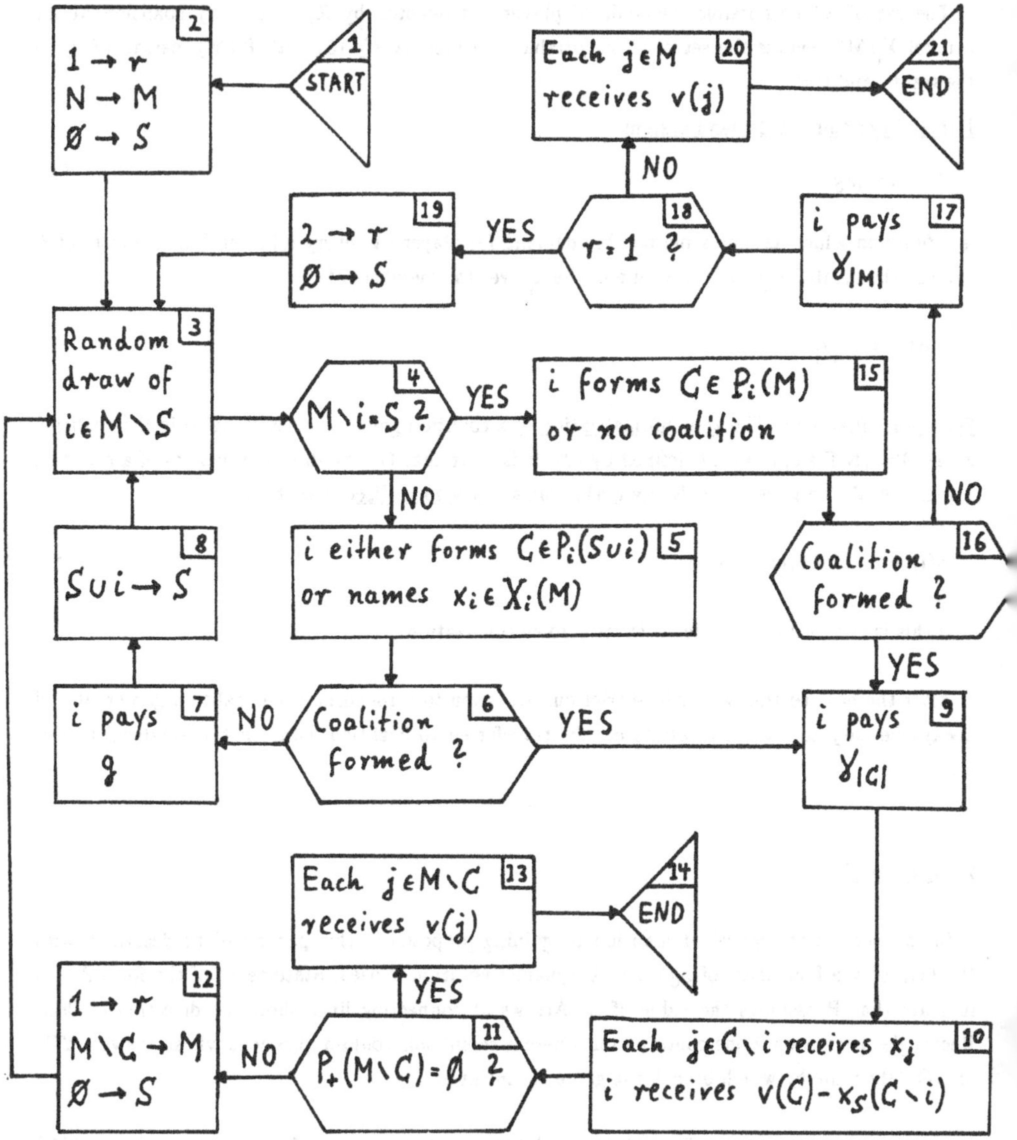

Figure 1: Flow chart of the model

flow chart will be explained in detail.

Rectangle 2: M is the set of active players, who are not yet in coalitions formed. At the beginning M is the player set N. The symbol r, the round number, indicates whether the current round is the first (r=1) or the second one (r=2) in its stage. The game begins with the first round of the first stage. The set S is the set of committed players. At the beginning, S is empty.

Rectangle 3: The term random draw means that every player $i \in M \setminus S$ has the same probability to be chosen as the next decision maker.

Rhomboid 4: It is important whether the decision maker i is the last uncommitted active player ($M \setminus i = S$) or not. The choice set is different in the two cases.

Rectangle 5: The decision maker is not the last uncommitted active player. He may either form a permissible coalition including himself whose other members – if there are any – are already committed to a payoff demand in the current round or he may name a payoff demand permissible with respect to the current round.

Rhomboid 6: It is now important whether the decision maker has formed a coalition or named a payoff demand.

Rectangle 7: A player who commits himself to a payoff demand incurs a cost of one bargaining unit.

Rectangle 8: Having named a payoff demand player i joins the ranks of committed players.

Rectangle 9: Player i has formed a coalition C and thereby incurs a cost of $\gamma |C|$

Rectangle 10: The committed members of C receive their payoff demands x_j as their basic payoffs and player i, who has formed C receives what is left over of the coalition value v(C).

Rhomboid 11: The game ends after the formation of a coalition, if the remaining players cannot form a genuine permissible coalition anymore. Otherwise the game enters its next stage.

Rectangle 12: The next stage begins with its first round. The remaining players form the new set of active players. All payoff demands are wiped out.

Rectangle 13: Players who at the end of the game are not in a coalition formed receive their one–person values as their basic payoffs.

Rectangle 15: The decision maker is the last uncommitted active player. He cannot name a payoff demand (since at the end of the round all payoff demands are wiped out this would make no sense). He can either form a permissible coalition including himself or he can refuse to do so.

Rhomboid 16: It is now important whether the decision maker has formed a coalition or not.

Rectangle 17: If a last uncommitted active player refuses to form a coalition he incurs a cost equal to the cost $\gamma_{|M|}$ of forming the coalition of all active players (if it is permissible).

Rhomboid 18: If no coalition has been formed at the end of a first round, the game goes on. After a second round without the formation of a coalition, the game ends.

Rectangle 19: The second round begins with the same set of active players. All demand commitments are wiped out.

Rectangle 20: No coalition has been formed in a second round. The game ends. All active players of this last round receive their one–person values as their basic payoffs.

3. Positions, histories, strategies, and equilibria

The flow chart of figure 1 contains all the information necessary for the construction of the extensive game $\Gamma(G)$ generated by the model for a coaliton game $G = (N,P,v,g,\gamma)$. However, precise mathematical statements of definitions and results require a more formal description of $\Gamma(G)$. For this purpose, we introduce the notion of a position. Roughly speaking a position summarizes the strategic information at a decision point as far as the possibilities of the future are concerned without giving a full account of the past.

A position is described by a quadruple (M,r,x_S,i) whose components have the following meaning: M is the set of active players, r is the number of the round within its stage, x_S is the system of payoff demands of the committed players and i is the decision maker. The set S of committed players does not appear separately in the description, but as a part of the information provided by x_S. Formally, a quadruple

$$u = (M,r,x_S,i) \tag{9}$$

is a position of $\Gamma(G)$ with $G = (N,P,v,g,\gamma)$ if the following conditions are satisfied:

$$i \in M \subseteq N \tag{10}$$

$$P_+(M) \neq \emptyset \tag{11}$$

$$r \in \{1,2\} \tag{12}$$

and

$$x_S = (x_i)_{i \in S} \tag{13}$$

with

$$S \subseteq M \backslash i \tag{14}$$

and

$$x_i \in X_i(M) \qquad \text{for each } i \in S \tag{15}$$

The set of all positions u is denoted by U.

Choice sets: The choice set A(u) at a position $u=(M,r,x_S,i)$ describes the set of all choices available at a decision point corresponding to u and is defined as follows:

$$A(u) = \begin{cases} P_i(S \cup i) \cup X_i(M) & \text{for } S \neq M \backslash i \\ P_i(M) \cup \{w\} & \text{for } S = M \backslash i \end{cases} \tag{16}$$

The symbol w stands for the choice of a last uncommitted active player of a round not to form a coalition.

Explanation of table 1: Under the heading NEXT POSITIONS the table shows which positions can follow a position u by a choice $a \in A(u)$ according to the conditions on u and $a \in A(u)$ shown in the other columns. The set of all positions which can follow u by $a \in A(u)$ is denoted by B(u,a). The entry END indicates that B(u,a) is empty. B(u) stands for the union of all B(u,a) with $a \in A(u)$. The symbol $x_{S \cup i}$ indicates the demand system obtained, if x_S is complemented by player i's demand x_i.

Histories: A history q of $\Gamma(G)$ is a sequence of positions:

$$q = (u_1, \ldots, u_T) \tag{17}$$

with u_1 of the form

$$u_1 = (N, 1, x_\emptyset, i) \tag{18}$$

and with

$$u_{t+1} \in B(u_t) \qquad \text{for } t = 1,\ldots,T-1 \tag{19}$$

The player who has to make a choice at u_t is called the <u>t-th decision maker</u> in q. The set of all histories q is denoted by Q.

<u>Comment:</u> The choice taken at u_t for $t=1,\ldots,T-1$ and the outcome of the random draw of the

CONDITIONS ON THE POSITION	CHOICES	CONDITIONS ON THE CHOICE	NEXT POSITIONS
	Coalition: $C \in P_i(S \cup i)$	$P_+(M \setminus C) \neq 0$	$(M \setminus C, r, x_\emptyset, j)$ with $j \in M \setminus C$
		$P_+(M \setminus C) = \emptyset$	END
$S \neq M \setminus i$	payoff demand: $x_i \in X_i(M)$		$(M, r, x_{S \cup i}, j)$ with $j \in M \setminus (S \cup i)$
$S = M \setminus i$ $r=1$	no coalition: w		$(M, 2, x_\emptyset, j)$ with $j \in M$
$S = M \setminus i$ $r=2$	no coalition: w		END

<u>Table 1:</u> Next positions

next decision maker need not be expressed explicitely by the formal definition of a history. This information is supplied by the next position u_{t+1}.

<u>Terminal choices:</u> A choice $a \in A(u)$ at a position u is called <u>terminal,</u> if $B(u,a)$ is empty. It is clear that a terminal choice at $u = (M,r,x_S,i)$ is either a coalition C with $P_+(M \setminus C) = \emptyset$ or the choice w for $r=2$ and $S=M\setminus i$. Not every position u permits a terminal choice. A position u with the

property, that $A(u)$ contains at least one terminal choice, is called <u>terminable.</u> The set of all terminable positions is denoted by U_0. A history $q = (u_1,...,u_T)$ is called <u>terminable,</u> if its last position is terminable. The set of all terminable histories is denoted by Q_0.

<u>Plays:</u> A <u>play,</u> z is defined as a terminable history $q=(u_1,...,u_T)$ together with a terminal choice $a_T \epsilon A(u_T)$ at its last position:

$$z = (u_1,...,u_T;a_T) \qquad (20)$$

The <u>t–th decision maker</u> in z is the t–th decision maker in $q=(u_1,...,u_T)$. The set of all plays z is denoted by Z.

<u>Payoffs:</u> The payoff $E_i(z)$ of player i at the end of a play $z \epsilon Z$ is determined as his basic payoff obtained in the play reduced by the costs attached to choices taken by him in the course of the play. Since this description is sufficiently clear we omit the definition of $E_i(z)$. In the laboratory context payoffs are in money. We shall assume that utilities are linear in money.

<u>Strategies:</u> A <u>local strategy</u> φ_u at a position u is a probability distribution over the choice set $A(u)$. The probability assigned by φ_u to $a \epsilon A(u)$ is expressed by $\varphi_u(a)$. The set of all local strategies at u is denoted by Φ_u. A local strategy φ_q at a history $q=(u_1,...,u_T)$ is a local strategy at its last position u_T. The set of all local strategies at q is also denoted by Φ_q and $A(q)$ is used for the set of all choices at the last position u_T of q. The probability assigned by φ_q to a choice $a \epsilon A(q)$ is also expressed by $\varphi_q(a)$.

A local strategy is called <u>pure,</u> if it assigns probability 1 to one choice and zero to all others; otherwise a local strategy is called <u>genuinely mixed.</u> We identify pure local strategies with the choices to whom they assign probability 1. By this identification the choice sets become subsets of the corresponding local strategy sets.

A <u>global strategy</u> φ is a function which assigns a local strategy φ_q to every history $q \epsilon Q$. The set of all global strategies is denoted by Φ. A global strategy φ is called <u>pure</u> if it assigns pure local strategies to all histories.

<u>Comment:</u> It is well known that subgame perfect equilibria in behavior strategies of games with perfect information are characterized by the property that at every decision point of a player his local strategy is locally optimal, where local optimality means optimality with respect to the expected payoff in the subgame at the decision point under the condition that later all players always play their equilibrium strategies. In view of this fact we can avoid the definition of strategies for individual players. It is sufficient to consider global strategies. The notion of a global strategy is equivalent to the usual notion of a combination of behavior strategies. The requirement of subgame perfectness can be expressed by local optimality conditions.

Local payoffs: Let $q=(u_1,...u_T)$ be a history with $u_T=(M,r,x_S,i)$ and let φ be a global strategy. Moreover, let ψ_q be a local strategy at q. Consider the decision point reached by q in $\Gamma(G)$ where player i has to take a choice at u_T. The subgame of $\Gamma(G)$ at this decision point is denoted by $\Gamma_q(G)$. The local payoff $E(q,\varphi,\psi_q)$ at q for φ and ψ_q is player i's expected payoff in $\Gamma_q(G)$ if he plays ψ_q at the origin of $\Gamma_q(G)$ and later all players play their local strategies prescribed by φ everywhere else in $\Gamma_q(G)$. Since the local payoff is always the payoff of the decision maker in the last position of q, our notation does not have to express by an index i, whose payoff it is.

Local best replies: We say that a local strategy ψ_q at a history q is a local best reply to a global strategy φ if we have:

$$E(q,\varphi,\psi_q) = \max_{a\in A(q)} E(q,\varphi,a) \tag{21}$$

On the right hand side, it is sufficient to maximize over the choice set A(q), since the maximum over all local strategies is always attained by a pure local strategy. ψ_q is called a strict local best reply to φ at q, if it is the only best reply to φ at q. It is clear that a strict local best reply must be a pure local strategy.

Subgame perfect equilibrium: A global strategy φ is a subgame perfect equilibrium of $\Gamma(G)$, if every local strategy φ_q assigned to a history $q\in Q$ by φ is a local best reply to φ at q. As has been pointed out above this definition is equivalent to the usual definition of a subgame perfect equilibrium.

The play distribution: For every global strategy $\varphi \in \Phi$ and every history $q=(u_1,...,u_T)$ let $p(q,\varphi)$ be the probability with which q occurs if φ is played, or, in other words, with which the origin of the subgame $\Gamma_q(G)$ is reached if φ is played. If we say that q is reached by φ we mean that this probability $p(q,\varphi)$ is positive. A position u is reached by φ if at least one history is reached by φ whose last position is u.

For every $\varphi \in \Phi$ and every $z \in Z$ let $p(z,\varphi)$ be the probability with which the game $\Gamma(G)$ ends at the endpoint corresponding to z, if φ is played. We say that z is reached by φ if $p(z,\varphi)$ is positive.

The probability distribution $p(\cdot,\varphi)$ over the set Z of all plays is called the play distribution of φ. It can be seen easily that the play distribution uniquely determines the probabilities $p(q,\varphi)$ for all $q \in Q$. The local strategies φ_q assigned by φ to histories q reached by φ are also uniquely determined by the play distribution $p(\cdot,\varphi)$. However, the play distribution contains no information on the local strategies at unreached histories.

A play $z \in Z$ is called a subgame perfect equilibrium play if it is reached by at least one subgame perfect equilibrium.

Determinateness: We say that the game $\Gamma(G)$ generated by the model for a coalition game G is determinate if all subgame perfect equilibria φ of $\Gamma(G)$ have the same play distribution $p(\cdot,\varphi)$.

Absolute optimality and non-optimality: Let ψ_u be a local strategy at a position u. We say that ψ_u is absolutely optimal if at every history q whose last position is u, the local strategy ψ_u is a local best reply to every subgame perfect equilibrium φ. Similarly ψ_u is called absolutely non-optimal, if ψ_u is not a local best reply to any subgame perfect equilibrium at any history q whose last position is u.

Remark: It is clear that a local strategy ψ_u at a position u is absolutely optimal if and only if every $a \in A(u)$ with $\psi_u(a)>0$ is absolutely optimal. If one of the choices $a \in A(u)$ with $\psi_u(a)>0$ is absolutely non-optimal, then ψ_u cannot be a local best reply to any subgame perfect equilibrium at any history whose last position is u.

Transfer of definitions to subgames: Notions defined in this section for $\Gamma(G)$ – histories, plays, local and global strategies, local payoffs, local best replies, subgame perfect equilibria, the play distribution, determinateness, absolute optimality, and absolute non-optimality – will be transferred in the obvious way to subgames of $\Gamma(G)$. Notational conventions will also be transferred.

Second round subgames: Consider a subgame Γ' of $\Gamma(G)$ which starts with the random draw of the first decision maker at the beginning of the second round of the first stage of $\Gamma(G)$. Such subgames are called second round subgames. The game $\Gamma(G)$ has many second round subgames. Apart from the payoffs all second round subgames have the same structure, in the sense that all of them have the same set of histories. The histories $q=(u_1,\ldots,u_T)$ of second round subgames begin with a position u_1 of the form.

$$u_1=(N,2,x_\emptyset i) \tag{22}$$

and satisfy (19). The set of all these histories is denoted by Q_2.

The payoffs at the endpoints of a second round subgame depend on the costs incurred by the choices taken in the first round. Obviously, these "sunk" costs do not matter for the analysis of a second round subgame. Therefore, for every endpoint of $\Gamma(G)$ we define second round payoffs for each player which consist of his basic payoffs reduced by costs incurred from the second round of the first stage on but not by costs incurred in the first round of the first stage.

If in a second round subgame Γ' of $\Gamma(G)$ all payoffs at all endpoints are replaced by the corresponding second round payoffs, one obtains a new game Γ''. The structure of this new game does not depend on the particular second round subgame from which it is derived. Histories, plays, payoffs, and strategies are the same in all cases. Therefore, it is natural in our framework to identify all these new games. By this identification we always obtain the same normalized second round subgame, denoted by $\Gamma^2(G)$.

The normalized second round subgame will be important in the analysis of the unanimity game and the 3–person–quota game without the grand coalition. In both cases the analysis of the normalized second round subgame will precede the analysis of the whole game.

4. Unanimity games

A coalition game $G=(N,P,v,g,\gamma)$ is a unanimity game, if N is the only genuine permissible coalition, i.e. if we have:

$$P_{+}(N)=\{N\} \tag{23}$$

A zero–one normalized unanimity game is a zero normalized unanimity game with the additional property:

$$v(N)=1 \tag{24}$$

In the following we shall restrict our attention to zero–one normalized unanimity games with

$$N=\{1,...,n\} \tag{25}$$

It will be easy to see that zero–one normalization together with (24) involves no loss of generality. In this section $G=(N,P,v,g,\gamma)$ will always be a zero–one normalized unanimity game with (25).

Explanation of table 2: All subgame perfect equilibria of a zero–one normalized unanimity game with $N=\{1,...,n\}$ have the common structure shown by table 2. The table distinguishes 8 types of histories by the properties of the last position $u=(M,r,x_S,i)$. The first column assigns a type number to each type of history. The round number r and the number $|S|$ of committed players are shown in the next two columns. The fourth column shows the choices which can be taken with positive probability in subgame perfect equilibrium. C stands for a coalition formed and x_i for a payoff demand named by the decision maker player i.

It will be the content of theorem 1 that all subgame perfect equilibria of zero–one normalized unanimity games have the properties shown by table 2. It is a consequence of table 2 that all subgame perfect equilibrium plays of a second round subgame have the following structure: the first decision maker names the payoff demand $1-(n-1)g$; all other players with the exception of the n–th decision maker name the payoff demand g and the n–th one forms the coalition N.

The auxiliary variable D can be interpreted as the maximum demand the decision maker player i can name if after the deduction of the demands of the players in $S \cup i$ from $v(N)=1$ at least g must be left over for each of the $n-|S|-1$ remaining uncommitted players.

It is a consequence of table 2 that a subgame perfect equilibrium play of the whole game $\Gamma(G)$ has the following structure: for $t=1,\dots,n-1$ the t-th decision maker names the payoff demand $1/n$ and the n-th decision maker forms N. – The auxiliary variable F can be interpreted as the maximum demand the decision maker player i can name if after the deduction of the demands of

TYPE OF HIS-TORY	ROUND r	NUMBER OF COMMIT-TED PLAYERS $\|S\|$	CONDITIONS ON $D=1-x_S(S)-(n-\|S\|-1)g$ and $F=1-x_S(S)-(n-\|S\|-1)\frac{1}{n}$	CHOICES POSSIBLY TAKEN WITH POSITIVE PROBABILITY
1	2	n–1	$D>0$	C=N
2	2	n–1	$D\leq 0$	C=i
3	2	<n–1	$D>0$	$x_i=D$
4	2	<n–1	$D\leq 0$	C=i
5	1	n–1	$F\geq\frac{1}{n}$	C=N
6	1	n–1	$F<\frac{1}{n}$	w
7	1	<n–1	$F\geq\frac{1}{n}$	$x_i=F$
8	1	<n–1	$F<\frac{1}{n}$	$x_i>F$

Table 2: Properties of subgame perfect equilibria of zero–one normalized unanimity games with $N=\{1,\dots,n\}$

the players in $S \cup i$ from $v(N)=1$ at least $1/n$ must be left over for each of the $n-|S|-1$ remaining uncommitted players.

Lemma 1: Let $G=(n,P,v,g,\gamma)$ be a zero–one normalized unanimity game with $N=\{1,...,n\}$ and let Γ' be a second round subgame of $\Gamma(G)$. A global strategy φ' for Γ' is a subgame perfect equilibrium of Γ' if and only if it satisfies the conditions imposed by table 2 on choices taken with positive probability at histories of types 1 to 4. At these histories the choices listed by table 2 are absolutely optimal and choices not listed there are absolutely non–optimal. Γ' is determinate. Every subgame perfect equilibrium play of Γ' has the following structure: The first decision maker names the payoff demand $1-(n-1)g$; for $t=2,...,n-1$ the t–th decision maker names the payoff demand g and the n–th decision maker forms N.

Proof: If a coalition is formed, regardless of whether this is the coalition N or a one–person coalition, the game ends. Therefore, for every position $(M,2,x_S,i)$ in Γ' we must have $M=N$. Therefore, every history of Γ' must be of one of the types 1 to 4. It is sufficient to prove the absolute optimality of the choices listed for these histories in table 2 and the absolute non–optimality of the choices not listed there. The assertion on subgame perfect equilibria is an immediate consequence. The structure of subgame perfect equilibrium plays and the determinateness of $\Gamma(G)$ follow by table 2.

We can replace the analysis of Γ' by the analysis of the normalized second round subgame $\Gamma^2(G)$. Consider the situation of a player i in the role of the n–th decision maker. He can either form N or $\{i\}$ or choose w. Coalition N yields a payoff of $D-\gamma_n$. His payoff is $-\gamma_1$ for $\{i\}$ and $-\gamma_n$ for w. In view of $\gamma_n>\gamma_1$ it never pays to choose w. It is clear that it is absolutely optimal to choose N for $D>0$ and $\{i\}$ for $D\leq 0$ and that in both cases everything else is absolutely non–optimal.

It remains to show the assertions on absolute optimality and non–optimality concerning histories of types 3 and 4. Define $k=n-|S|-1$. We show the assertions on absolute optimality and non–optimality concerning histories of types 3 and 4 by induction on k. In the case $k=1$, it follows by what has been proved on histories of types 1 and 2 that a payoff demand of $x_i \leq D$ leads to the formation of N whereas a demand of $x_i > D$ has the consequence that the n–th decision maker forms his one–person coalition. Therefore, $x_i \leq D$ yields a payoff of $x_i - g$ and $x_i > g$ yields a payoff of $-g$ whereas $\{i\}$ yields $-\gamma_1$. Therefore, $x_i = D$ is absolutely optimal for $D>0$ and $\{i\}$ is absolutely optimal for $D \leq 0$. In both cases nothing else is optimal.

Assume that the assertions hold for $k=1,...,m$ and consider the situation of a decision maker with $k=m+1$. Here, too, a payoff demand of $x_i \leq D$ leads to the formation of N whereas a demand of $x_i > D$ has the consequence that the next decision maker forms his one–person coalition. The same payoff comparisons as above show that the assertions hold for k=m+1.

The second round value: In view of lemma 1, the expected payoff of a player i=1,...,n for a subgame perfect equilibrium φ^2 of the normalized second round subgame $\Gamma^2(G)$ is the same for each player and for every subgame perfect equilibrium of $\Gamma^2(G)$. We use the symbol V for this payoff and refer to it as the second round value. We have:

$$V=\frac{1}{n}-\frac{n-1}{n}g-\frac{1}{n}\gamma_n \qquad \text{for } i=1,\dots,n \tag{26}$$

This is due to the fact that in subgame perfect equilibrium play a player has a probability of (n–1)/n to become one of the first n–1 decision makers and a probability of 1/n to become the n–th decision maker. In the first case, his cost is g and in the second one, it is γ_n. We have

$$\frac{1}{n}-g<V<\frac{1}{n} \tag{27}$$

in view of $0<\gamma_n<g$.

Theorem 1: Let $G=(N,P,v,\gamma)$ be a zero–one normalized unanimity game with $N=\{1,\dots,n\}$. A global strategy φ for $\Gamma(G)$ is a subgame perfect equilibrium of $\Gamma(G)$ if and only if it satisfies the conditions imposed by table 2 on choices taken with positive probability. For each of the 8 history types the choices listed by table 2 are absolutely optimal and those not listed by table 2 are absolutely non–optimal. $\Gamma(G)$ is determinate. Every subgame perfect equilibrium play of $\Gamma(G)$ has the following structure: For t=1,...,n–1 the t–th decision maker names the payoff demand 1/n and the n–th decision maker forms N.

Proof: Since in $\Gamma(G)$ all positions are of the form (N,r,x_S,i) every history in $\Gamma(G)$ must be of one of the 8 types of table 2. As in the case of lemma 1, it is sufficient to prove the assertions on absolute optimality and non–optimality.

Consider the situation of a player i in the role of the n–th decision maker. He can either form N or {i} or choose w. Coalition N yields $F-\gamma_n$ and {i} yields $-\gamma$. The choice of w yields $V-\gamma_n$. In view of (3), (4), and (27), we have

$$V-\gamma_n>\frac{1}{n}-g-\gamma_n\geq g-\gamma_n>0 \tag{28}$$

Therefore, w is better for player i than {i}. Since F is negotiable, it follows by (27) that F is greater than V for $F\geq 1/n$ and smaller than V for $F<1/n$. It is now clear that at histories of types 5 and 6 the choices listed in table 2 are absolutely optimal and everything else is absolutely non–optimal.

It remains to show the assertions on absolute optimality and absolute non-optimality concerning histories of types 7 and 8. As in the proof of lemma 1, this will be shown by induction on $k=n-|S|-1$. In the case $k=1$, it follows by what has been proved on histories of types 5 and 6 that a payoff demand of $x_i \leq F$ leads to the formation of N whereas a demand of $x_i > F$ has the consequence that the n-th decision maker chooses w. Therefore, $x_i \leq F$ yields a payoff of $x_i - g$ and $x_i > F$ yields a payoff of $V-g$. The choice of {i} yields a payoff of $-\gamma_1$. In view of (3) and (27), we have

$$V - g > \frac{1}{n} - 2g > 0 \tag{29}$$

Since F is greater than V for $F \geq 1/n$ and smaller than V for $F < 1/n$, it follows that $x_i = F$ is absolutely optimal for $F \geq 1/n$ and every x_i with $x_i > F$ is absolutely optimal for $F < 1/n$. In both cases everything else is absolutely non-optimal.

Assume that the assertions on optimality and non-optimality for histories of types 7 and 8 are satisfied for $k=1,...,m$. Consider the situation of a decision maker with $k=m+1$. Here, too, a payoff demand of $x_i \leq F$ leads to the formation of N whereas a demand of $x_i > F$ has the consequence that the next decision maker names a demand greater than the new F left over for him which eventually leads to the choice of w by the n-th decision maker. The same payoff comparisons as above show that the assertions hold for $k=m+1$.

Remark: It follows by theorem 1 that for $i=1,...,n$ and every subgame perfect equilibrium φ of $\Gamma(G)$ the expected payoff $E_i(\varphi)$ is V. In this sense we say that V is the value of $\Gamma(G)$.

5. 3-person quota games without the grand coalition

For the purposes of this paper, a 3-person quota game is defined as a coalition game $G=(N,P,v,g,\gamma)$ with 3 players in which all 2-person coaltions are permissible with the additional property that

$$v(ij)+v(ik)>v(jk)+4g \tag{30}$$

holds for every player i where j and k are the two other players. The number

$$q_i=\frac{v(ij)+v(ik)-v(jk)}{2} \tag{31}$$

is called the quota of player i. The quotas have the property

$$q_i+q_j=v(ij) \tag{32}$$

for $i \neq j$. It follows by the divisibility condition (see section 2) that the quotas are negotiable. In view of (30), each of the quotas is greater than $2g$. In this respect, our definition of a 3-person quota game is a little more restrictive than the usual one. Thereby, we exclude uninteresting border cases. The usual definition requires non-negative quotas, but not more.

A 3-person quota game, in which the 3-person coalition is not permissible is called a 3-person quota game <u>without the grand coalition.</u> In this section $G=(N,P,v,g,\gamma)$ will always be a zero-normalized 3-person quota game without the grand coalition.

<u>Explanation of Table 3:</u> Consider a second round subgame Γ' of the extensive game $\Gamma(G)$ generated by the model for a 3-person quota game G without the grand coalition. It will be the content of lemma 2 that all subgame perfect equilibrium plays of Γ' have the following structure: The first decision maker names his quota as his payoff demand and the second decision maker

DECISION MAKER			NECESSARY CONDITION IN SUBGAME PERFECT EQUILIBRIUM	BASIC PAYOFF OF THE FIRST DECISION MAKER
FIRST: i	SECOND: j	THIRD: k		
PAYOFF DEMAND	POSSIBLE PURE LOCAL BEST REPLIES	POSSIBLE PURE LOCAL BEST REPLIES		
$x_i \leq q_i$	$C=ij$	–		x_i
$v(ik)>x_i>q_i$	undercutting $x_j=q_j+x_i-q_i-g$	$C=jk$		0
	matching $x_j=q_j+x_i-q_i$	$C=ik$ $C=jk$	PROBABILITY of $C=ik$ at most $\frac{g}{q_j+x_i-q_i}$	$<q_i$
$x_i \geq v(ik)$	leaving g to k $x_j=v(jk)-g$	$C=jk$		0
	pushing k out: $x_j \geq v(jk)$	$C=k$	$\frac{q_i-q_j}{2}>q_k-g$	$\frac{v(ij)}{2}$

<u>Table 3:</u> Consequences of first demands x_i in $\Gamma^2(G)$, where G is a 3-person quota game without the grand coalition.

forms the 2–person coalition with the first decision maker. It is sufficient to show this for the normalized second round subgame $\Gamma^2(G)$. In order to prove lemma 2, it is necessary to investigate the consequences of all possible payoff demands of the first decision maker, if later all players play a subgame perfect equilibrium.

Table 3 provides an overview over these consequences. The symbols i, j, and k denote the first, the second and the potential third decision maker, respectively. This notational convention will always be used in this section. The first column distinguishes three cases with respect to the payoff demand x_i of the first decision maker. The second and third column show possible pure local best replies to subgame perfect equilibria for the second and third decision maker, respectively. The symbol C indicates a coalition and x_j or x_k ,respectively a payoff demand of j and k. Choices of j and k not listed by table 3 for one of the cases with respect to previous decisions are absolutely non–optimal for this case. The assertions expressed by table 2 will be shown in the proof of lemma 2.

For $x_i \leq q_i$ it is absolutely optimal for player j to form the coalition ij. In the case $v(ik) > x_i > q_1$, there are two types of possible best replies: undercutting and matching. Undercutting means that player j offers player k one bargaining unit more than player i does. Matching means that player j offers player k the same amount as i does. The only best reply of player k to undercutting is C=jk. There are two local best replies to matching, C=ik and C=jk. However, matching is not a local best reply for player j unless player k's probability for ik is sufficiently low as required by the condition in the fourth column. This condition will be derived by a comparison of player j's payoff for undercutting and matching.

In the case $x_i \geq v(ik)$ there are two possible pure local best replies for player j: leaving g to k and pushing k out. Leaving g to k means that j offers k just one bargaining unit which is enough to induce k to form jk. Pushing k out means to demand at least v(jk) in order to make it optimal for k to form his one–person coalition. Pushing k out may be a local best reply; a necessary condition for this is shown in the fourth column.

The last column of table 3 shows consequences for player i's basic payoff. These consequences will permit the conclusion that it is absolutely optimal for the first decision maker to demand his quota and that everything else is absolutely non–optimal for him.

Lemma 2: Let G be a 3–person quota game without the grand coalition and let Γ' be a second round subgame of the extensive game Γ(G) generated by the model. Γ' is determinate. Every subgame perfect equilibrium play of Γ' has the following structure: The first decision maker names his quota as his payoff demand and the second decision maker forms the 2–person coalition with the first decision maker.

Proof: It is sufficient to prove the lemma for the normalized second round subgame $\Gamma^2(G)$ instead of Γ'. It is our plan to show the assertions expressed by table 3 in order to derive the statements of the lemma.

The case $x_i \leq q_i$: Assume that the first decision maker names a demand $x_i \leq q_i$. We shall show that it is absolutely optimal for player j to form ij and that everything else is absolutely non–optimal for him. If player j forms ij he receives the payoff $q_j+q_i-x_i-\gamma_2$. If he names a demand $x_j \leq q_j+q_i-x_i$ he receives less than that. Assume that he demands $x_j>q_j+q_i-x_i$. Then player k receives the basic payoffs $q_k+q_i-x_i$ for ik and $q_k+q_j-x_j$ for jk. In view of $x_i \leq q_i$ we have $x_j>q_j$. It follows that the formation of ik is more advantageous for k than the formation of jk. In view of $q_k>0$ and $x_i \leq q_i$ the remaining choices $\{k\}$ and w are less profitable for k than ik. Therefore, it is the only local best reply of k to form ik. This means that j receives the payoff $-g$. It follows that it is absolutely optimal for j to form ij and that everything else is absolutely non–optimal for him. Therefore, i receives the basic payoff x_i for a demand $x_i \leq q_i$.

The case $v(ik)>x_i>q_i$: Now assume $v(ik)>x_i>q_i$. Suppose that j demands $x_j>q_j+x_i-q_i$. Then ik is more advantageous for k than jk. Since $v(ik)-x_i-\gamma_2$ is positive ik is k's only local best reply. This means that j receives a payoff of $-g$. He can get more by the formation of his one–person coalition. Therefore, a demand $x_j>q_j+x_i-q_i$ is absolutely non–optimal for j.

Suppose that j demands $x_j<q_j+x_i-q_i$. Then for similar reasons it becomes absolutely optimal for k to form jk. Obviously, player j's best demand of this kind is the demand connected to undercutting. Consequently, undercutting and matching are the only candidates for local best replies to subgame perfect equilibria.

Suppose that j responds to $v(ik)>x_i>q_i$ by matching. Then both ik and jk become local best replies of k. Let α be the probability with which k chooses ik in this situation. If matching is a local best reply than player j's local payoff for matching must be at least as great as his local payoff for undercutting. This yields the following condition:

$$(1-\alpha)(q_j+x_i-q_i) \geq q_j+x_i-q_i-g \tag{33}$$

The costs of g are the same for both payoff demands and therefore cancel out. As we have seen jk becomes absolutely optimal for k in the case of undercutting. Inequality (33) is equivalent to the following condition:

$$\alpha \leq \frac{g}{q_j+x_i-q_i} \tag{34}$$

This condition is shown in the fourth column of table 3. In the case of undercutting player i receives the basic payoff zero. Therefore, player i receives the basic payoff α g for a demand x_i

with $v(ik)>x_i>q_i$. We shall now show that we have:

$$\alpha x_i<q_i \tag{35}$$

Inequality (34) yields

$$\alpha x_i \leq \frac{g x_i}{q_j+x_i-q_i} \tag{36}$$

Therefore, (35) is implied by

$$g x_i<q_i(q_j+x_i-q_i) \tag{37}$$

This is equivalent to

$$q_i(q_j-q_i)+x_i(q_i-g)>0 \tag{38}$$

In view of $x_i>q_i$ inequality (38) is implied by

$$q_i(q_j-g)>0 \tag{39}$$

Inequality holds since our definition of a 3–person quota game requires that quotas are greater than g. Therefore, (35) holds. As shown by the last column of table 3 player i's expected basic payoff for a demand x_i with $v(ik)>x_i>q_i$ is less than q_i if later a subgame perfect equilibrium is played.

<u>The case $x_i \geq v(ik)$:</u> Assume that the first decision maker names a demand $x_i \geq v(ik)$ and that later a subgame perfect equilibrium is played. If player j names a demand $x_j<v(jk)$ it becomes absolutely optimal for player k to form jk; obviously, player j's best choice among these demands is leaving g to k. However, pushing k out may lead to a better payoff or an equally high payoff. A demand $x_j \geq v(jk)$ makes it absolutely optimal for player k to form his one–person coalition, since w or a 2–person coalition with basic payoffs of zero incur higher cost. This has the consequence that in stage 2 a subgame is played whose structure is that of the extensive game generated by the model for the zero normalized unanimity game with the player set ij and with the value v(ij) for the grand coalition. The expected basic payoff for this subgame is v(ij)/2 for each of both players i and j.

Since additional costs arise in a second stage pushing k out cannot be at least as good for player j as leaving g to k unless the following inequality holds:

$$\frac{v(ij)}{2} > v(jk)-g \tag{40}$$

The right hand side shows j's basic payoff for leaving g to k. Inequality (40) can be rewritten as follows:

$$\frac{q_i+q_j}{2} > q_j+q_k-g \tag{41}$$

$$\frac{q_i-q_j}{2} > q_k-g \tag{42}$$

This is the condition shown by the fourth column of table 3. In view of $q_k \geq g$ inequality (42) implies $q_i > q_j$. This has the consequence that q_i is greater than $v(ij)/2$. Therefore, after a demand $x_i \geq v(ik)$ player i's expected payoff is less than q_i if later a subgame perfect equilibrium is played.

Completion of the proof: It is now clear that player i obtains the basic payoff q_i for $x_i=q_i$ and less for any other demand, if later a subgame perfect equilibrium is played. Other demands cause the same costs in stage 1 and possibly additional costs in stage 2. Player 1's one–person coalition yields $-\gamma_1$ which in view of $q_i \geq 3g$ is less than the payoff q_i-g for $x_i=q_i$. Therefore, $x_i=q_i$ is absolutely optimal for player i and everything else is absolutely non–optimal for him. Faced with $x_i=q_i$ it is absolutely optimal for player j to form ij and everything else is absolutely non–optimal for him. This proves the statement on the structure of subgame perfect equilibrium plays. The determinateness of Γ' is an immediate consequence.

Second round values: The second round value of a player $h \in N$ for a zero–normalized 3–person quota game G without the grand coalition is his expected payoff in $\Gamma^2(G)$ if a subgame perfect equilibrium is played. In view of the determinateness of $\Gamma^2(G)$ this value, denoted by V_h, does not depend on the particular subgame perfect equilibrium. V_h is as follows:

$$V_h=\frac{2}{3}q_h-\frac{1}{3}g-\frac{1}{3}\gamma_2 \tag{43}$$

This follows by the randomness of the order, in which the players become decision makers. A player h receives his quota with probability 2/3 and has to pay g with probability 1/3 as the first decision maker and γ_2 with probability 1/3 as the second decision maker.

The critical level: Consider the situation of the third decision maker, player k in a play of Γ(G) if later a subgame perfect equilibrium is played. Under which conditions is at least one of the two coalitions ik and jk a local best reply of player k? As we shall see, this is the case, if and only if we have:

$$\max[v(ik)-x_i, v(jk)-x_j]-\gamma_2 \geq V_k-\gamma_3 \tag{44}$$

The left hand side shows what can be obtained by a 2–person coalition and the right hand side is the local payoff for player k's choice w. We have to show that this local payoff is greater than the payoff $-\gamma_1$ for k's one-person coalition. (43) yields:

$$V_k-\gamma_3 = \frac{1}{3}(2q_k-g-\gamma_2-3\gamma_3) \tag{45}$$

Since our definition of a 3-person quota game requires $q_k \geq 3g$ it follows by $\gamma_2<g$ and $\gamma_3<g$ that $V_k-\gamma_3$ is positive. This shows that player k's one-person coalition is absolutely non-optimal for him. Define

$$y_k = V_k-\gamma_3+\gamma_2 = \frac{1}{3}(2q_k-g+2\gamma_2-3\gamma_3) \tag{46}$$

We call y_k player k's <u>critical level.</u> In view of $q_k \geq 3g$ and $0<\gamma_1<\gamma_3<g$ the following inequality holds:

$$0<y_k<q_k \tag{47}$$

One of the two coalitions ik or jk is a local best reply of player k, if and only if we have:

$$\max\,[v(ik)-x_i, v(jk)-x_j] \geq y_k \tag{48}$$

and w is a local best reply of player k, if and only if the reverse inequality

$$\max\,[v(ik)-x_i, v(jk)-x_j] \leq y_k \tag{49}$$

holds. The only other choice of player k, his one-person coalition is absolutely non-optimal. These facts will be important in the proof of theorem 1.

<u>The rounding function:</u> For every real number η the smallest negotiable amount not smaller than η is denoted by $\rho(\eta)$. In this way we define a <u>rounding function</u> ρ which maps a non-negotiable amount to the next higher negotiable amount and leaves negotiable amounts unchanged.

<u>Explanation of tables 4 and 5:</u> These tables are analogous to table 3 with the difference that they refer to the first round of $\Gamma(G)$. The same notational conventions are used. Table 4 is valid for cases in which the critical level y_k of the third decision maker is non-negotiable and table 5 concerns cases in which y_k is negotiable (in the same game y_k may be negotiable for some players but not for others). Choices not listed as possible pure best replies are absolutely non-optimal under the conditions indicated by the tables.

In both tables the case $x_i \leq q_i$ is analogous to table 3. The case $v(ik)-y_k>x_i>q_i$ in tables 4 and 5 is analogous to the case $v(ik)>x_i>q_i$ in table 3. The case $v(ik)-y_k \geq x_i>q_i$ in table 5 is slightly

DECISION MAKER			NECESSARY CONDITIONS IN SUBGAME PERFECT EQUILIBRIUM	BASIC PAYOFF OF PLAYER i
FIRST: i	SECOND: j	THIRD: k		
PAYOFF DEMAND	POSSIBLE PURE LOCAL BEST REPLIES	POSSIBLE PURE LOCAL BEST REPLIES		
$x_i \le q_i$	C=ij	–		x_i
$v(ik)-y_k>x_i>q_i$	undercutting $x_j=q_j+x_i-q_i-g$	C=jk		0
	matching $x_j=q_j+x_i-q_i$	C=ik C=jk	Probability of C=ik at most $\frac{g}{q_j+x_i-q_i}$	$<q_i$
$x_i>v(ik)-y_k$	pushing k close to the critical level $x_j=v(jk)-\rho(y_k)$	C=jk		0

Table 4: Consequences of first demands in $\Gamma(G)$ where G is a 3–person quota game without the grand coalition under the condition that y_k is non–negotiable.

different since here w appears as an additional pure best reply of player k after a matching response of the second decision maker, but only in the special case $x_i= v(ik)-y_k$.

The case $x_i>v(ik)-y_k$ in tables 4 and 5 is different from the case $x_i \geq v(ik)$ in table 3 since leaving g to k and pushing k out are replaced by pushing k close to the critical level and pushing k to the critical level. Pushing k close to the critical level means that just enough is left over for player k, to make it unattractive for him to choose w. Pushing k to the critical level means to make k indifferent between jk and w. This is impossible if y_k is non–negotiable. The assertions expressed by tables 4 and 5 will be shown in the proof of theorem 2.

Theorem 2: Let G be a 3–person quota game without the grand coalition. $\Gamma(G)$ is determinate. Every subgame perfect equilibrium play of $\Gamma(G)$ has the following structure: The first decision

DECISION MAKER			NECESSARY CONDITION IN SUBGAME PERFECT EQUILIBRIUM	BASIC PAYOFF OF PLAYER i
FIRST: i	SECOND: j	THIRD: k		
PAYOFF DEMAND	POSSIBLE PURE LOCAL BEST REPLIES	POSSIBLE PURE LOCAL BEST REPLIES		
$x_i \leq q_i$	C=ij	–		x_i
$v(ik)-y_k \geq x_i > q_i$	undercutting $x_j=q_j+x_i-q_i-g$	C=jk		0
	matching $x_j=q_j+x_i-q_i$	C=ik C=jk w*)	Probability of C=ik or w at most $\frac{g}{q_j+x_i-q_i}$	$<q_i$
$x_i > v(ik)-y_k$	pushing k close to the critical level $x_j=v(jk)-y_k-g$	C=jk		0
	pushing k to the critical level $x_j=v(jk)-y_k$	C=jk w		$<\frac{2}{3}q_i$

Table 5: Consequences of first demands in $\Gamma(G)$ where G is a 3–person quota game without the grand coalition under the condition that y_k is negotiable.

*) for $x_i=v(ik)-y_k$ only

maker names his quota as his payoff demand and the second decision maker forms the 2–person coalition with the first decision maker.

Proof: It is our plan to show the assertions expressed by tables 4 and 5 in order to derive the structure of subgame perfect equilibrium plays as the final conclusions.

<u>The case $x_i \leq q_i$ in tables 4 and 5:</u> The proof is analogous to the proof for the case $x_i \leq q_i$ in table 3, with the only difference that one must consider the fact that w is now a more attractive alternative for player k than before. However, in view of (48) and $\gamma_3 > \gamma_2$ it is clear that w cannot be more attractive for k than ik.

<u>The case $v(ik)-y_k > x_i > q_i$ in table 4:</u> The proof is analogous to the proof for the case $v(ik) > x_i > q_i$ in table 3. The condition $v(ik)-y_k > x_i$ excludes the possibility that w is a local best reply of player k after the matching response of player j. We cannot have $v(ik)-y_k = x_i$ since y_k is non–negotiable.

<u>The case $v(ik)-y_k \geq x_i > q_i$ in table 5:</u> For the subcase of undercutting the proof is analogous to the proof for the case $v(ik) > x_i > q_i$ in table 3. The possibility that w is a local best reply of player k is excluded by $v(ik)-y_k \geq x_i$.

Consider the subcase of matching. As long as $v(ik)-y_k$ is greater than x_i, the choice w cannot be a local best reply of player k to a subgame perfect equilibrium, but this is different in the special case $x_i = v(ik)-y_k$. Assume that player i has named this demand. It follows by what has been shown in connection with (48) and (49) that in this case w and the two 2–person coalitions ik and jk are absolutely optimal for player k and that his one–person coalition is absolutely non–optimal for him.

Regardless of whether player i's demand is $v(ik)-y_k$ or lower let α be the probability with which player k chooses ik and let β be the probability with which he chooses w. Matching cannot be a local best reply of player j unless it yields at least as much as undercutting. We must have:

$$(1-\alpha-\beta)(q_j+x_i-q_i)+\beta V_j \geq q_j+x_i-q_i \tag{50}$$

Since βV_j is non–negative, (50) permits the conclusion:

$$\alpha+\beta \leq \frac{g}{q_j+x_i-q_i} \tag{51}$$

Player i's expected basic payoff is $\alpha x_i + \beta V_i$. In view of $V_i < x_i$ this payoff is at most $(a+\beta)x_i$. In the same way as (35) has been derived in the proof of lemma 2, it follows that we have:

$$(\alpha+\beta)x_i < q_i \tag{52}$$

Therefore player i's expected basic payoff in the case under consideration is less than q_i as indicated by table 5.

<u>The case $x_i \geq v(ik)-y_k$ in table 4:</u> Consider the situation of the second decision maker, player j, and assume that after his choice a subgame perfect equilibrium is played. Suppose that j names a payoff demand $x_j > v(jk)-\rho(y_k)$. It follows by the non–negotiability of y_k that $v(jk)-x_j$ is smaller

than y_k. Therefore w becomes the only local best reply for player k. This means that player j's local payoff for $x_j > v(jk) - p(y_k)$ is $V_j - g$.

Suppose that j names a demand $x_j \leq v(jk) - \rho(y_k)$. In this case it is the only best reply of player k to form jk. Player j receives the local payoff $x_j - g$ for a demand of this kind. Obviously, pushing k close to the critical level specifies the most advantageous demand of this kind. Player j's local payoff for pushing k close to the critical level is $v(jk) - \rho(y_k) - g$. In view of (47) we have $\rho(y_k) \leq q_k$. Therefore pushing k close to the critical level yields at least $q_j - g$ which is more than $V_j - g$. Consequently, pushing k close to the critical level is player j's only local best reply. Therefore, player i receives the basic payoff zero if he names a demand $x_i > v(ik) - y_k$.

<u>The case $x_i \geq v(ik) - y_k$ in table 5:</u> Consider the situation of player j and assume that after his choice a subgame perfect equilibrium is played. The effects of pushing k close to the critical level and lower payoff demands need not be discussed in detail, since they are the same as in the corresponding case of table 4.

Whether pushing k close to the critical level is a local best reply or not depends on player k's local strategy after pushing k to the critical level. The demand $x_j = v(jk) - y_k$ makes k indifferent between jk and w. Both choices become absolutely optimal for him. Pushing k to the critical level is a local best reply if k's prabability for jk after $x_j = v(jk) - y_k$ is sufficiently high and pushing k close to the critical level is a local best reply if this probability is sufficiently low. In the subcase of pushing k close to the critical level player i receives the basic payoff zero and in the subcase of pushing k to the critical level his basic payoff is smaller than $2q_i/3$.

<u>Completion of the proof:</u> As in the case of table 3 it follows by the last columns of tables 4 and 5 that it is absolutely optimal for player i to commit himself to his quota q_i as his payoff demand; everything else is absolutely non–optimal for him. Faced with this demand it is absolutely optimal for player j to form ij and everything else is absolutely non–optimal for him. This proves the statement on the structure of subgame perfect equilibrium plays. The determinateness of $\Gamma(G)$ is an immediate consequence.

<u>Remark:</u> For every $h \in N$ the second round values V_h defined by (43) are also the expected payoffs $E_h(\varphi)$ for every subgame perfect equilibrium φ of $\Gamma(G)$.

6. Strategic choice of the next decision maker

In this section we shall consider a modified model in which for $m=1,\dots,|M|-1$ the m–th decision maker, if he names a payoff demand, simultaneously has to choose the (m+1)–th decision maker among the remaining uncommitted players still in the game. Otherwise the structure of the model of section 1 remains unchanged. This means that figure 1 is changed as follows: In rectangle 5 player i either forms a coalition $C \in P_i(S \cup i)$ or names a pair (x_i, j) with $x_i \in X_i(M)$ and $j \in M \setminus S$. In rectangle 8 after $S \cup i \to S$ the additional operation $j \to i$ has to be performed. The arrow from rectangle 8 to rectangle 3 is deleted and replaced by an arrow from rectangle 9 to rhomboid 4.

For $m=|M|-1$ the choice of the next decision maker is purely formal, since there is only one player which can be chosen. In the case of a 3–person game only the first decision maker in a round with $M=N$ has a real choice with respect to the next decision maker. Positions, histories, and plays are the same as in section 1, even if the strategy sets are larger.

In the following we shall look at a numerical zero–normalized 3–person quota game $G=(N,P,v,g,\gamma)$ without the grand coalition. The player set is $N=\{1,2,3\}$. The characteristic function and the bargaining unit are specified as follows:

$$v(1)=v(2)=v(3)=0 \tag{53}$$
$$v(12)=v(13)=150 \tag{54}$$
$$v(23)=60 \tag{55}$$
$$g=1 \tag{56}$$

The cost structure γ will not be numerically specified. The game has the following quotas:

$$q_1=120 \tag{57}$$
$$q_2=q_3=30 \tag{58}$$

Let Γ be the extensive game generated by the modified model for this coalition game G. We shall construct a subgame perfect equilibrium φ for Γ with the property that outcomes far away from quota divisions are reached if one of the players 2 or 3 is the first decision maker.

The definitions of a second round subgame and the normalized second round subgame are transferred to the present context in the obvious way. The normalized second round subgame of Γ is denoted by Γ^2. As before a subgame perfect equilibrium of Γ^2 is also a subgame perfect equilibrium of every second round subgame Γ' of Γ. Only two different subgame perfect equilibria φ^2 and ψ^2 of Γ^2 are induced by φ on second round subgames of Γ. As we shall see the game ends in the first round, if φ is played. Therefore one of the players must have deviated from φ if a second round subgame is reached, and he must have deviated by taking a choice which is not taken with positive probability, if φ is played. We call such a choice an <u>observable deviation.</u> The player who was the earliest one to choose an observable deviation is called the <u>first deviator.</u> φ^2 is the subgame perfect equilibrium induced by φ on second round subgames reached by a history with player 1 as the first deviator; ψ^2 is induced on all other second round subgames.

We shall first describe the local strategies of φ, φ^2, and ψ^2 at histories reached in Γ and Γ^2, respectively. On this basis, we shall later complete the construciton of φ. Finally, we shall argue by the sketch of a proof, that φ is a subgame perfect equilibrium.

Explanation of table 6: This table shows the equilibrium plays reached by φ and the local strategies prescribed by φ at histories reached by φ. The first column shows the demand of the first decision maker and the second one indicates whom he chooses as the next decision maker, "prob. 1/2" means that the concerning choice is taken with probability 1/2. The two last columns show the choices taken by the second and the third decision maker. x_i indicates a demand and C a coalition formed.

FIRST DECISION MAKER		SECOND DECISION MAKER	THIRD DECISION MAKER
DEMAND	CHOICE OF NEXT DECISION MAKER		
x_1=120	j=2 prob. $\frac{1}{2}$	C=12	–
	j=3 prob. $\frac{1}{3}$	C=13	–
x_2=100	j=3	x_3=100	C=12 prob. $\frac{1}{2}$
			C=13 prob. $\frac{1}{2}$
x_3=100	j=2	x_2=100	C=13 prob. $\frac{1}{2}$
			C=12 prob. $\frac{1}{2}$

Table 6: Equilibrium plays reached by φ and local strategies prescribed by φ at histories of Γ reached by φ.

Player 1 as the first decision maker demands his quota q_1=120 and chooses each of both other players with equal probabilities as the next decision maker who then forms the 2–person coalition with player 1. Here a quota division results. However, player 2 or 3 as the first decision maker demands 100 and choose the other one of these players as the next decision maker, who then also demands 100. Player 1 then forms 12 and 13 with equal probabilities. Obviously, the payoffs obtained by the coalition members, namely 50 for player 1 and 100 for the other member are far away from their quotas 120 and 30, respectively.

Explanation of tables 7 and 8: The notational conventions are the same as in table 6. The subgame perfect equilibrium φ^2 is induced by φ on second round subgames reached by histories, in which player 1 is the first deviator. Table 7 shows the local strategies prescribed by φ^2 at histories of Γ^2 reached by φ^2. Player 1 as the first decision maker behaves like in table 6; the second decision maker then forms the 2–person coalition with player 1. Player 2 or 3 as the first decision maker demands his quota and chooses the other one of these two players as the next decision maker, who then forms 23.

Table 8 shows the plays reached by the subgame perfect equilibrium ψ^2 induced by φ on second round subgames reached by histories in which either player 2 or player 3 is the first deviator. Plays starting with player 1 as the first decision maker are like those in tables 6 and 7. Player 2 or 3 as the first decision maker demands his quota as in table 7, but he chooses player 1 as the next decision maker; then 12 or 13, respectively, is formed.

Construction of φ^2 and ψ^2: The subgames of Γ^2 reached after the choice of the first decision maker agree with the corresponding subgames of the normalized second round subgame $\Gamma^2(G)$ for the unmodified model. Let π^2 be an arbitrary fixed subgame perfect equilibrium of $\Gamma^2(G)$. It can be seen easily that we obtain subgame perfect equilibria for Γ^2 if we complement the prescriptions of tables 7 and 8 by the local strategies of π^2 at histories not reached by the plays listed there. In This way we construct φ^2 and ψ^2.

The second round expectations for φ^2 and ψ^2: For every player i let Y_i and Z_i be the expected payoffs in Γ^2 if φ^2 and Ψ^2, respectively, are played. In view of tables 7 and 8 we have:

$$Y_1 = 50 - \frac{1}{3}g \tag{59}$$

$$Y_2 = Y_3 = 25 - \frac{1}{3}g - \frac{1}{2}\gamma_2 \tag{60}$$

$$Z_1 = 120 - \frac{1}{3}g - \frac{1}{2}\gamma_2 \tag{61}$$

$$Z_2 = Z_3 = 15 - \frac{1}{3}g - \frac{1}{6}\gamma_2 \tag{62}$$

FIRST DECISION MAKER		SECOND DECISION MAKER
DEMAND	CHOICE OF NEXT DECISION MAKER	
x_1=120	j=2 prob. $\frac{1}{2}$	C=12
	j=3 prob. $\frac{1}{2}$	C=13
x_2= 30	j=3	C=23
x_3= 30	j=2	C=23

Table 7: Equilibrium plays reached by φ^2 and local strategies prescribed by φ^2 at histories of Γ^2 reached by φ^2.

FIRST DECISION MAKER		SECOND DECISION MAKER
DEMAND	CHOICE OF NEXT DECISION MAKER	
x_1=120	j=2 prob. $\frac{1}{2}$	C=12
	j=3 prob. $\frac{1}{2}$	C=13
x_2= 30	j=1	C=12
x_3= 30	j=1	C=13

Table 8: Equilibrium plays reached by ψ^2 and local strategies prescribed by ψ^2 at histories of Γ^2 reached by ψ^2.

Player 1 as the first deviator is punished by not beeing chosen by player 2 and 3. This gives him a probability of only 1/3 to be in the coalition formed. Player 2 or 3 as the first deviator is punished by not being chosen by the other one of these two players. This gives him a probability of only 1/2 to be in the coalition formed.

Construction of φ: Let Γ^1 be the game which results from Γ if every second round subgame after a history with player 1 as the first deviator is replaced by the payoff vector $(Y_1,Y_2,Y_3.)$ and all other second round subgames are replaced by the payoff vector (Z_1,Z_2,Z_3). We call Γ^1 the truncated game. We now construct a global strategy φ^1 for Γ^1. As we shall argue later in the sketch of a proof, φ^1 is a subgame perfect equilibrium of Γ^1. Let π^1 be an arbitrary fixed subgame perfect equilibrium of Γ^1. The global strategy φ^1 is as follows: At histories of Γ^1 at which table 8 specifies local strategies the local strategies prescribed by φ^1 are those specified there; everywhere else φ^1 agrees with π_1.

The global strategy φ for Γ is defined by the requirement that it induces φ^1 on Γ^1, and as been said already, φ^2 on second round subgames after histories for which player 1 is the first deviator and ψ^2 on all other second round subgames.

φ is a subgame perfect equilibrium – sketch of a proof: In order to show that φ is a subgame perfect equilibrium of Γ, it is sufficient to show that φ^1 is a subgame perfect equilibrium of the truncated game Γ^1. We shall prove this by showing that the following statement is true: An observable deviation of a first deviator cannot be a local best reply to φ. It is sufficient to show this.

It will be convenient to say that a choice is optimal at a history of Γ^1 if at this history it is a local best reply to φ and that it is non-optimal otherwise. We shall refer to player 1 as the strong player and to players 2 and 3 as the weak player. We shall distinguish the five cases (i)–(v) with respect to the situation in which a first deviator can be. In view of the symmetry of Γ^1 with respect to the weak players, we can restrict our attention to players 1 and 2 as first deviators.

case (i): Player 1 as the third decision maker faced with $x_2=x_3=100$

case (ii): Player 2 as the second decision maker faced with $x_3=100$

case (iii): Player 2 as the second decision maker faced with $x_1=120$

case (iv): Player 1 as the first decision maker

case (v): Player 2 as the first decision maker.

In the treatment of these cases, the terms undercutting, matching, pushing k close to the critical level, and pushing k to the critical level will be used analogously to section 4 (see tables 4 and 5). However, the critical level now depends on who is the first deviator. The critical level of player i in case (i) and (iv) is $Y_1-\gamma_3+\gamma_2$ and in all other cases it is $Z_i-\gamma_3+\gamma_2$.

The case (i): Obviously, forming his one-person coalition is not optimal for player 1. The only other possible observable deviation is w which yields $Y_1-\gamma_3$ which is less than $50-\gamma_2$.

The case (ii): If player 2 does not deviate he receives the expected payoff 50–g. Obviously, naming a demand x_2 different from 100 does not pay unless we have $x_2 \geq 50$. If he does this, w becomes the only optimal choice of player 1 since $Z_1-\gamma_3$ is greater than 100. In view of $Z_2 < 15$ this is disadvantageous for player 2. Obviously, {2} is not optimal.

The case (iii): If player 2 does not deviate he receives $30-\gamma_2$. Therefore a demand x_2 does not pay unless we have $x_2 \geq 31$. After this demand it becomes the only optimal choice of player 3 to form 13. This is disadvantageous for player 2. Obviously, {2} is not optimal.

The case (iv): If player 1 does not deviate he receives 120–g. A demand x_1 different from 120 cannot pay unless we have $x_1 \geq 121$. Depending on the local strategies of the third decision maker and on how high x_1 is, it becomes optimal for the second decision maker to undercut, to match to push k close to the critical level or to push k to the critical level. As in connection with tables 4 and 5, it can be seen that each of these alternatives is disadvantageous. With the exception of "matching" these responses have the consequence that the third decision maker chooses w or forms the coalition 23. In view of $Y_1-\gamma_3+\gamma_2 < 50$ this is disadvantageous for player 1. As in connection with tables 3,4, and 5 it can be shown that "matching" does not offer a high enough chance for 13 or w to make $x_1 \geq 121$ optimal for player 1.

The case (v): A demand x_2 different from 100 cannot be optimal unless we have $x_2 \geq 50$. It is clear that a demand of this kind has the consequence that the coalition 23 becomes non-optimal for player 3 regardless of whether he is the second or third decision. Since Z_1 is far greater than 100 an end of the game by the choice w of the third decision maker is better for player 1 than coalition 12 which yields at most $100-\gamma_2$ for him. Regardless who the second decision maker is, he responds to $x_2 \geq 50$ by pushing the third decision maker close to the critical level or to the critical level. Nothing else can be optimal for him. It becomes then optimal for the third decision maker to form 13 or to choose w. Therefore it is not optimal for player 2 to deviate from $x_2=100$.

7. Final remarks

We have shown that the non-cooperative analysis of the model of demand commitment bargaining presented in this paper yields clear results in the two basic cases of unanimity games and 3-person quota games without the grand coalition. The analysis has been based on subgame perfectness alone without additional selection criteria. In the case of the zero-normalized unanimity game one obtains the equal payoff division as the unique equilibrium outcome. In the case of the 3-person quota game without the grand coalition and with quotas of at least three bargaining units one obtains the three quota divisions, each of them with probability 1/3.

In section 5 we have looked at a modified model, in which each decision maker with the exception of the last one in a round strategically determines the next decision maker. In the case of the 3-person quota game subgame perfect equilibrium outcomes far away from quota divisions become possible in the modified model.

Another modification of the model suggests itself. Instead of only two rounds one could permit any fixed number of rounds in a stage, before the game ends in conflict. With the help of the results of sections 3 and 4, it can be seen without difficulty that this modification of the model would not change the conclusions reached by the non-cooperative analysis in the two basic cases.

We have analysed the model on the basis of the assumption that utilities are linear in money. The nature of our results suggests the conjecture that the conclusions would remain the same for concave utilities in money.

It is doubtful whether utility theory has any behavioral relevance. The author has strongly expressed his opinion that Bayesianism as a descriptive theory is refuted (Selten, 1989). However, it is possible that under appropriate conditions, learning converges to the maximization of expected money payoffs. As long as elaborate behavioral theories are not available this hypothesis can serve as a reasonable point of departure in the evaluation of experiments. If the hypothesis turns out to be wrong one can search for the reasons in the hope to improve our understanding of human economic behavior.

References

Bennett, E. (1991). "Folk theorems for the proposal-making model", in: R. Selten (ed.), *Game equilibrium models, III, strategic bargaining*, Springer-Verlag

Bennett, E., and E. van Damme (1990). "Demand commitment bargaining – the case of apex games", Discussion paper, CentER, Tilburg University

Binmore, K.G. (1985). "Bargaining and coalitions", in: A. Roth (ed.), *Game-theoretic models of bargaining*, Cambridge University Press, Cambridge

Chatterjee, K., B. Dutta, D. Ray, and K. Sengupta (1990). "A non-cooperative theory of coalitional bargaining", unpublished

Gul, F. (1989). "Bargaining foundations of Shapley value", *Econometrica 57*, 81–95

Harsanyi, J.C. (1974). "An Equilibrium point interpretation of stable sets and a proposed alternative definition", *Management Science*, *20*, 1422–1495

Harsanyi, J., and R. Selten (1988). "A general theory of equilibrium selection in games", The MIT Press

Kahan, J.P., and A. Rapoport (1984). "Theories of coalition formation", Lawrence Erlbaum Associates, Publishers, Hellsdale, N.J., - London

Maschler, M. (1978). "Playing an n-person game - an experiment", in: H. Sauermann (ed.), Coalition forming behavior, 2331-328, Beiträge zur experimentellen Wirtschaftsforschung 8, J.C.B. Mohr (Paul Siebeck), Tübingen

Moldovanu, B. (1991). "Competition and bargaining in games and markets", dissertation, University of Bonn

Rockenbach, B., and G.R. Uhlich. "The negotiation agreement area: An experimental analysis of two-person characteristic function games", Discussion Paper SFB No. B-126, University of Bonn, October 1989

Selten, R. (1981). "A non-cooperative model of characteristic function bargaining", in: V. Böhm and H. Nachtkamp (eds.), Essays in honor of Oskar Morgenstern, Bibliographisches Institut, Mannheim

Selten, R. (1987). "Equity and coalition bargaining in experimental three-person games", in: A. Roth (ed.), Laboratory experimentation in economics, Cambridge University Press

Selten, R. (1989). "Evolution, Learning, and Economic Behavior", Nancy L. Schwartz Lecture, Kellogg, Northwestern University

Uhlich, G.R. (1989). "Descriptive theories of bargaining", Lecture Notes in Economics and Mathematical Sciences 341, Springer Verlag

The Reconciliation of Micro and Macro Economics

M. Shubik
Cowles Foundation for Research in Economics, Yale University
New Haven, CT 06520

Abstract: It is suggested that the appropriate structure for the reconciliation of micro and macroeconomics is an infinite horizon overlapping generations (OLG) model with many finitely lived natural persons and one infinitely lived strategic player without preferences whose choice rule is determined by the periodic political choice of the finitely lived players who are alive and politically strategically active at the time of choice. This player may be interpreted as government. There may also be a class of corporations with infinite lives and a decision rule.

In the steps from the finite horizon general equilibrium (GE) model to the overlapping generations model to a government guided overlapping generations model (GGOLG) it is suggested that even without exogenous uncertainty, if economic efficiency is to be attained it is logically and technologically necessary to introduce government, government money, credit, bankruptcy and inheritance conditions. Firms as corporate entities are legal but not natural persons. Their behavior may be determined by a game within the game.

1. Introduction

Macroeconomic models of the economy are intrinsically dynamic. They are also frequently often considered to be *ad hoc* in some aspects. They are implicitly, if not explicitly institutional, they have at least consumer/workers, producer/firms and government as agents. Government money and bank or private credit are often explicit. The models are implicitly open in time. Although not always spelled out, there appears to always have been a yesterday and a tomorrow is expected.

In contrast with the macroeconomic models the standard general equilibrium (GE) model is preinstitutional. It is suspended in a context free static world of production and exchange over a finite time horizon by individual agents in a moneyless institution free world. All individuals face an isolated one time optimization problem which can be viewed as selecting an optimal strategy in an appropriately related game in strategic form.

Yet in spite of the extreme simplicity and parsimony of the GE model it has considerable appeal for several reasons. In contrast with the macroeconomic models (abbreviated henceforth as MM) the GE models are high dimensional, handling many types of agents and goods and are logically consistent and complete offering an attractive picture of decentralized optimization.

It is clear that the Arrow-Debreu model has many features missing which are needed before it could be regarded as a means for providing the basis for macroeconomic theory. In the remainder of this article the steps needed to enlarge the economic analysis of GE to encompass the politico-economic concerns of the MM, are described.

There are five different types of enlargement of the GE model which are called for in our approach to macroeconomics. They are listed then discussed. They are displayed in Figure 1: (1) A process model of exchange and of production and exchange (SMG). (2) A GE model with international trade. (3) A GE model with an infinite horizon (GE∞). (4) A GE model with public goods. (5) A GE model with incomplete markets (GEI).

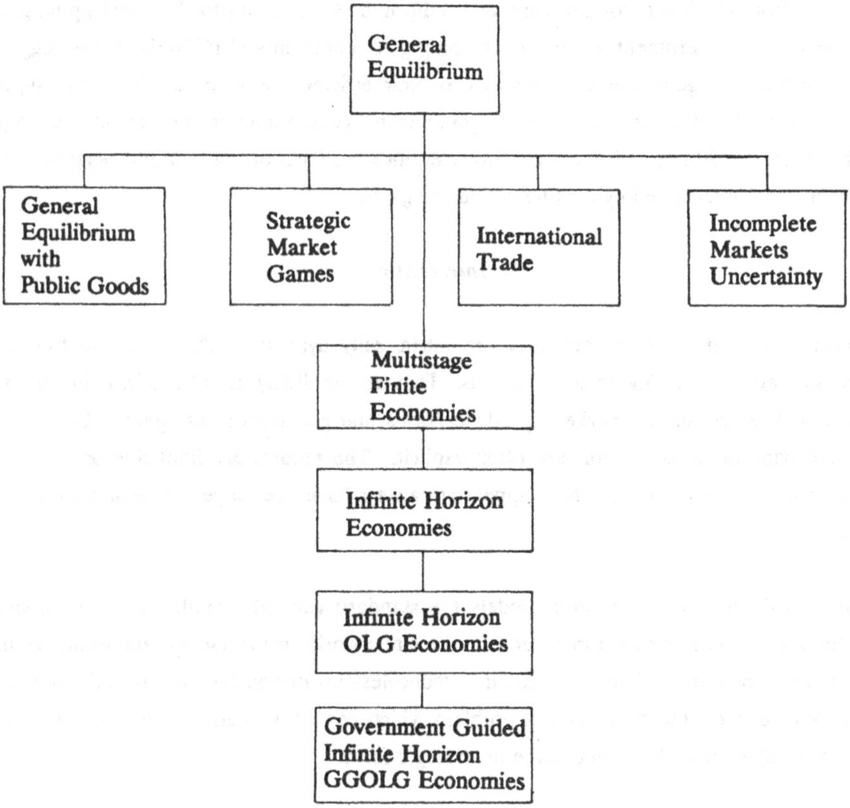

FIGURE 1

1) The approach of Debreu (1959) and most subsequent authors working on GE models was to present an existence proof for efficient prices, not to specify any price formation mechanism or to even indicate the nature of a playable game.

The act of attempting to convert the GE model into a playable strategic market game (SMG) raises explicitly basic problems concerning the role of money and the meaning of "enough money" in an economy. These features have been discussed extensively elsewhere (Shubik [9]).

2) The control of the economy as a whole calls for the provision of a host of different public goods. However even without considering the usual array of goods and services provided by government the running of an OLG economy may require government and a financial system for efficiency and a society might wish to consider land as a public good. In keeping with a belief that if difficulties can be separated and studied individually it is usually desirable to do so; we deal only with government as a financial control system. After the structure of the GGLOG economies have been exposited then features such as public goods may be integrated into the framework.

International trade calls for a consideration of at least two governments and the possibility for two or more national monies. Similarly the introduction of uncertainty with incomplete markets raises new conceptual problems in the enlargement of the scope of general equilibrium. However these too may be considered separately from the key question of how to extend in a logically satisfying manner the preinstitutional economic theory of general equilibrium to the political economy of macroeconomics.

We pursue the path which adds only consideration of time and dynamics. The extension of general equilibrium to many, but finite periods causes no extra conceptual problems if complete markets are considered. If exchange and production are viewed as a game of strategy with a finite number of players then new problems involving oligopolistic inefficiency and the complexity of the strategies of each individual appear.

If we go to the infinite horizon then new problems appear even for the extended general equilibrium model assuming a continuum of agents. In particular how do we define the utility functions or payoffs for both consumers and firms if both live forever? The usual answer is to introduce a discount rate where the firms maximize their expected discounted profit

$$\Pi_j = \sum_{t=1}^{\infty} \left(\frac{1}{1+r}\right)^{t-1} z_t$$

where z_t is the net revenue in period t. The consumers are assumed to maximize

$\sum_{t=0}^{\infty} \beta^{t-1}\varphi(x_t)$ where x_t is the vector of goods consumed by the individual during t.

All agents are assumed to live forever. Logically there is nothing wrong with the assumption, that individuals live forever. Factually it appears to be a poor assumption about natural persons. Also the need to "balance the budget only at infinity" leads to problems with the possibility of "Ponzi financing" where debts are continually rolled over and never paid back.

A model that is closer to reality has overlapping generations of finitely lived agents, and this is where the potential for the construction of models for political economy which yield an understanding of the role of institutions, begins.

2. Overlapping Generaltion: Natural and Legal Persons

2.1. Allais and Samuelson

In their seminal works on overlapping generations both Allais [1] and Samuelson [7] suggest a role for government. In Samuelson's work verbal reference is made to both government and the role of money, but no formal effort is made to model government as an active agent or player. In contrast in a lengthy appendix (around 115 pages) Allais carries out a calculation with old and young agents, firms and government all introduced explicitly.

A possibly unsatisfactory aspect of this trailblazing exercise is that the justification for the goals of the two sets of potentially infinitely lived players, the firms and the government is not made clear.

Since the advent of the pioneering work the literature on overlapping generations models has proliferated. The survey by Geanakoplos [4] covers much of it. When a government is considered the opportunity appears to construct models and raise questions concerning the national debt, social security and a host of other welfare and macroeconomic problem.

2.2. Infinitely Lived Agents

At a formal level Muller and Woodford [6] investigate a mixed model with a set of agents with finite lives and a set of players with infinite lives.

They suggest utility functions of the form: $\sum_{t=0}^{\infty} \beta_j^t \varphi(x_t^j)$ for the infinite players. There are three interpretations of the infinite players. They are the government, corporations and dynasties.

Both the government and corporations are legal persons with infinite lives. Dynasties can be offered as a justification for this utility function where one might argue that a "superplayer" is composed of a set of finitely lived individuals each of whom regards the utility functions of future generations to be part of his own.

There is a modeling problem with the dynasty view of an infinitely discounted utility function. When generation t shows a concern for generation $t+1$, this concern may be manifested in a desire to leave an inheritance for $t+1$ such as $10,000 or the house or a family heirloom. It might also involve generation t's concern for the utility or values obtained by $t+1$. An empirically and logically easier and more satisfactory view of intergenerational concern is that the bequestor is concerned with his desires to make the bequest and not its utility to the recipient. If this view is adopted then we must abandon the dynasty interpretation of the infinite utility function.

If we interpret the two types of infinite agents as firms and government we must raise the question as to why we wish to ascribe to them utility functions of the form:

$$\Phi_j = \sum_{t=0}^{\infty} \beta_j^t \varphi(x_t^j) \text{ with } 0 \leq \beta_j < 1. \tag{1}$$

The first answer might be mathematical convenience. This form gives us bounded values. But, upon reflection, we observe that they are not "natural persons" and there is a question as to whether the attribution of a utility function to these entities is called for or is necessary.

A player in game of strategy is characterized by a strategy set and a payoff function. A solution of a game involves the specification of some sort of choice rule usually based on some operation by the player on his strategy set. This operation may involve the player's attempt to maximize his payoff function. But in order to describe a game of strategy it is not necessary that an individual who is a strategic player has a utility function. Instead it could be equipped with a *choice rule* which instructs it as to what to do in all circumstances. Even though the player may have an infinite life the decision rule whereby it advances into the future may be based on a finite set of arguments. It is suggested here that an appropriate way to incorporate government and corporations into an overlapping generations model is to introduce them as infinitely lived strategic players without utility functions but with a decision or choice rule determined by the real persons. For simplicity in the discussion firms will be considered later and our remarks are concentrated on the modeling of government which is controlled by the real persons. Figure 2 suggests the scheme for an overlapping generations economy containing three types of agents. They are: (1) the young, (2) the old and (3) the government. The first two, the real or natural persons, can be regarded as many in number, possibly most conveniently represented by a continuum of agents and the third is a single large atom.

At point of time $t = 0$ there are two generations alive, those born at time $t = -1$ and $t = 0$. Government as well as the live agents must select its move at $t = 0$. A reasonable restriction on its choice rule is that it depends at most on all of the decisions of all of the natural persons alive at $t = 0$. The idea of a government plan will be discussed later.

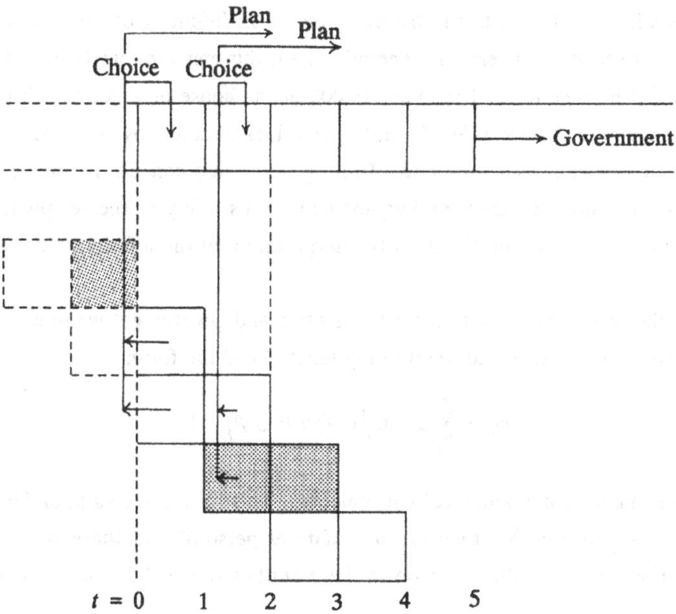

Control of Government

FIGURE 2

At time t there are two generations alive, those born at $t-1$ and those born at t. They are concerned with maximizing their welfare where an individual born at time t is assumed to have a utility function of the form:

$$u_t = \varphi(x_t^t, x_{t+1}^t) \tag{2}$$

where x_t^t may be interpreted as a vector of resources obtained by a representative individual born at time t.

We may consider that an individual born at t has a set of strategies S_t. Part of any strategy is an action, message or instruction that he sends to the government each period.

Heuristically the actions of the government at time t might depend on cultural, historical, social, political and economic factors. If (both for ease and parsimony) we choose to model the agents as nonsocial, nonhistorical individualistic local maximizers we then have the most unfavorable and simplest set of conditions to examine the possibility that there may be structures where the needs of future generations could be served by the selfish behavior of the living.

One way of trying to explain cooperative behavior and institutional structure is to try to start with the individuals alone without any societal context whatsoever. For example we might start with an analysis of repeated plays of an abstract 2 × 2 matrix game. We do not follow this approach here as there is a way to split the study of institutions into two parts and to consider them sequentially. The first step may be regarded as a comparative analysis of institutions in the sense that one takes *as given* an institution called government whose strategy set and decision rule are given. We then consider the outcome of a game played *in the context* of these rules. Having considered politico-economic or economic efficiency within the context of the given institution we may then wish to ask a separate question. That is how did the institution evolve in the first place and how is its structure modified.

Even if one is studying equilibrium conditions in an abstract game one must be careful to consider how general and how relevant and accurate is the abstraction of the biological, anthropological, cultural, historical, sociological, legal, political and economic context of the game. A way to avoid spurious or dubious generality is to specify the institutional structure within which the game is to be played.

At equilibrium much of the institutional structure may not be apparent because much of its function is to be a carrier of process and this is best seen out of equilibrium.

2.3. Government Policy and Choice: Social Welfare and Strategy

In formal game theory it is important to keep in mind the distinction between move and strategy. A move is a choice made at a specific information set. A strategy is a plan which describes contingent choices over a sequence of moves. When finite games are studied in strategic form it is assumed that players select strategies and their commitments are firm.

Government behavior can be considered in terms of both moves and strategy, but the more appropriate words are choice or action and policy or plan. We may regard a government at any particular point in time as carrying out two activities. It sets taxes, subsidies, fixes the money supply, controls the central bank rate of interest and manages the national debt, as well as taking action on many items of social welfare entailing public goods and services. The government also announces or otherwise indicates policy. Thus it indicates that it has a program or plan over the next few years to raise taxes and to lower the public debt.

Empirically we may observe that most governments, at any point in time, appear to have some form of plan, but the plan is continually updated, modified or otherwise changed. There is often a considerable skepticism held by the public concerning the plausibility that the stated plans of the government at time t will be consistent with its actions at time $t+1$. Yet in human affairs, even if policy changes in the future there is a perceived need that it be announced in the present, if only to provide some indication of the gap between promise and fulfillment.

When one looks at governmental planning and policy, at least at a level of explicitness it does not appear to be particularly long term. At most one suspects that it stretches no longer than between administrations. In a democratic society comprised of selfish individuals it is reasonable to accept as an upper bound on the length of any announced plan, the lifetime of the youngest strategically active individual.

Figure 3, which contains more detail than we will be using in the remainder of this discussion suggests roughly the divisions of political and economic activity of a natural person assumed to live to 75.

Discretionary spending of the young may be economically significant even before they start to earn. The political franchise of an individual lasts even after economically productive activity.

Referring back to Figure 2 we may become somewhat more formal as follows.

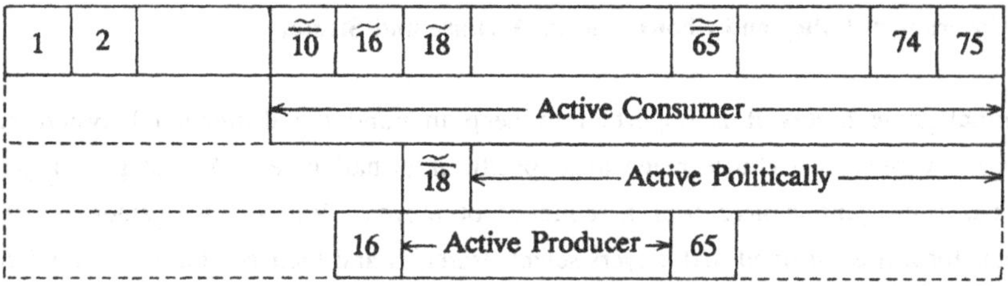

FIGURE 3

Let the set of actions available to the government at time t be A_t, a specific action is $a_t \in A_t$. Let P_t be a set of plans one of which announced by the government at time t $p_t \in P_t$ is a specific plan.

Both the action and the announced plan of the government at time t may be regarded as functions of the moves of the old and young natural persons at time t. Thus

$$a_t = f_1(m_t^{t-1}, m_t^t), \quad p_t = f_2(m_t^{t-1}, m_t^t).$$

We might require that a_t and p_t be consistent, i.e. at least your current actions are consistent with your current plan. We might wish to impose consistency between overlapping plans but for an evolutionary theory that may not be a logical necessity.

If the government is an artificial player and all of the natural persons are small relative to the society as a whole then the strategy of each individual will be relatively simple and consist of a sequence of moves dependent upon the state of the system at the end of each previous period.

A review of Figure 2 shows one cross-hatched cell which indicates the young who will be born in the future when those who are alive and young today will still be alive. Thus in the game that is played in the space of the lifetimes of those who are alive today there will be three natural persons as well as government. There are two ways to take into account the influence of the newborn who will be of strategic relevance to those currently alive, one can make a general behavioral assumption replacing them by an expectation on their behavior, or one can make a special assumption which tries to deal with the full structure of the potential infinite regression in the strategic interlinkage of the overlapping generations. We return to this in Section 2.5.

Returning to the consideration of government as an artificial player. It has been suggested that we may regard its choices as based on the choices of the real persons. A simpler assumption is that they are based on *the characteristics* of the real persons rather than their actions. This in essence provides for a society with a government guided by a *social welfare function.*

2.4. The Game within the Game: Artificial Players and Politicians

In the model described in Section 2.3 government is described as an artificial player. There is a somewhat different and more complex mode for modeling government as a game within the game (see Downs, 1957, Shubik, 1984, Ch. 22). For example we might consider the strategy set of what government can do to be given by some form of constitution, but the selection of a move by government to be the result of a game played by a set of fiduciaries, politicians or bureaucrats who have been elected or appointed to act or play in the political and bureaucratic game which in turn determines the move of government.

A model of representative government or government by bureaucrats suggests government as more than a direct artificial player, but a device whose moves are the outcome of a game of strategy among an elite. We note, but do not attempt to deal with this level of complexity at this time.

2.5. Expectations

Figure 2 and the example below are utilized to attempt to specify the need for the introduction of expectations if a forward solution to the progression of the political economy is attempted.

Let the overall set of strategies of an individual born at time t be S_t. For simplicity we assume that a government announces only its moves and that at time t a move by government is determined by $f(s_{t-1}, s_t)$.

We now describe the payoffs to three live players but include, in part, strategic inputs from five. Consider the game at $t = 0$

$$
\begin{aligned}
\pi^{-1} &= \Pi^{-1}(\hat{s}_{-2}, s_{-1}, s_0, f(\hat{s}_{-2}, s_{-1}), f(s_{-1}, s_0)) \\
\pi^{0} &= \Pi^{0}(s_{-1}, s_0, s_1, f(s_{-1}, s_0), f(s_0, s_1)) \\
\pi^{1} &= \Pi^{1}(s_0, s_1, \tilde{s}_2, f(s_0, s_1), f(s_1, \tilde{s}_2)).
\end{aligned}
\tag{3}
$$

Referring back to Figure 2 there are two boxes, one with square hatching involving generation -2 and one with cross hatching for generation 2. If we wish to consider the game played today including all those alive in the span of a generation as strategic players then we must supply two sets of information, one historical and the other involving expectations. We need to regard as datum what player -2 did in period 1 when he was old. This is relevant to the payoff of player -1. This is denoted by $\hat{s}^{-2}$.

Player 1 will live only half of his life in the game at $t = 1$. His actions will be influenced by his expectation of the actions of player 2 at time 2 thus we need as datum to be able to fill in the information to account for the cross-hatched square. That is the relevant part of the strategy for player 2. This is noted by $\tilde{s}_2$.

The assumptions made in defining the three player game are that the payoff characteristics of player 1 are known to -1 and 0 before he is born and the strategic choice of player 2 is known to player 1 before player 2 is born.

Where these assumptions come from must be accounted for by the invocation of parsimony, a normative argument or a theory of the formation of expectations possibly supported by observation or as an implication of a psychological or social-psychological theory.

3. OPTIMIZING AND BEHAVIORAL MODELS

3.1. Stationarity, Optimization and the Disappearance of Time

An aesthetically pleasing theory of multigenerational economic behavior might call for consistency of expectations combined with some criterion of optimality.

If we make the assumption that all generations are identical and that all enter with the same endowments as their ancestors, without trying to patch up a local optimization process with history and expectations assumed exogenously we might search for stationary solutions which yield consistent expectations over the infinite horizon. The assumption of the existence of a stationary (or cyclical) solution is tantamount to the eradication of time from the problem. The expectation of what future generations will do is born out. The optimization problem involving a countable infinity of strategic variables is replaced by a finite cycle.

We need to contemplate the full knowledge and information requirements of this view of overall optimization. Table 1 suggests several of the relevant factors.

The assumptions that an individual knows his own utility function and strategy are relatively strong but will not be challenged in detail here, beyond observing that a better picture for even those individuals who have been schooled in normative theory is that at most the planning horizon is a few years and the plan is updated in a year or even sooner.

The knowledge of the strategies of other contemporaries can be high in long term one on one competition; thus in markets with two or three long lived firms they may try to form a picture of each other's strategies from a history of moves and verbal statements. In mass societies, however, at best, individuals form views of aggregates. The aggregate form may be justified or rationalized by introspection ("they are like me") or by some conception of "the average."

In essence individuals may see the moves of others or some composite outcome (such as the stock market closing prices), but it is difficult, if not impossible to deduce strategies of others in mass markets.

Invoking some form of continuity and genetics together with symmetry one might argue that children or grandchildren might be assumed to be relatively similar to their parents. It is even feasible for five generations to be alive simultaneously, possibly six or even seven might be biologically feasible (a birth at 17-20 and an age of 102-120) but this is suggestive of an extreme upper bound on any individual's conscious concern for further generations and their actions.

	Plaus-ible	Weakly Plaus-ible	Un-likely	Comment
knowledge of own utility function		√		assumed as best of workable assumptions
knowledge of utility function of other contemporaries such as		√		modified by use of introspection aggregation and proxy variables excess demand
knowledge of own strategy	√	√		need to define optimization but doubtful over more than a few years
knowledge of strategies of other contemporaries			√	highly implausible except in long term oligopolistic competition oligopolistic competition
knowledge of moves of others				empirically reasonable in some aggregate form
knowledge of goals of future generations			√	obtained by summetry assumptions: "They are like us"
knowledge of strategies of future generations			√	can be obtained by symmetry assumptions together with simplifications due to looking at stationarity
knowledge of size and power government	√			

TABLE 1

If we limit, by assumption, the ability of live natural players to include in their plans knowledge or foreknowledge beyond their own lives then we only allow optimization equations of type (3) which involve a bound to the number of strategic variables and an explicit need for expectations in contrast with processes where the argument for each payoff function involves a countable infinity of variables.

It appears that an attractive reason for assuming the similarity of future generations and concentrating on steady state, cyclical repetition of states or smooth growth is that not only is the mathematics more tractable than otherwise, but in essence a problem involving an infinite number of strategic variables is mapped into one with a finite number.

The need to plug in expectations as is indicated in Figure 2 and equations (3) is satisfied by the assumption that like agents in like circumstances will act the same way. Infinite time disappears and is replaced by a stationary cycle when similar circumstances are postulated.

The last item in Table 1 is government. It is reasonable to assume that most individuals concerned with the economy more or less know of the government's actions. Furthermore if government is large and influential (for example: taxes and subsidies might be bigger than 15 percent of GNP) a key factor in the dynamics of the economy will be government.

3.2. Expectations and Evolution

There is a literature showing how many interactions or trades between two partially informed agents will lead to consistent estimations (see Geanakoplos and Polemarchakis, 1982). The stationary or cyclical state OLG or GGOLG solutions are also characterized by consistent expectations among all agents. But this does not rule out other solutions. An individual agent may die having found out that his expectations were not met but it is too late to correct his previous actions. Thus invocation of the convergence of expectations may not be relevant if the period is not long enough.

As long as there is some rule of behavior given to supply the expectations indicated in Figure 2 then each generation has enough for a forward decision process. The infinite horizon solution with intergenerational consistent expectations is only one of many finite behavioral solutions available. Because of the assumptions of regularity concerning population and resources the aggregation from an infinite to a finite number of variables is straightforward and the chances for a stationary state or cycle consistent with the need for individuals to plan for no more than a lifetime or two,[1] are good.

The dynamics of the system has optimizing natural persons who, by their strategies constantly set the parameters of their own environment. But to *even* well define individual optimization for a forward solution even for a proposed rational decisionmaking criterion requires an assumption concerning expectations.

Government at its simplest is a set of rules; at a more complex level is a game within a game. Yet we observe societies evolve and decay and this involves the change of government structure, not merely a change in the value of parameters. A way to model this is to build several levels of games within the game. Thus, for example, although the election of representatives might be by simple majority vote every two years, there may be a constitutional amendment requiring a ninety percent majority which is considered every ten years.

In essence the formulation of the OLG model as a playable game requires the specification of rules which among other features specify the structure of the artificial players. These set parameters such as the money supply a rate of interest. But we may also contemplate a game with rules governing how to change the rules.

[1]Two lifetimes (not two generations) cover all those who can interact strategically at any point in time. The interaction of the oldest and youngest is characterized by $k-1$ periods of history and $k-1$ periods of expectations relevant to now for the living.

3.3. An Aside on Population

Economics and demography are highly related but different topics. In particular demographic considerations are called for to supply several of the basic gaps implicit in the assumptions concerning economic growth models, especially when OLG models are considered. Undoubtedly there are economic components to intergenerational transfers of assets and to conscious decisions by prospective parents as to how many children they wish to have. But biology and society, instinct, nature and nurture all play their roles. An adequate model of a political economy with overlapping generations needs to reflect the explicit assumptions about the motivation for new population and the factors determining resource transfer between the old and young.

The easiest assumptions are of a constant or constantly growing population given exogenously with no explanation or suggestion of motivation. These assumptions have been made here. The easiest assumptions concerning inheritance and other intergenerational transfer is that of perfectly selfish individuals who choose to give away nothing. This minimizes the biological and societal links postulated and enables us to concentrate on the efficacy of governmental actions in providing for interlinkage.

It is important to note a distinction between general equilibrium type of modeling and game theoretic modeling even where the mathematics utilized to examine equilibrium is virtually identical. In particular even if we assume no inheritance, in the game, in disequilibrium an individual might die before arranging for a net worth of zero at death. In order to well-define the game there must be a rule specified for the disposal of accidentally left over assets.

Over the course of history many governments have attempted to influence population growth by economic means including subsidies for or taxes on the number of children. The study of growth calls for an explicit description of the economic aspects and some assessment as to their relative importance or unimportance in economic development.

4. Corportations as Artificial Players

4.1. Business and Not-for-Profit Institutions

Most modern societies have three distinguishable types of corporate institutions they are government, business firms and not-for-profit organizations such as churches, universities, charities and various cultural organizations. We have discussed some of the aspects of government. The consideration of not-for-profit organizations is postponed; here we concentrate on the business firm.

The individually owned firm poses no problem in modeling. We may assume that the owner attempts to maximize his utility over his lifetime. But it is fairly obvious that for technological reasons vast aggregates of physical and financial capital and individuals form jointly owned corporations of considerable size. Unlike government or not-for-profit organizations the suggested purpose of the business corporation is explicitly economic and the nature of its strategic variables are for the most part economic. In particular the firm in contrast with natural persons does not act as an ultimate consumer and lives forever. It buys, manufactures or otherwise transforms its inputs and it sells most of its outputs. Most (but not all) of its transactions are monetary. Two key questions that must be answered are what is meant by profits and do business corporations attempt to maximize profits?

The share an individual has in government is more or less implicit, the shares held of a corporation are explicitly defined. The corporation is an artificial person whose decision is meant to be some selection from its choice set determined by the actions of its current stockholders as interpreted or influenced by management. A cynical; but easy way to model both government and corporations is to consider that citizens and stockholders have no influence. This is suggestive of models of oligopoly and plutocracy. The institutions are games within the overall game run by small sets of natural persons who are the real players accountable to none other. In contrast we might wish to specify the conditions under which stockholder control results in some appropriate form of profit maximization.

We may wish to avoid considering oligopolistic behavior by postulating a continuum of firms, possibly of several types. We may imagine that the stocks of the same type are sold from a mutual.

A firm is assumed to buy inputs, produce, sell outputs, borrow and deposit and pay dividends.

4.2. Time and Production

In economic models with OLG and production, time plays a critical role. A taxonomy of goods is called for to stress that there is a theoretical basis for many casual business distinctions. Table 2 is suggestive of some of the basic distinctions.

	Producible	Storable	Moveable	Production Time	Length of Life	Deprec-iation
Consumer good	2	2	2	3	4	4
Producer/consumer	2	2	2	3	4	4
Production good	2	2	2	3	4	4

TABLE 2

Leaving aside factors such as indivisibility, there are 3 basic types of goods from the view point of production and consumption and 384 different categories for each. They are

Producible	yes	no
Storable	yes	no
Moveable	yes	no
Production Time	less than, equal to, greater than a life span	
Length of Life	less than, equal to, greater than a life span or ∞	
Depreciation	(1) None; (2) one "hoss" shay; (3) linear; (4) exponential	

Empirically as a good first order approximation all but a few of the categories exist. However, fortunately, probably only a few dozen of the over 1000 possibilities are of major concern to a general theory of government guided overlapping generations.

Gold and land emerge as the two infinite lived nondepreciating assets. Labor is a production/ consumption good with many special properties. Land and existing works of art are essentially nonreproducible. Large machines are essentially producer goods with no direct consumption value. Many food items such as salt or milk may be used for final consumption or as productive inputs.

The length of a productive process is often difficult to define and measure. For example many cathedrals have taken several hundreds of years to build. Yet the delays appear to have been primarily financial, social and political rather than technological. The great wall of China appears to have been built in fits and starts for such the same reasons. Yet when one includes major buildings, dams, machines it is difficult to find examples of production processes requiring more than a generation, let alone a lifetime. But the length of time it takes to manufacture an item and the length of time it lasts are critical items in forging intergenerational links in a world with selfish individuals. If you "cannot take it with you," in one way or another the next generations will get your assets. If it takes longer than a lifetime to build, the only motivation for nonaltruistic individuals to start to produce a capital good would be an expectation that they can successfully sell intermediate product.

In actuality there are buildings and other assets which last longer than a lifetime (although when maintenance and part replacement are accounted for definitional problems appear--"this old axe has lasted for a hundred years with five new handles and seven new heads").

Regardless of whether production processes and the length of life of assets last more or less than an individual life the possibility for a host of cyclical equilibria appears where the length of the cycle may be related to the relative primes of human life, length of the production process and length of life of capital goods.

4.3. Income and Capital Accounts and Profit

A standard chestnut of economic theory and accounting involves the distinction between capital and income accounts. Fisher stressed this distinction, but not in the context of a formal closed model of the economy. This distinction emerges as a "rules of the game" necessity if we are to well define the concept of profit. If a firm maximizes profit, how is it defined? In general equilibrium theory with a finite horizon the definition is simple. It is the difference between the value of all outputs and inputs. In essence there is no time hence no distinction between long and short term profits.

A natural way to define short term profits is to consider the value of outputs minus inputs plus the value of ending stocks minus starting stocks. The starting value of stocks including physical and financial assets.

4.4. Control: Artificial Players and Real Fiduciaries

The most that the choice rule of the corporation depends upon is a set of actions by current stockholders if we view the corporation as purely a mechanism. If in contrast we believe that a better model is a game within a game, then the choices of the firm are influenced by the strategic behavior of management whose strategy sets are constrained by the stockholders.

5. Macroeconomics, Institutions and Dynamics

5.1. Institutions and the Rules of the Game

In order to model the politico-economy as a playable game the rules of the game must be sufficiently explicit to be able to describe the state of the system under all circumstances. Thus in essence the rules are the carriers of process and can be regarded as defining institutions and instruments.

There may well be a deep philosophical question as to how the first institutions emerged from the prehistory of savages, but for the purposes of understanding the functioning of a modern political economy the institutions are best taken as given. Institutions evolve and laws change, but as already noted one can construct rules for the change of rules. The sequence of games within games provides for an evolving system driven by finitely lived natural persons who may be characterized as following optimizing behavior.

Questions of economic optimality and equity can be well defined in the context of the institutions assumed. Once time is a feature of the analysis and process must be described there is no institution free political economy. From the viewpoint of macroeconomics comparative studies of different rules of the game or different institutional structures around the economy are worth investigating. But we can also attempt to select a minimal set of rules basing our criteria on logical requirements, efficiency and parsimony. thus we may even attempt to well define the minimal set of instruments required for government control of the economy. Consideration of optimality requires the invention of outside money in an OLG model to complete the double coincidence of wants. But if there is any fluctuation in the economy the amount of outside money must be varied thus there must be a device for increasing and decreasing this supply. Taxes and subsidies alone would suffice. The presence of a national debt manipulated by the buying and selling of bonds is another alternative, or one could contemplate a mixture of both.

A clear feature in the type of models postulated here is that although there may be economic factors which decide political control, there is a clear separation between economic and political control mechanisms. In particular it is apparent that the use of a price system has little if anything to do with the motivation of the political structure.

5.2. Macroeconomics and Mathematical Institutional Economics

Money and financial institutions arise from a combination of logical, technological and optimizing conditions required for efficient trade. Once the instruments and institutions have been invented they take on a life of their own. In particular they provide a government with the weapons to control the economy.

Macroeconomics has been regarded as somewhat *ad hoc* and theoretical as compared with general equilibrium theory. It is suggested here that when *process models* of exchange and production are considered financial instruments must be invented. But when overlapping generations are considered *outside money* is required for optimality and this in turn calls for the description of government as a logical necessity. But even if we limit ourselves to an economy without public goods the control of production and exchange via the money supply and interest rate may be in the self interest of the natural agents. Thus the concept of macroeconomics emerges, not as institutional *ad hocery* but as a necessity to provide control and direction to an overlapping generations economy. General equilibrium provides only a small part of this more general schema.

There is no way that assumptions concerning expectations can be avoided in attempting to solve an OLG model, thus there is no pure optimization. At the least a behavioral assumption such as rational expectations must be made in order to well define the individual optimization.

The government as an artificial player is in reality part of the rules of the game. Society and the polity in the course of time may change the rules. But at any time there is only a finite number of finitely lived individuals acting. Their efforts to optimize in the environment they perceive, results in the evolution of the institutions.

Institutional economics is often viewed in contrast with mathematical economics. The first is held to the historically oriented and *ad hoc*. Yet it is argued here that the comparison is misleading. A mathematical economics that deals with process invents rudimentary institutions. When it deals with infinite horizons the rigid distinction between maximization and evolution is difficult to maintain. Although the short term explicit economic game may best be explained by economic optimization, the rules may be changed by a Russian doll sequence of games within the game within the game.

General equilibrium economics in its stress on preinstitutional nonprocess optimization was a valuable but incomplete step in the reconciliation of micro and macroeconomics. A mathematical institutional economics with a stress on process provides the means for justifying the existence of institutions and the basis for a macroeconomics with a logical justification at least as compelling as that of general equilibrium.

References

[1] Allais, M. 1947. *Economie et interet*. Paris: Imprimerie Nationale.

[2] Debreu, G. (1959). *The Theory of Value*. New York: Wiley.

[3] Downs, A. (1957). *An Economic Theory of Democracy*. New York: Harper and Row.

[4] Geanakoplos, J. D. (1987). "Overlapping Generations: Survey Article," *New Palgrave Dictionary of Economics*.

[5] Geanakoplos, J. D. and H. M. Polemarchakis (1982). "We Can't Disagree Forever," *Journal of Economic Theory*, 28.

[6] Muller, J. W. and M. Woodford (1988). "Determinacy of Equilibrium in Stationary Economies with Both Finite and Infinite Lived Consumers," *Journal of Economic Theory*, 46: 255-290.

[7] Samuelson, P. A. (1958). "An Exact Consumption-loan Model of Interest with or without the Social Contrivance of Money," *Journal of Political Economy*, 66: 467-482.

[8] Shubik, M. (1984). *A Game Theoretic Approach to Political Economy*. Cambridge, MA: M.I.T. Press.

[9] Shubik, M. (1989). "Money and Financial Institutions: A Game Theoretic Approach," *Handbook of Monetary Economics*, forthcoming.

V. Utilitarianism and Related Topics

Harsanyi's Utilitarian Theorem: A Simpler Proof and Some Ethical Connotations

PETER J. HAMMOND, Department of Economics
European University Institute, Badia Fiesolana
50016 S. Domenico di Fiesole (FI), Italy;
and Stanford University, CA 94305–6072, U.S.A.

ABSTRACT

Harsanyi's utilitarian theorem states that the social welfare function is the weighted sum of individuals' utility functions if: (i) society maximizes expected social welfare; (ii) individuals maximize expected utility; (iii) society is indifferent between two probability distributions over social states whenever all individuals are. After giving a simpler proof, an alternative axiomatic foundation for Vickrey-Harsanyi utilitarianism is provided. By making using an extended version of Harsanyi's concept of a player's "type" in the theory of games with incomplete information, the problem of forming social objectives when there is incomplete information can also be resolved, at least in principle.

1. Introduction

Gabriel Cramer (1728) and then Daniel Bernouilli (1738, 1954) first proposed as a decision criterion the maximization of expected utility rather than of expected wealth. Much later, in an appendix to their classic work, von Neumann and Morgenstern (1943) set out for the first time an axiomatic justification for this criterion. They were also the first to appreciate fully how their theory provided a cardinal concept of utility — i.e., one that is unique up to linear (or affine) transformations of the utility function. This was in contrast to the ordinal concept of utility which is unique up to general increasing transformations.

Very shortly thereafter, Lerner (1944) showed that expected social welfare could be maximized, under certain symmetry conditions, by equalizing income — see Sen (1969, 1973) for futher discussion of this result. Perhaps more important, however, was Vickrey's (1945) realization that this von Neumann-Morgenstern cardinalization could be used to measure marginal utility in a way that relates to statements about what redistributions of income would be desirable. Of course, for any one individual's cardinal utility function, it is true that measures of that individual's marginal utility for different levels of income are all uniquely determined up to a single multiplicative constant. This use of the cardinal utility function was contested by Friedman and Savage (1952), which led in turn to Harsanyi's (1953) comment on their paper. It was in this comment that Harsanyi first enunciated his idea of "impersonality," according to which ethical decisions should be based upon the interests of persons who have had all personal biases removed by being put in a situation of complete uncertainty about their true identity.

Also in the early 1950's, a paper by Fleming (1952) appeared which advocated that the social welfare function should be additively separable over individuals, over time periods, and also over uncertain states of the world. This prompted Harsanyi (1955) to expand his idea of impersonality into a complete new theory. In part, Harsanyi adapted Lerner's idea significantly and considered, several years before Rawls (1959, 1971), an "original position" in which all individuals are supposed to choose the social state they prefer without knowing which members of the society they will become upon emerging from behind what Rawls so aptly called the "veil of ignorance." Unlike Rawls, however, Harsanyi has always stuck to the orthodox theory of choice under uncertainty — namely, the existence of subjective probabilities, and then the maximization of expected utility. Of course, in the original position, symmetry was postulated, so that there was an equal probability of becoming each individual in the society. Later, Harsanyi (1975a, 1975b) emphasized this crucial difference from what Rawls called the "difference principle." These articles by Harsanyi appeared after Rawls' theory had acquired its great popularity — which it fully deserved, though mostly for reasons having little to do with the difference principle *per se*. Indeed, the version of Harsanyi (1975a) which is reprinted in Harsanyi (1976) contains an additional section responding to Rawls' (1974) more thorough discussion of his reasons for using the difference principle rather than an expected social welfare criterion.

A key step in Harsanyi's (1955) argument was the claim that expected social welfare would be the weighted sum of expected individual utility functions, assuming that whenever all individuals are indifferent between any two probability distributions over social states, then so is society. Strictly speaking, Harsanyi's justification for this claim relied on some implicit assumptions concerning possible variations in individuals' expected utility levels — assumptions similar to those which were also made in Hammond (1983). This was first pointed out by Domator (1979), it seems, who, along with a number of other authors more recently, have given rigorous proofs without such additional assumptions — see especially Border (1985), Coulhon and Mongin (1989), and also Broome (1990). Section 2 below will present what I believe to be an equally rigorous, but rather simple proof. For the case of a finite number of social states, this proof uses an elementary result in linear algebra which can be found, for instance, in Gale (1960). The idea of using this kind of result is due to Border (1981), which was a privately circulated precursor to Border (1985). Very similar proofs for this special case can also be found in Selinger (1986) and Weymark (1990). For the general case of an infinite number of social states, the proof presented here relies only on the finite intersection property of compact sets.

For too long a time Harsanyi's approach was not very widely appreciated, and even today remains controversial. Fleming (1957), Diamond (1967), and Pattanaik (1968) made relatively early criticisms. Diamond's criticism, which Sen (1970) also expressed, and to which Harsanyi (1975b) contains a response, was that maximizing expected social welfare could produce unacceptable inequalities of utility. Yet it is not clear what these inequalities really signify until we give "utility" some concrete meaning; once we do, the criticism essentially loses its force, as Broome (1989) in particular has pointed out.

Pattanaik's concern was more with Harsanyi's original position argument, and the claim that a better understanding of individual psychology was likely to bring us closer to a social welfare function that all could agree to. In fact, despite Harsanyi's serious attempts to argue otherwise, it seems all too likely that different individuals, even with a perfect understanding of psychology, and even behind an apparently common veil of ignorance, would still retain their different views about what other individuals' attitudes to risk are likely to be, and about how to weight the von Neumann-Morgenstern utility functions of different individuals which represent these attitudes to risk. Oddly enough, a similar debate surrounds the assumption of Harsanyi (1967–8) and Aumann (1987) regarding the existence of common prior beliefs in game theory.

Given these and other problems with original position arguments, Section 3 suggests a procedure for side-stepping the issue entirely. The argument is actually no more than a summary, and perhaps a clearer presentation, of ideas discussed more extensively in Hammond (1987). Indeed, those ideas build on or relate to Hammond (1983, 1986, 1988a, b, c) and the realization that a new "consequentialist" framework, based on analysing behaviour in decision trees, could also help to justify the axioms behind conventional expected utility theory. This is really the reason why I find Harsanyi's fundamental work so relevant to ethical decision making.

Finally, Section 4 discusses a natural extension of the previous formulation to societies in which there is incomplete information about individuals' true utilities and other features relevant to a proper ethical decision. Adapting Harsanyi's (1967–8) key insight regarding games of incomplete information, it becomes clear that one needs to consider not just social states in the usual sense, but *contingent* social states which depend on different individuals' types. These are closely related to the "game forms" which Sugden (1985, 1986) has argued, in opposition to Sen, are the right way of modelling individual rights — see also Gaertner, Pattanaik and Suzumura (1988) and Riley (1989, 1990).

2. Proof of Harsanyi's Theorem

Let X be the space of social states, which is assumed to be a (Borel) measurable set with σ-algebra $\mathcal{X}$. Let $\mathcal{M}(X)$ be the set of probability measures on X with this σ-algebra. Suppose that each individual i in the membership M (a finite set) has a welfare ordering $\succsim_i$ on $\mathcal{M}(X)$ represented by the expected value $\mathbf{E}_\mu v_i(x)$ with respect to $\mu \in \mathcal{M}(X)$ of the von Neumann-Morgenstern utility function (NMUF) $v_i : X \to \Re$. Suppose too that there is a social ordering $\succsim$ on $\mathcal{M}(X)$ which is represented by the expected value $\mathbf{E}_\mu w(x)$ of a "von Neumann-Morgenstern Bergson social welfare function" $w : X \to \Re$. Finally, suppose that Pareto indifference is satisfied — i.e., that

$$[\forall i \in M : \mu_i \sim_i \nu_i] \Longrightarrow \mu \sim \nu,$$

or equivalently, that

$$[\forall i \in M : \mathbf{E}_\mu v_i = \mathbf{E}_\nu v_i] \Longrightarrow \mathbf{E}_\mu w = \mathbf{E}_\nu w$$

for all pairs $\mu, \nu \in \mathcal{M}(X)$. Then *Harsanyi's theorem* is the implication that there exist *welfare weights* ω_i ($i \in M$) and an additive constant α such that

$$w(x) \equiv \alpha + \sum_{i \in M} \omega_i \, v_i(x) \tag{1}$$

on X. This theorem was proved by Harsanyi (1955). As pointed out in the introduction, however, there were a number of unnecessary implicit assumptions concerning how possible variations in the social state x could lead to entirely independent variations in the value of each individual's utility $v_i(x)$.

The following proof considers first the case when $X = A$, a finite set consisting of $\#A$ members. Then the argument uses ideas similar to those in Border (1981). The (new) proof for a general measurable space $(X, \mathcal{X})$ follows later.

Proof (when $X = A$, a finite set).

Let $\eta(x)$ $(x \in A)$ be any set of $\#A$ real numbers satisfying

$$\sum_{x \in A} \eta(x) = 0 \quad \text{and} \quad \sum_{x \in A} \eta(x)\, v_i(x) = 0 \quad (\text{all } i \in M). \tag{2}$$

For all $x \in A$, define

$$\mu(x) := (1/\#A) + \lambda\, \eta(x); \quad \nu(x) := (1/\#A) - \lambda\, \eta(x) \tag{3}$$

where $\lambda > 0$ is small enough to ensure that $\mu(x), \nu(x) \geq 0$ for all $x \in A$. Then $\mu, \nu \in \mathcal{M}(A)$ and also

$$\mu(x) - \nu(x) = 2\lambda\, \eta(x) \quad (\text{all } x \in A). \tag{4}$$

It follows that

$$\mathbf{E}_\mu\, v_i(x) - \mathbf{E}_\nu\, v_i(x) = \sum_{x \in A} [\mu(x) - \nu(x)]\, v_i(x) = 2\lambda \sum_{x \in A} \eta(x)\, v_i(x) = 0 \tag{5}$$

for every $i \in M$. So, by the assumption of Pareto indifference,

$$0 = \mathbf{E}_\mu\, w(x) - \mathbf{E}_\nu\, w(x) = \sum_{x \in A} [\mu(x) - \nu(x)] w(x) = 2\lambda \sum_{x \in A} \eta(x) w(x). \tag{6}$$

Thus any $\#A$-vector $\eta(x)$ $(x \in A)$ satisfying (2) above also satisfies

$$\sum_{x \in A} \eta(x)\, w(x) = 0. \tag{7}$$

Consider now any row vector η with elements $\eta(x)$ $(x \in A)$. Suppose that η lies in the null space of the $\#A \times (\#M + 1)$ matrix V whose elements in the first column are all ones and whose elements in the last $\#M$ columns are $v_i(x)$ $(x \in A,\ i \in M)$. Then $\eta\, V = 0$, which implies that η satisfies (2) above, and so we have just shown that it must also satisfy (7). From the *Solvability Theorem* due to Gale (1960, p. 41), it follows that the column vector w with components $w(x)$ $(x \in A)$ is spanned by the columns of V matrix. So there must exist constants α and ω_i $(i \in M)$ for which $w = V\binom{\alpha}{\omega_M}$ or

$$w(x) \equiv \alpha + \sum_{i \in M} \omega_i\, v_i(x) \quad (\text{all } x \in A) \tag{1'}$$

as required. ▮

The above result can now be used to prove Harsanyi's theorem for a general (possibly infinite) measurable set X as follows.

Proof (when X is an infinite set).

First, reduce M if necessary to a subset M^* with the linear independence property that the only solution to the equation

$$0 \equiv \alpha + \sum_{i \in M^*} \omega_i \, v_i(x) \quad (\text{all } x \in X) \tag{8}$$

in the unknown constants α and ω_i ($i \in M^*$) is the trivial solution with $\alpha = \omega_i = 0$ (all $i \in M^*$). This is possible because, if the identity

$$0 \equiv \alpha + \sum_{i \in M} \omega_i \, v_i(x) \quad (\text{all } x \in X) \tag{9}$$

has a non-trivial solution in α and ω_i ($i \in M$), then there must be at least one $j \in M$ for which $\omega_j \neq 0$. But this implies that the corresponding function $v_j(x)$ can be expressed as the linear combination

$$v_j(x) \equiv \left(-\frac{\alpha}{\omega_j}\right) + \sum_{i \in M \setminus \{j\}} \left(-\frac{\omega_i}{\omega_j}\right) v_i(x) \quad (\text{all } x \in X) \tag{10}$$

of a constant term and of all the other functions $v_i(x)$ ($i \neq j$). If the linear independence property (8) is still not satisfied even by the reduced set $M^* := M \setminus \{j\}$, then M^* can be reduced still further in this way as many times as necessary, until M^* does finally satisfy the linear independence property. After this reduction has been completed, for every $j \in M \setminus M^*$ it will be true that the function $v_j(x)$ can be expressed as a linear combination

$$v_j(x) \equiv \alpha'_j + \sum_{i \in M^*} \omega'_{ji} \, v_i(x) \quad (\text{all } x \in X) \tag{11}$$

for some constants α'_j and ω'_{ji} ($i \in M^*$).

For any finite $A \subset X$, define the set

$$\Omega(A) := \{ (\lambda, \alpha, \omega^{M^*}) \in \Re^{\#M^*+2} \mid \forall x \in A : \lambda \, w(x) \equiv \alpha + \sum_{i \in M^*} \omega_i \, v_i(x) \text{ and } \lambda^2 + \alpha^2 + \sum_{i \in M^*} \omega_i^2 = 1 \}. \tag{12}$$

Since Harsanyi's theorem has just been proved for the case when A is finite, there certainly exist both an $\#M$-dimensional vector $\tilde{\omega}^M \in \Re^M$ and a constant $\tilde{\alpha} \in \Re$ such that

$$w(x) \equiv \tilde{\alpha} + \sum_{i \in M} \tilde{\omega}_i \, v_i(x) \quad (\text{all } x \in A). \tag{13}$$

But then (11) implies that in fact there must exist a different $\#M^*$-dimensional vector $\omega^{M^*} \in \Re^{M^*}$ and a constant $\alpha \in \Re$ such that

$$w(x) \equiv \alpha + \sum_{i \in M^*} \omega_i \, v_i(x) \quad (\text{all } x \in A). \tag{14}$$

Now just multiply each side of (14) by $\lambda := \left(1 + \alpha^2 + \sum_{i \in M^*} \omega_i^2\right)^{-\frac{1}{2}}$ in order to ensure that the new normalized coefficients lie on the surface of the unit sphere: the result makes it evident that the set $\Omega(A)$ is non-empty for every finite $A \subset X$.

Notice how, for $A \subset X$, the set $\Omega(A) = \cap_{x \in A} \Omega(\{x\})$. Now, for each $x \in X$, the set $\Omega(\{x\})$ is evidently compact because it is a closed subset of the surface of the unit sphere in $\Re^{\#M^*+2}$. But when A is finite, it has just been shown that the intersection $\Omega(A) = \cap_{x \in A} \Omega(\{x\})$ must be non-empty. By the finite intersection property for arbitrary families of compact sets, it follows that the intersection $\cap_{x \in X} \Omega(\{x\})$ is non-empty. So there exists some combination

$$(\lambda, \alpha, \omega^{M^*}) \in \cap_{x \in X} \Omega(\{x\}) \subset \Re^{\#M^*+2}. \tag{15}$$

By definition of $\Omega(\{x\})$, this combination must satisfy

$$\lambda\, w(x) \equiv \alpha + \sum\nolimits_{i \in M^*} \omega_i\, v_i(x) \quad (\text{all } x \in X) \tag{16}$$

and also

$$\lambda^2 + \alpha^2 + \sum\nolimits_{i \in M^*} \omega_i^2 = 1. \tag{17}$$

Now, if it were true that $\lambda = 0$, then (8) would be satisfied for some constants satisfying $\alpha^2 + \sum_{i \in M^*} \omega_i^2 = 1$, which therefore cannot all be zero. This would contradict the construction of the reduced set M^* which has to satisfy the linear independence property that (8) has only a trivial solution. So $\lambda \neq 0$. One can therefore divide each side of (16) by λ in order to obtain

$$w(x) \equiv \bar{\alpha} + \sum\nolimits_{i \in M^*} \bar{\omega}_i\, v_i(x) \equiv \bar{\alpha} + \sum\nolimits_{i \in M} \bar{\omega}_i\, v_i(x) \quad (\text{all } x \in X) \tag{18}$$

for the newly defined constants

$$\bar{\alpha} := \alpha/\lambda \quad \text{and} \quad \bar{\omega}_i := \begin{cases} \omega_i/\lambda & \text{if } i \in M^*; \\ 0 & \text{if } i \in M \setminus M^*. \end{cases} \tag{19}$$

This completes the proof. ∎

3. Social Welfare, Personal Issues, and Individual Welfare

This section will provide an alternative motivation to that of Harsanyi's original position for his form of utilitarianism. Indeed, it will be claimed that his form of utilitiarianism is a logical implication of the standard "Bayesian" approach to decision-making under uncertainty, when this is combined with a plausible formalization of the key idea in individualistic ethics — namely, that actions should be judged by whether they produce desirable consequences for individuals.

The argument will rely upon the fiction that a different personalized version x_i of the social state $x \in X$ is possible for each different individual $i \in M$. This idea is very similar to the notion of personalized public goods, as exploited by Milleron (1972) and many successors. So the extended Cartesian product domain $X^M := \prod_{i \in M} X_i$ of possible *personalized social states* will be considered, in which each component space X_i ($i \in M$) is a copy of the space X. Of course, the constraint that $x_i = x$ for all $i \in M$ and for some single social state $x \in X$ will usually have to be observed in practice, just as all individuals are generally required to have the same bundle of public goods. Nevertheless, in thinking about society's objectives, it will be useful to contemplate what would be possible in the absence of this constraint.

Ethical decisions whose consequences lie in this space of personalized social states are assumed to satisfy the standard *Bayesian rationality* postulate. Thus it will be assumed that:

ASSUMPTION 1. *There exists some* **social welfare ordering** $\succsim$ *on the space* $\mathcal{M}(X^M)$ *which represents the relative ethical desirabilities of different uncertain social consequences. Moreover, there is a* **von Neumann-Morgenstern social welfare function** $W : X^M \to \Re$ *with the property that, for every pair of probability distributions* $\mu, \nu \in \mathcal{M}(X^M)$, *one has* $\mu \succsim \nu$ *if and only if the expected levels of welfare satisfy* $\mathbf{E}_\mu W \geq \mathbf{E}_\nu W$.

So far, $\mathbf{E}_\mu W$ could be an expected welfare function representing a purely collectivist ethic, paying no attention to individuals whatsoever. To capture the idea that it is only each individual's welfare which matters, two other assumptions will be made. Before they can be stated, however, some additional notation is needed. For any given joint probability distribution $\mu \in \mathcal{M}(X^M)$ over all the different individuals' personalized social states $x^M = \langle x_i \rangle_{i \in M}$, and any given individual $i \in M$, let $\mu_i \in \mathcal{M}(X_i)$ denote the marginal distribution $\text{marg}_{X_i} \mu$ of i's personalized social state x_i.

It will now be assumed that any stochastic dependence between the personalized consequences of different individuals is irrelevant to the welfare ordering, so that the social ordering depends only on the marginal distribution over the personalized consequences of each individual. Formally:

ASSUMPTION 2. *For all* $\mu, \nu \in \mathcal{M}(X^M)$, *assume that if* $\mu_i = \nu_i$ *for all* $i \in M$, *then* $\mu \sim \nu$.

Let $\mu^M := \prod_{i \in M} \mu_i$ denote the joint probability distribution of the vector x^M having the property that the distributions of the different individuals' personalized social states x_i ($i \in M$) are all independent, and equal to the appropriate marginal distributions μ_i. As an implication of Assumption 2 notice that $\mu \sim \mu^M$ for all $\mu \in \mathcal{M}(X^M)$. Indeed, it is always sufficient to consider the collection μ_i ($i \in M$) of marginal distributions, without worrying at all about any interdependence within the joint distribution μ of different personalized social states x_i ($i \in M$).

Finally, it will also be assumed that any ethically relevant external effects which are of consequence to individual i have already been included within the personalized social state x_i. In particular, it will be assumed that the ethically proper choice of the marginal distribution $\mu_i \in \mathcal{M}(X_i)$ of i's personalized social state x_i is completely unaffected by the joint distribution $\mu_{-i} \in \mathcal{M}(X^{M \setminus \{i\}})$ of the profile $x_{-i} \in X^{M \setminus \{i\}} := \prod_{j \in M \setminus \{i\}} X_j$ consisting of all the other individuals' personalized social states. Formally:

ASSUMPTION 3. *For each individual* $i \in M$, *there exists an ordering* $\succsim_i$, *called* **individual** *i's* **ethical welfare ordering**, *such that for every fixed distribution* $\bar{\mu}_{-i} \in \mathcal{M}(X^{M \setminus \{i\}})$ *and all pairs* $\mu_i, \nu_i \in \mathcal{M}(X_i)$, *one has*

$$\mu_i \succsim_i \nu_i \iff (\mu_i, \bar{\mu}_{-i}) \succsim (\nu_i, \bar{\mu}_{-i}).$$

Note how each individual's welfare ordering has essentially been derived from the social ordering over personal issues, rather than the social ordering being derived from all the different individuals' welfare orderings. Moreover, it is not explicitly assumed that each individual's welfare ordering can be represented by the expected value of some cardinal individual welfare function, even though this will turn out to be an implication of all the three assumptions together. This indirect approach to the definition of an individual's welfare ordering is similar to that of Broome (1987). At first it may seem excessively paternalistic. Yet if each individual's "ethical welfare" preferences or "interests" get properly respected in constructing the social ordering $\succsim$ and so each individual i's welfare ordering $\succsim_i$, then this is entirely consistent with what was called "ethical liberalism" in Hammond (1987).

Because the social welfare ordering $\succsim$ on $\mathcal{M}(X^M)$ is represented by the expected value of the social welfare function $W(x^M)$, it follows that, for each fixed $\bar{x}_{-i} \in X^{M\setminus\{i\}}$, the individual welfare ordering $\succsim_i$ on $\mathcal{M}(X_i)$ is represented by the expected value of $W(x_i, \bar{x}_{-i})$. So, for each $i \in M$, there exists a unique cardinal equivalence class of individual welfare functions $w_i(x_i)$ with the property that

$$w_i(x_i) \equiv \alpha_i(\bar{x}_{-i}) + \beta_i(\bar{x}_{-i})\, W(x_i, \bar{x}_{-i})$$

for all $x_i \in X_i$ and all $\bar{x}_{-i} \in X^{M\setminus\{i\}}$, where $\alpha_i(\bar{x}_{-i})$ is an arbitrary real-valued function of $\bar{x}_{-i}$, and $\beta_i(\bar{x}_{-i})$ is a positive real-valued function of $\bar{x}_{-i}$. In future $w_i(x_i)$ will be used to denote any particular member of this equivalence class.

Assumptions 2 and 3 above now imply that, for all pairs $\mu, \nu \in \mathcal{M}(X^M)$ satisfying $\mu_i \sim_i \nu_i$ for all $i \in M$, it must be true that

$$(\mu^J, \nu^{M\setminus J}) = (\mu^{J\setminus\{j\}}, \mu_j, \nu^{M\setminus J}) \sim (\mu^{J\setminus\{j\}}, \nu_j, \nu^{M\setminus J}) = (\mu^{J\setminus\{j\}}, \nu^{M\setminus[J\setminus\{j\}]})$$

whenever $j \in J \subset M$. From this it follows easily by induction on the number of members in the set J that $\mu_i \sim_i \nu_i$ (all $i \in M$) implies $\mu \sim \nu$. We have therefore confirmed that

$$[\forall i \in M : \mathbf{E}_\mu w_i = \mathbf{E}_\nu w_i] \Longrightarrow \mathbf{E}_\mu W = \mathbf{E}_\nu W.$$

So, by Harsanyi's theorem which was proved in the last section, there must exist welfare weights ω_i ($i \in M$) and an additive constant α such that

$$W(x^M) \equiv \alpha + \sum\nolimits_{i \in M} \omega_i\, w_i(x_i).$$

Indeed, for the special case being considered in this section, the device of personalized social states allows sufficient independent variations in different individuals' utilities for Harsanyi's (1955) original proof to be used.

Moreover, Assumption 3 in particular allows a stronger conclusion in this case. This is because, for every fixed $\bar{\mu}_{-i} \in \prod_{j \in M\setminus\{i\}} \mathcal{M}(X_j)$ and for all pairs $\mu_i, \nu_i \in \mathcal{M}(X_i)$, the weak preference $(\nu_i, \bar{\mu}_{-i}) \succsim (\mu_i, \bar{\mu}_{-i})$ would imply that $\nu_i \succsim_i \mu_i$ because of Assumption 3. Therefore, for all $i \in M$ it must actually be true that

$$\mu_i \succ_i \nu_i \Longrightarrow (\mu_i, \bar{\mu}_{-i}) \succ (\nu_i, \bar{\mu}_{-i}),$$

implying that all the welfare weights ω_i are strictly positive. Now, however, the cardinal individual welfare functions $w_i(x_i)$ can be re-normalized so that they become $\tilde{w}_i(x_i) := \omega_i \, w_i(x_i)$, and the social welfare function $W(x^M)$ can be replaced by $\tilde{W}(x^M) := W(x^M) - \alpha$. Accordingly, one has

$$\tilde{W}(x^M) \equiv \sum\nolimits_{i \in M} \tilde{w}_i(x_i).$$

This has therefore become a version of classical utilitarianism, but with a much more general interpretation of individual utility or welfare. Each individual i's function $w_i(x_i)$ represents the ethical value or "goodness" of i's personalized consequence x_i. For Benthamites, goodness corresponded to pleasure minus pain. But much wider and less naïve interpretations of individual utility or welfare functions are certainly allowed. The possibilities are rich enough, in fact, to embrace almost any individualistic ethical theory.

Of course, all the constructions presented here rely on interpersonal comparisons of utility. These have been controversial among economists. Nevertheless, they can be interpreted as ethical preferences for different kinds of people — an idea which is expounded at some length in Hammond (1991b), so I shall not repeat the discussion here. Nor will I repeat here the possible reformulations of Arrow's independence of irrelevant alternatives condition so that it becomes consistent with Harsanyi's form of utilitarianism — on this topic, see Hammond (1991a).

4. Societies with Incomplete Information and Personal Rights

The previous sections have considered only societies in which each individual $i \in M$ was assumed to have a known welfare ordering. Yet real ethical decisions often have to be taken which affect individuals whose interests may be known only very imperfectly. In addition, the ethical decision maker is usually not the only person taking decisions. The individuals in the set M may also be making their own choices. These may affect each other both directly, and also indirectly through their effect on what possibilities remain open to the ethical decision maker.

Such complicated interactions are, of course, the subject of game theory. It is as though some ethical "principal" were confronting a set of individual "agents," each having their own personal objective. Moreover, it is natural to think of the society having to be described by a game of incomplete information, in the sense which Harsanyi (1967–8) was the first to set out formally. After all, both the agents and the ethical principal are players who are likely to be imperfectly informed about one another's objectives, beliefs, etc. In addition, as I have tried to argue in Hammond (1990b), the (ethical) principal will not be able to make Bayesian rational decisons unless they are analysed within some complete game model of the society in which they are all living. It is then necessary as well to form appropriate prior beliefs about what strategies the various agents will choose in that model. This is entirely in accord with Bernheim (1984, 1986) and Pearce's (1984) work on rationalizable strategies, as well as that of Aumann (1987) and others on correlated equilibrium.

A framework is needed to describe the ethical principal's incomplete information, including incomplete information about the incomplete information of other individuals, and about their incomplete information regarding the incomplete information of others, etc. Such a framework is provided by Harsanyi's (1967–8) notion of the "type" of a player in the game. It is actually a far from trivial issue whether a big enough space of possible types to accommodate this infinite regress of incomplete information could ever be constructed. Yet an affirmative answer, at least in principle, has now been provided by, for example, Mertens and Zamir (1985) or Tan and Werlang (1988, pp. 373–5). They use ideas that were, however, pioneered earlier by Böge and his associates — see Armbruster and Böge (1979), Böge and Eisele (1979), and the earlier unpublished work cited therein.

In game theory, a player's type should include everything relevant for determining that player's: (i) payoff function; (ii) beliefs about all the other players' types; (iii) rule for selecting a particular strategy that maximizes expected utility, whenever there is more than one strategy that does so. In ethics, it is also necessary to include: (iv) everything relevant to determining that individual's ethical welfare ordering (which is generally different from the individual's own payoff function).

A society with incomplete information can now be formulated. It will consist of a set of individuals whose "names" or labels lie in the finite set N. For each $j \in N$ there is a set T_j of potential types for person j. Each possible type $t_j \in T_j$ of each individual $j \in N$ will be regarded as a separate "contingent individual," described by the pair (j, t_j). The set of contingent individuals who actually exist is effectively random, since it is unknown to the ethical decision maker. Thus society must be thought of as consisting of *all* possible contingent individuals, since all have potential interests which can be affected by the ethical decision. The membership of the society is therefore given by the set

$$M := \bigcup_{j \in N} (\{j\} \times T_j)$$

of all possible pairs (j, t_j) satisfying $j \in N$ and $t_j \in T_j$. In the special case when $T_j = T$ for all $j \in N$, then M can be expressed more simply as the product space $N \times T$. A serious complication, of course, is that if any of the type spaces T_j is infinite then the set M will also be infinite — even though N itself is only finite. This gives rise to analytical complexities such as the need to replace sums by integrals and to discuss measurability issues. In order to avoid these, it will simply be assumed here that, as in Harsanyi's original formulation of games of incomplete information, each player j has a finite type space T_j.

As in Section 2, the ethical decision maker is assumed to be concerned about the social states or consequences in some domain X. Only now it has to be recognized that, even if full control over the social state really were posible, it might well not be ethically desirable. Instead, the social state should probably respond to changes in individuals' types, because part of each individual i's type describes i's own ethical welfare ordering. Changes in this ordering should often give rise to changes in the social state. In economics, for example, where the social state is the entire allocation of resources within an economy, maintaining efficiency requires responding appropriately to changes in each individual's wants and needs. Moreover, such changes may be inevitable because the ethical decision maker has limited ability to prevent individuals from choosing certain aspects of the social state as they please.

To represent this dependence of the social state on individuals' types, an extended notion of *type-contingent social state* becomes necessary. First, let $t^N \in T^N := \prod_{j \in N} T_j$ denote a typical *type profile*, with one type for each named individual $j \in N$. Responsiveness to types obviously requires that the social state x should be a function $t^N \mapsto x = \xi(t^N)$ of the variable type profile, where $\xi : T^N \to X$. The space of all possible type-contingent social states is then

$$\Xi := X^{T^N} := \prod_{t^N \in T^N} X(t^N) = \{ \xi : T^N \to X \},$$

where $X(t^N)$ denotes the set of social states that can occur when the type profile is t^N. Such type-contingent social states are effectively the same as the "game forms" used by Sugden (1985, 1986) and others in their discussion of rights. Individuals' types are equivalent to strategies in such a game form, and the outcome of the game is what I am calling a social state.

All the previous arguments of Harsanyi and of this paper could now be applied to the society with membership M and space of type-contingent social states Ξ. They suggest that an appropriate ethical objective is the maximization of the expected value of some cardinal social welfare function having the additive form $W(\xi) \equiv \sum_{i \in M} W_i(\xi)$ for suitable cardinal individual welfare functions W_i $(i \in M)$. Now, however, there is much more structure. Really, the only reason why the ethical decision maker needs to consider all different possible type profiles $t^N \in T^N$ is because of uncertainty about which is the right one. Along with Harsanyi, I continue to impose full Bayesian rationality, and so claim that this uncertainty should be described by some subjective probability distribution $\pi(t^N)$ $(t^N \in T^N)$. Then the ethical decision maker should be maximizing the expected value with respect to π of some welfare function $W(x; t^N)$ which would be the appropriate one if the type profile were known to be t^N. Moreover, the previous arguments can be used yet again to claim that, for each $t^N \in T^N$, the social welfare function $W(x; t^N)$ should have the additive form $W(x; t^N) \equiv \sum_{j \in N} w_j(x; t_j)$ for suitable type-dependent cardinal individual welfare functions $w_j(\cdot; t_j)$ $(j \in N;\ t_j \in T_j)$. Note how it is being assumed that j's welfare function $w_j(\cdot; t_j)$ depends only on j's own type t_j; this is natural on the understanding that all kinds of external effects should be included within the definition of any social state or consequence $x \in X$.

When the type-contingent social state is $\xi : T^N \to X$, putting these different functions $W(x; t^N)$ together gives

$$W(\xi) \equiv \sum_{t^N \in T^N} \pi(t^N) \sum_{j \in N} w_j(\xi(t^N); t_j)$$

as the appropriate measure of expected social welfare, after allowing for uncertainty about the type profile $t^N \in T^N$. Generally this uncertainty forces the ethical decision maker to trade off decisions leading to good consequences for different type profiles, and to do so in a way that is sensitive to the subjective probability assessments $\pi(t^N)$ $(t^N \in T^N)$. Only by discovering the true type profile can this uncertainty be avoided. There are, however, serious obstacles in the way of doing so; this is the final topic to be discussed in this paper.

Indeed, the question of how to elicit private information has been the topic of much work since the early 1970's. There is no space here to discuss properly what has become an enormous literature. Nor is there any need, since the basic point can be made very briefly. It is that, in order to be able to have the social state adapt to changes in individuals' private information, it is necessary that the right incentives be created. I have already suggested that one can usefully think of a society as composed of individuals who interact within some enormously complicated game of incomplete information — see also Hammond (1990a). The issue becomes one of how the ethical decision maker should behave in such a game. The ethical objectives have already been discussed. They are the topic of social choice theory, and all the previous part of this paper has been devoted to giving reasons why the most suitable objective is likely to be some version of Harsanyi's form of utilitarianism. The need to provide incentives, however, arises because the ethical decision maker does not have full control over the process which determines the social state — other individuals' actions are also important, and will usually be much more so. Thus incentives are concerned with constraints on what the ethical decision maker can achieve, rather than with desirable social objectives. For this reason, they really lie beyond the scope of this paper.

Nevertheless, it is already possible to see that the issue of individual rights arises much more naturally when society is modelled as having incomplete information. Dasgupta (1982) was probably the first to point this out with any kind of formal argument. An increased respect for rights also emerges if we take into account the limited control that the ethical decision maker has over individual actions which affect the social state. Both incomplete information and moral hazard give individuals powers to affect their own destinies in ways which would meet with the approval of libertarians. An ethical decision maker can only control individuals' actions if some enforcement mechanism is in place. Most of the time, such enforcement mechanisms are inevitably both intrusive and costly in other ways. Then, however, a proper ethical utilitarian calculation of the costs and benefits of having an enforcement mechanism is quite likely to decide that effective enforcement is not worthwhile. The same is true of incomplete information about something which, if known, could improve the quality of some important ethical decision. In order that something which is known only privately can be discovered and the information used, individuals have to be provided with incentives to reveal what they alone know. These can be normal incentives, such as those a shop-keeper provides to encourage customers to reveal what they want to buy and what is an upper bound on the price which they are willing to pay. Alternatively, the ethical decision maker may be able to coerce the information out of the individual in some unpleasant way. In the latter case, however, the coercion itself imposes enormous costs which are very rarely going to be outweighed by any ethical benefits which the knowledge might yield. The point is that a full description of the type-contingent social state has to specify what means will be used to enforce certain kinds of behaviour and to encourage certain kinds of private information to be revealed. When this is done properly, many attempts to infringe what people see as their rights, even if such attempts would otherwise be ethically valid, are likely to be evaluated as ethically unacceptable when all kinds of enforcement cost are taken into account.

REFERENCES

W. ARMBRUSTER AND W. BÖGE (1979), "Bayesian Game Theory," in *Game Theory and Related Topics* edited by O. Moeschlin and D. Pallaschke (Amsterdam: North-Holland), pp. 17–28.

R.J. AUMANN (1987), "Correlated Equilibrium as an Expression of Bayesian Rationality," *Econometrica*, **55**: 1–18.

B.D. BERNHEIM (1984), "Rationalizable Strategic Behavior," *Econometrica*, **52**: 1007–1028.

B.D. BERNHEIM (1986), "Axiomatic Characterizations of Rational Choice in Strategic Environments," *Scandinavian Journal of Economics*, **88**: 473–488.

D. BERNOUILLI (1738; translated by L. Sommer, 1954) "Specimen theoriae novae de mensura sortis" [title of translation: 'Exposition of a New Theory on the Measurement of Risk'], *Econometrica*, **22**: 23–36.

W. BÖGE AND T. EISELE (1979), "On Solutions of Bayesian Games," *International Journal of Game Theory*, **8**: 193–215.

K.C. BORDER (1981), "Notes on von Neumann-Morgenstern Social Welfare Functions," unpublished preprint, California Institute of Technology.

K.C. BORDER (1985), "More on Harsanyi's Utilititarian Cardinal Welfare Theorem," *Social Choice and Welfare*, **1**: 279–281.

J. BROOME (1987), "Utilitarianism and Expected Utility," *Journal of Philosophy*, **84**: 405–422.

J. BROOME (1989), "Should Social Preferences Be Consistent?" *Economics and Philosophy*, **5**: 7–17.

J. BROOME (1990), "Bolker-Jeffrey Expected Utility Theory and Axiomatic Utilitarianism," *Review of Economic Studies*, **57**: 477–502.

T. COULHON AND P. MONGIN (1989), "Social Choice Theory in the Case of Von Neumann-Morgenstern Utilities," *Social Choice and Welfare*, **6**: 175–187.

G. CRAMER (1728), Letter to Nicolas Bernouilli; extracts printed in Bernouilli (1738) and in Sommer's (1954) translation.

P.S. DASGUPTA (1982), "Utility, Information, and Rights," in *Utilitarianism and Beyond* edited by A.K. Sen and B. Williams (Cambridge: Cambridge University Press), pp. 199–218.

P.A. DIAMOND (1967), "Cardinal Welfare, Individualistic Ethics, and Interpersonal Comparisons of Utility: Comment," *Journal of Political Economy*, **75**: 765–766.

Z. DOMOTOR (1979), "Ordered Sum and Tensor Product of Linear Utility Structures," *Theory and Decision*, **11**: 375–399.

M. FLEMING (1952), "A Cardinal Concept of Welfare," *Quarterly Journal of Economics*, **66**: 366–384.

M. FLEMING (1957), "Cardinal Welfare and Individualistic Ethics: A Comment," *Journal of Political Economy*, **65**: 355–357.

M. FRIEDMAN AND L.J. SAVAGE (1952), "The Expected Utility Hypothesis and Measurement of Utility," *Journal of Political Economy*, **60**: 463–474.

W. GAERTNER AND P.K. PATTANAIK (EDS.) (1988), *Distributive Justice and Inequality*. Berlin: Springer Verlag.

W. GAERTNER, P.K. PATTANAIK AND K. SUZUMURA (1988), "Individual Rights Revisited," Fachbereich Wirtschaftswissenschaft, Universität Osnabrück; to appear in *Economica*.

D. GALE (1960), *Theory of Linear Economic Models*. New York: McGraw-Hill.

P.J. HAMMOND (1983), "Ex-Post Optimality as a Dynamically Consistent Objective for Collective Choice under Uncertainty," in *Social Choice and Welfare* edited by P.K. Pattanaik and M. Salles (Amsterdam: North Holland), ch. 10, pp. 175–205.

P.J. HAMMOND (1986), "Consequentialist Social Norms for Public Decisions," in *Social Choice and Public Decision Making: Essays in Honor of Kenneth J. Arrow, Vol. 1* edited by W. P. Heller, R. M. Starr and D. A. Starrett (Cambridge: Cambridge University Press), ch. 1, pp. 3–27.

P.J. HAMMOND (1987), "On Reconciling Arrow's Theory of Social Choice with Harsanyi's Fundamental Utilitarianism," in *Arrow and the Foundations of the Theory of Economic Policy* edited by G.R. Feiwel (London and New York: Macmillan and New York University Press), ch. 4, pp. 179–222.

P.J. HAMMOND (1988a), "Consequentialism and the Independence Axiom," in *Risk, Decision and Rationality* edited by B.R. Munier (Dordrecht: D. Reidel), pp. 503–516.

P.J. HAMMOND (1988b), "Consequentialist Foundations for Expected Utility," *Theory and Decision*, **25**: 25–78.

P.J. HAMMOND (1988c), "Consequentialist Demographic Norms and Parenting Rights," *Social Choice and Welfare*, **5**: 127–145; reprinted in Gaertner and Pattanaik.

P.J. HAMMOND (1990a), "Incentives and Allocation Mechanisms," in *Advanced Lectures in Quantitative Economics* edited by R. van der Ploeg (New York: Academic Press), ch. 6, pp. 213–248.

P.J. HAMMOND (1990b), "A Revelation Principle for (Boundedly) Bayesian Rationalizable Strategies," European University Institute, Working Paper ECO 90/4; to appear in R.P. Gilles and P.H.M. Ruys (eds.), *Economic Behaviour in an Imperfect Environment* (Amsterdam: North-Holland).

P.J. HAMMOND (1991a), "Independence of Irrelevant Interpersonal Comparisons," European University Institute, Working Paper ECO 90/5; *Social Choice and Welfare* (in press).

P.J. HAMMOND (1991b), "Interpersonal Comparisons of Utility: Why and How They Are and Should Be Made," European University Institute, Working Paper ECO 90/3; to appear in J. Elster and J. Roemer (eds.), *Interpersonal Comparisons of Well-Being* (New York: Cambridge University Press).

J.C. HARSANYI (1953), "Cardinal Utility in Welfare Economics and in the Theory of Risk-Taking," *Journal of Political Economy*, **61**: 434–5; reprinted in Harsanyi (1976).

J.C. HARSANYI (1955), "Cardinal Welfare, Individualistic Ethics, and Interpersonal Comparisons of Utility," *Journal of Political Economy*, **63**: 309–321; reprinted in Phelps (1973) and in Harsanyi (1976).

J.C. HARSANYI (1967–8), "Games with Incomplete Information Played by 'Bayesian' Players, I–III," *Management Science*, **14**: 159–182, 320–334, 486–502.

J.C. HARSANYI (1975a), "Nonlinear Social Welfare Functions: Do Welfare Economists Have a Special Exemption from Bayesian Rationality?," *Theory and Decision*, **6**: 311–32; reprinted in Harsanyi (1976).

J.C. HARSANYI (1975b), "Can the Maximin Principle Serve as a Basis for Morality? A Critique of John Rawls's Theory," *American Political Science Review*, **69**: 594–606; reprinted in Harsanyi (1976).

J.C. HARSANYI (1976), *Essays on Ethics, Social Behavior, and Scientific Explanation*. Dordrecht: D. Reidel.

J.C. HARSANYI (1978), "Bayesian Decision Theory and Utilitarian Ethics," *American Economic Review (Papers and Proceedings)*, **68**: 223–8.

A.P. LERNER (1944), *The Economics of Control*. London: Macmillan.

J.-F. MERTENS AND S. ZAMIR (1985), "Formalization of Bayesian Analysis of Games with Incomplete Information," *International Journal of Game Theory*, **14**: 1–29.

J.C. MILLERON (1972), "Theory of Value with Public Goods: A Survey Article," *Journal of Economic Theory*, **5**: 419–477.

J. VON NEUMANN AND O. MORGENSTERN (1943; 3rd edn., 1953), *Theory of Games and Economic Behavior.* Princeton: Princeton University Press.

P.K. PATTANAIK (1968), "Risk, Impersonality, and the Social Welfare Function," *Journal of Political Economy,* **76**: 1152–1169; reprinted in Phelps (1973).

D. PEARCE (1984), "Rationalizable Strategic Behavior and the Problem of Perfection," *Econometrica,* **52**: 1029–1050.

E.S. PHELPS (ED.) (1973), *Economic Justice.* Harmondsworth: Penguin Books.

J. RAWLS (1959), "Justice as Fairness," *Philosophical Review,* **67**: 164–94.

J. RAWLS (1971), *A Theory of Justice.* Cambridge, Mass: Harvard University Press.

J. RAWLS (1974), "Some Reasons for the Maximin Criterion," *American Economic Review (Papers and Proceedings),* **64**: 141–6.

J. RILEY (1989, 1990), "Rights to Liberty in Purely Private Matters: Parts I and II," *Economics and Philosophy,* **5**: 121–166 and **6**: 27–64.

S. SELINGER (1986), "Harsanyi's Aggregation Theorem without Selfish Prefences," *Theory and Decision,* **20**: 53–62.

A.K. SEN (1969), "Planner's Preferences: Optimality, Distribution and Social Welfare," in *Public Economics* edited by J. Margolis and H. Guitton (London: Macmillan), ch. 8, pp. 201–221.

A.K. SEN (1970), *Collective Choice and Social Welfare.* San Francisco: Holden-Day.

A.K. SEN (1973), "On Ignorance and Equal Distribution," *American Economic Review,* **63**: 1022–4; reprinted in Sen (1982).

A.K. SEN (1982), *Choice, Welfare and Measurement.* Oxford: Basil Blackwell, and Cambridge, Mass.: MIT Press).

R. SUGDEN (1985), "Liberty, Preference and Choice," *Economics and Philosophy,* **1**: 213–229.

R. SUGDEN (1986), *The Economics of Rights, Cooperation and Welfare.* New York: Basil Blackwell.

T.C.-C. TAN AND S.R. DA C. WERLANG (1988), "The Bayesian Foundations of Solution Concepts of Games," *Journal of Economic Theory,* **45**: 370–391.

W.S. VICKREY (1945), "Measuring Marginal Utility by Reactions to Risk," *Econometrica,* **13**: 319–33.

J.A. WEYMARK (1990), "Harsanyi's Social Aggregation Theorem with Alternative Pareto Principles," University of British Columbia, Department of Economics, Discussion Paper No.: 90-28; presented to the Fifth Karlsruhe Seminar on *Models and Measurement of Welfare and Inequality.*

Evolution and Convention: Social Contract IV

Ken Binmore
Economics Department
University of Michigan
Ann Arbor, MI 48109

Two men, who pull the oars of a boat, do it by an agreement or convention, tho' they have never given promises to each other.

Hume [1739, p.490]

1 Introduction.

This paper may be read in two different ways. It can be seen as a contribution to the literature on the evolution of bargaining conventions. (See, for example, Young [11].) Those who are interested only in this aspect of the paper will wish to read only Section 2. However, the chief motivation for writing the paper is to round off a philosophical point about the evolution of utilitarianism raised in an earlier paper (Binmore [4]).

Harsanyi [5,6,7,9] has been an unwavering and persuasive champion of utilitarianism for many years. Unlike most other advocates, he is very precise about the assumptions on which his arguments in favor of utilitarianism are based. It is therefore possible to obtain a clear picture of what human nature would need to be like for a Harsanyi–style utilitarian society to be viable. One can then ask whether Harsanyi's assumptions really do reflect the "nature of human nature". It seems to me that the question boils down to whether *homo sapiens* is more like *homo economicus* or *homo behavioralis*. The latter is a hominid who is programmed directly with behavior, like a chocolate dispensing machine.

The distinction between *homo economicus* and *homo behavioralis* to be emphasized here is the manner in which they get to equilibrium. I have made the virtually tautologous point elsewhere (Binmore [2]) that the equilibrium that gets selected will be a function of the equilibriating process, the *libration*, that selects it. A distinction was made between *evolutive* and *eductive* librations. *Homo behavioralis* gets to equilibrium via an evolutive libration: a slow trial-and-error process in which Nature tries out various strategies, and eliminates those which are "less successful". *Homo economicus* gets to equilibrium via an eductive libration. He *thinks* his way to an equilibrium strategy using a classification tâtonnement argument.

Why should evolution generate *homo economicus* when *homo behavioralis* is available? Brain power is, after all, an expensive investment. Moreover, *homo economicus* is somewhat of a nuisance from Nature's[1] point of view. *Homo behavioralis* can be manipulated directly. His behavior can be tailored precisely to the environment by providing him with a suitable program. *Homo economicus* can only be manipulated via his preferences. Nature cannot get directly at his behavior.

In economics, the "principal-agent problem" involves a principal with certain aims who can only act through an agent who may have aims of his own. The principal therefore seeks to design an incentive scheme which minimizes the distortions resulting from having to delegate to the agent.

[1]In such assertions, of course, Nature is used in a metaphorical sense. No teleological interpretation is intended.

As a principal, Nature is blind, but her agent *homo economicus* can see his environment. Thus, although Nature *loses* fine control in being unable to modify his behavior directly, she *gains* access to information that would otherwise have been unavailable to her. If the environment changes sufficiently rapidly, the gains will outweigh the losses. *Homo behavioralis* is as blind as his mistress. She can only learn about his environment indirectly by observing which types of *homo behavioralis* are reproductively successful. But this learning process is very slow. In brief, *homo economicus* adapts quickly but *homo behavioralis* does not.

In Binmore [4], the Nash demand game was taken as a paradigm[2] for the human "coordination problem"—the problem of selecting an equilibrium when many are available. It was shown that biological evolution can lead a variety of the species *homo behavioralis* to a utilitarian outcome in the Nash demand game. Nature's hidden hand can therefore engineer a utilitarian's "first-best" result.

Unlike *homo behavioralis*, *homo economicus* can adapt quickly to a changing environment, but one can argue that his societies must pay a price in having to live with outcomes that are "second-best" from the welfarist's viewpoint. One might express the difference by saying that *homo behavioralis* can behave *as though* motivated by a single "common interest", but *homo economicus* can only sustain "common understandings" that provide the basis for the *coordinating conventions* used by his societies to select an equilibrium from those available. It is the commonly held system of principles built into the coordinating conventions used in selecting equilibria that I follow many others in believing should be identified with the *social contract*. Such a view denies the existence of *a priori* "natural laws", and sees a social contract as an artefact of a society's history. As such, it must be seen as being shaped, not by biological evolution, but by *social* evolution.

This paper seeks to explore the viewpoint expounded in the previous paragraph by following the evolution of a "common understanding" in a population of specimens of *homo economicus*. The mechanism is entirely orthodox. What the players know and believe about each other will evolve over time as a consequence of the players' updating their beliefs using Bayes' rule. Conditions are studied under which the players' resulting behavior in the Nash demand game converges on the Nash bargaining solution.

Two warnings are necessary. The primary purpose of the paper is philosophical, and it is intended for a wider audience than that for which mathematical economists commonly write. Although mathematical, the exposition is therefore *informal*. The second warning concerns the conditions under which results are derived. It would not be easy to verify whether or not they are satisfied in situations of any complexity.

2 Nash Demand Game

In the Nash demand game, two bargainers, Adam and Eve, make simultaneous demands. Adam demands a Von Neumann and Morgenstern utility a. Eve demands b. If these demands are compatible, both receive their demands. Otherwise, each gets a disagreement payoff that will always be normalized to zero in this paper. The two demands are compatible if and only if the pair (a, b) lies in a given feasible set X defined by

$$X = \{(x_A, x_E) : x_E \leq \varphi(x_A)\},$$

where $\varphi : \mathbb{R} \to \mathbb{R}$ is strictly decreasing, concave and continuously differentiable. The concavity is not essential but is included so that X is convex (as well as bounded above, closed and comprehensive) since this is a standard assumption.

As in Binmore [4], Adam and Eve are drawn at random from a large population. In this previous paper, the population consisted of specimens of *homo behavioralis* and the conclusion was

[2] And reasons were given for rejecting the Prisoners' Dilemma for such purposes.

that, with suitable assumptions, biological evolution can generate a utilitarian outcome. *Homo behavioralis* then acts as though moved by a "common interest". In the current analysis, the population will consist of specimens of *homo economicus*, who are capable only of sustaining a "common understanding".

In the model to be described here, social evolution, acting on this "common understanding", engineers the *Nash bargaining solution* as the final outcome. This is the point at which the product $n_A n_E$ is maximized subject to the constraints $n \geq O$ and $n \in X$. It is *not* claimed that social evolution will *inevitably* generate the Nash bargaining solution. Substantive assumptions are needed for this purpose whose plausibility requires evaluation. However, when these assumptions are satisfied, *homo economicus* may be said to be paying for his individual talents by being forced to live with a "second best" social outcome.

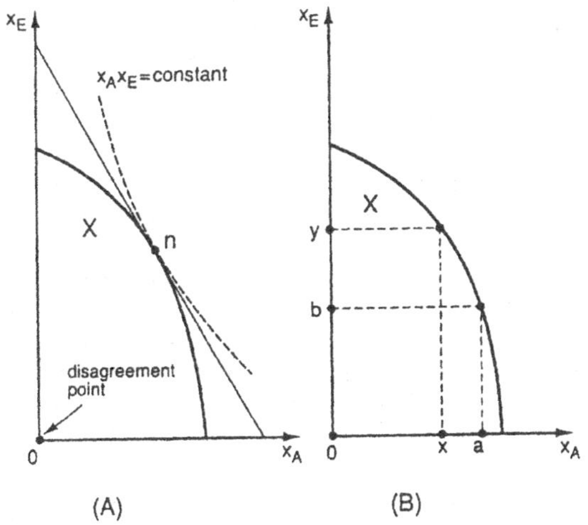

Figure 1: Nash demands

Some comment on modeling questions is necessary in replacing *homo behavioralis*, as used in Binmore [4], by *homo economicus* in the context of the Nash demand game. First, it should be noted that, while the Nash demand game perhaps has some claim to be considered a natural bargaining habitat for *homo behavioralis*, the same cannot be claimed for *homo economicus* because of his inability to make commitments. His natural bargaining habitat is rather some iterated version of the Nash demand game, or a Rubinstein alternating offers game. The Nash demand game is not studied here because it is thought to be an interesting *bargaining* model for *homo economicus*. It is proposed as a paradigm for the human *coordination problem*: the problem of selecting an equilibrium when many are available.

It is interesting in this role because it is a very simple game but has a continuum of Nash equilibria. Each pair (a, b) with $a \geq 0$, $b \geq 0$ and $b = \varphi(a)$ is a Nash equilibrium. That is to say, any pair of Pareto efficient, "individually rational" demands is an equilibrium. These might be called the "cooperative equilibria". There are also trivial or "non-cooperative equilibria" which result in each player getting a payoff of zero. For example, both might demand so much that no demand the other could make would be compatible.

The question is: how is one of these equilibria selected? Nash [10] proposed what would nowadays be regarded as a refinement of Nash equilibrium to deal with this problem. He introduced some *exogenous* uncertainty into his model. When the amount of uncertainty is small, in an appropriate sense, all non-trivial Nash equilibria in the modified model approximate the Nash bargaining solution in the unmodified model. Nash's idea of introducing exogenous uncertainty is important for the study of *homo behavioralis* in Binmore [4]. But its use in the context of *homo economicus* would finesse the very problems on which this inquiry is intended to focus. One of the principle distinctions between *homo behavioralis* and *homo economicus* is that the latter can costlessly explore his environment. He should therefore be assumed to be able to locate the frontier of X to as close a degree of approximation as he chooses which, at the level of abstraction of this paper, means without any error at all.[3] Any uncertainty ought to appear *endogenously* in the form of doubts that he may have about what the other player is *thinking*.

Thus the assumption will be that specimens of *homo economicus* know exactly what X is. What they will not know for sure is what might be going on in the heads of other members of the population. I hasten to insist that they will have no doubts about the fact that the population consists entirely of rational optimizers. Each player will believe that it is certain that, were he in another's shoes, with her information, he would reason precisely as she reasons. His difficulty will be that he does not know *precisely* what information others have about the population as a whole.

This remark may confuse laymen, since the game to be studied will be one of *complete* information. This means that the players' preferences *and beliefs* are common knowledge, along with the rules of the game. It is important to be clear about *when* common knowledge of beliefs is assumed in this statement. The beliefs that are common knowledge are those the players have *before* the game begins. In a game of *perfect* information, like Chess or Backgammon, everything that happens during the game is *public* and so, although the players may update their beliefs as the game proceeds, their beliefs will always remain common knowledge. But, in a game of *imperfect* information, the players will receive *private* information. For example, in Poker, it is reasonable to assume that the beliefs everybody has about what is likely to be dealt are common knowledge *before the deal*. However, *after the deal* the players will have different beliefs about what has been dealt and the game would be pointless if those updated beliefs were common knowledge. Thus although Poker, as normally analyzed, is a game *of complete* information, it is very definitely not a game of *perfect* information.

The players' *a priori* beliefs in a game of complete information are common knowledge. In this paper, their beliefs are about the *kinds*[4] of player who make up the population of which society is formed. A somewhat more general formulation will be required shortly, but the following simplified story might be a helpful introduction.

Suppose that the members of the population correspond to the real numbers θ satisfying $0 \leq \theta \leq 1$. Each of these potential players of the Nash demand game is dealt a card at random from an infinite deck containing only jokers and knaves. The fraction of the population dealt a joker is p and the fraction dealt a knave is $q = 1 - p$. It is important that nobody knows the value of p. That is to say, there is uncertainty about the distribution of jokers and knaves in the population. What is assumed to be common knowledge is the probability distribution from which the number p is drawn.[5] One must therefore think of the rules of the game as requiring an initial chance move that selects a deck of cards, followed by a second chance move comprising the deal.

For simplicity, assume that p is uniformly distributed on $[0, 1]$. After learning what kind of person they are, each player will update his *a priori* belief about the value of p (i.e., the structure of the population) using Bayes' rule. Thus, a joker's posterior density $f_J : [0,1] \to \mathbb{R}$ will be given

[3] I know that locating the frontier of X requires knowing the preferences of others as well as one knows one's own preferences. But this lack of realism has to be swallowed if one accepts that it is premature to attempt an analysis of incomplete information games.

[4] I would like to write "type" to indicate that the idea is borrowed from Harsanyi's [8] theory of incomplete information, but there is a risk of confusion with its technical usage in Harsanyi's formulation.

[5] This assumption of common knowledge of a common prior is the "Harsanyi doctrine" criticized in Binmore [3], but used in a more harmless context here.

by $f_J(p) = 2p$ and a knave's by $f_K(p) = 2q$. Jokers and knaves will therefore have different beliefs about the population of which they are a part. A joker will believe that a randomly chosen member of the population will be another joker with probability of

$$\int_0^1 p f_J(p) dp = \tfrac{2}{3}.$$

A knave will attach precisely the same probability of $\frac{2}{3}$ to the event that a knave will be chosen.

The next step is to attribute behavior to jokers and knaves in the event that they are called upon to occupy the role of Adam or Eve in the Nash demand game. It is necessary not to lose sight of the fact that jokers and knaves are specimens of *homo economicus* who optimize given their beliefs. Of course, their optimization problem in the Nash demand game is not a one-person decision problem. They have to *think* their way to an *equilibrium*. The process by means of which they do this is more than a little problematic (Binmore [2]). But this paper will proceed in the standard way by supposing that the process achieves the end for which it is designed without enquiring into the question of precisely what the process is.

In discussing evolution in Binmore [4], a distinction was made between *nominal* players and *actual* players. The distinction is not so significant in the current context but it will be made here also in an attempt to clarify the terminology used in describing equilibria. My own preference is to regard the set of actual players as consisting of the population as a whole. The chance move which selects actual players, or actors, to fill the roles of Adam and Eve then needs to be modeled as part of the formal game. Each actor decides in advance what strategy he will choose if called upon to play. One need only require of these strategies that they constitute a *Nash equilibrium*—that is, each actor's strategy decision is optimal given the choices of the others.

An alternative and equivalent approach is to treat Adam and Eve, not as nominal players as in the discussion above, but as the actual players. The game they play is simply the one-shot Nash demand game. The price paid for these simplifications is that it becomes necessary to introduce the idea of a *correlated equilibrium* in the sense of Aumann [1]. Although this approach is not used in this paper, it is worth bearing in mind that everything which is said can be reinterpreted in these terms.

Suppose that each joker plans to demand $a > 0$ if he finds himself in the role of Adam, and to demand $b > 0$ in the role of Eve. Each knave plans to demand $x > 0$ in the role of Adam and $y > 0$ in the role of Eve. Assume that (a, b) and (x, y) are Pareto-efficient points of X and that $a > x$, as in Figure 1B. Then, given the beliefs attributed to jokers and knaves, their planned behavior constitutes a Nash equilibrium if and only if

$$\left.\begin{array}{ccccc} \frac{3}{2}x & \leq & a & \leq & 3x \\ \frac{3}{2}b & \leq & y & \leq & 3b \end{array}\right\} \tag{1}$$

The inequality $\frac{2}{3}a \geq x$ means that a joker would not want to imitate a knave in the role of Adam. The inequality $x \geq \frac{1}{3}a$ means that a knave would not want to imitate a joker in the role of Adam. The second of the displayed inequalities follows from similar considerations with Adam replaced by Eve.

Recall that the focus of interest is on how "common understandings" provide a basis for different specimens of *homo economicus* zeroing in on the same equilibrium. In the model under discussion, the "common understanding" includes what kinds of people there are, what they believe, how they behave and, most importantly, how they are distributed in the population as a whole. The latter information is derived from the *a priori* common knowledge that p is uniformly distributed on $[0, 1]$, and that each actor has learned only what kind of person he is himself, and nothing else. The question is how social evolution will shape such "common understandings". Only the crudest type of social evolution will be captured in the story that follows. What will evolve is the pool of what is commonly known as the players take on board new data. The engine of evolutionary change will therefore simply be Bayesian updating.

A suitable evolutionary story could be told as follows. First, the deal described above takes place at time 0. As time passes, each actor randomly encounters other actors in the population from whose behavior, it will be assumed, he is able to deduce the entirety of their experience from time 0 onwards. At time t, each actor will therefore know the values of N of the cards dealt, and not just the value of his own card. There will therefore no longer be just two kinds of person: jokers and knaves. Instead, $N+1$ different kinds of person will be distinguished, corresponding to the $N+1$ hands containing N cards. Each such hand will be assigned a label $\lambda\,(0 \leq \lambda \leq 1)$ by taking λN to be the number of jokers the hand contains.

An actor holding hand λ will be called Larry. His posterior probability density $f_\lambda : [0,1] \to \mathbb{R}$ for p is given by

$$f_\lambda(p) = (N+1)\binom{N}{\lambda N} p^{\lambda N}(1-p)^{(1-\lambda)N} \qquad (0 \leq p \leq 1).$$

Larry may be contrasted with another actor called Michael who holds the hand μ. It is an algebraic exercise for Larry to calculate the probability $\text{prob}(\mu|\lambda)$ that a randomly drawn opponent is of the same kind as Michael.[6] In fact,

$$\begin{aligned} N\text{prob}(\mu|\lambda) &= \frac{N(N+1)}{2N+1}\binom{N}{\mu N}\binom{N}{\lambda N}\Big/\binom{2N}{(\mu+\lambda)N} \\ &\sim \sqrt{\tfrac{1}{2}\pi N}\left(\frac{W^2_{(\lambda+\mu)/2}}{W_\lambda W_\mu}\right)^N, \end{aligned}$$

where $W_\alpha = \alpha^\alpha(1-\alpha)^{1-\alpha}$.

If N is large, Larry's beliefs about the distribution of other kinds of persons in the population can be approximated by a probability distribution function $G(\bullet|\lambda) : [0,1] \to [0,1]$ that has a continuous density function $g(\bullet|\lambda) : [0,1] \to \mathbb{R}$ satisfying $g(\mu|\lambda) \sim N\text{prob}(\mu|\lambda)$ $(\mu = 0, 1/N, 2/N, \ldots, 1)$. Figure 2 illustrates such density functions for Larry and two different Michaels. After contemplating these graphs, the preceding algebra may be forgotten. What is important is that $g(\bullet|\lambda)$ is closely concentrated at λ and $g(\bullet|\mu)$ is closely concentrated at μ.

How has what is "commonly understood" evolved at time t? The important point is that it is now common knowledge that everybody has very sharp beliefs compared with their beliefs at time 0. Larry believes that the probability that a randomly chosen Michael is holding a hand which is approximately the same as Larry's is nearly one. Since each actor's beliefs are a well-behaved function of the hand they are holding, Larry therefore believes that all but a tiny fraction of the population believe nearly the same things as he does. (In Figure 2, Larry thinks that nearly all Michaels are holding a hand like μ_1.) Of course, Larry knows of the existence of what he believes to be oddball Michaels (like those holding hand μ_2 in Figure 2). Larry knows that such oddballs believe that all but a tiny fraction of the population are like themselves and it is people like Larry who are the oddballs. But this does not lead Larry to change his views which are based on good firm evidence: namely, the sample of N cards that Larry knows were dealt in the original deal.

The story of the deal at time 0 has no further role in what follows except insofar as it may serve to motivate some of the assumptions to be made. Instead a more abstract framework will be adopted. At any time t, an actor will be characterized by a label λ drawn from an *open*[7] interval Λ of real numbers. As before, an actor characterized by λ will be called Larry. Larry has beliefs about the different kinds of people in the population. These beliefs are specified by a *continuous*

[6] Although an actor, Larry is a specimen of *homo economicus* and hence knows Stirling's formula: $n! \sim \sqrt{2\pi n}\, n^n e^{-n}$. He also knows that

$$\text{prob}(\mu \mid \lambda) = \int_0^1 \binom{N}{\mu N} p^{\mu N}(1-p)^{(1-\mu)N} f_\lambda(p)\,dp.$$

[7] A smallest or a largest label is being ruled out.

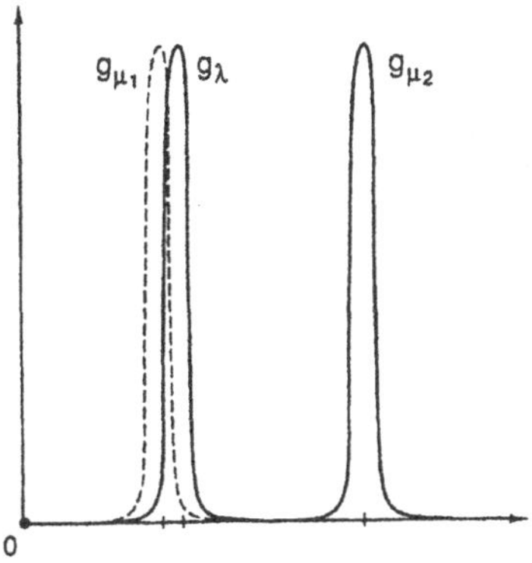

Figure 2: Larry and Michael's beliefs

probability density function $g_t(\bullet|\lambda) : \Lambda \to \mathbb{R}$. At time 0, it is common knowledge that all actors share the same beliefs about what kinds of people there will be at time t. The beliefs are specified by a continuous density function $h_t : \Lambda \to \mathbb{R}$. Bayes' rule may then be expressed in the form

$$g_t(\mu|\lambda)h_t(\lambda) = g_t(\lambda|\mu)h_t(\mu). \tag{2}$$

Having discussed the actors' beliefs at time t, it is now necessary to discuss their planned behavior should they be called upon to play the Nash demand game, either in the role of Adam or in the role of Eve. For this purpose, two functions $\alpha_t : \Lambda \to \mathbb{R}$ and $\beta_t : \Lambda \to \mathbb{R}$ are introduced. The number $a = \alpha_t(\lambda)$ is the demand Larry plans to make if cast as Adam. The number $b = \beta_t(\lambda)$ is the demand he plans to make if cast as Eve. The numbers $x = \alpha_t(\mu)$ and $y = \beta_t(\mu)$ are the corresponding demands for Michael.

The function α_t is assumed to be strictly increasing and continuously differentiable on I. The function β_t is assumed to be strictly decreasing and continuously differentiable.[8] A function $\Psi_t : A_t \to B_t$ is defined so that $\Psi_t(a) = b$ if and only if $(a,\, b) = (\alpha_t(\lambda),\, \beta_t(\lambda))$ for some $\lambda \in \Lambda$. Thus $\Psi_t = \beta_t \circ \alpha_t^{-1}$. Figure 3A may serve to clarify these definitions.

The chief aim of the analysis is to show that, if sufficient assumptions are made, then, for each $\lambda \in \Lambda$,

$$\alpha_t(\lambda) \to n_A \text{ and } \beta_t(\lambda) \to n_B \quad \text{as} \quad t \to \infty,$$

where (n_A, n_B) is the Nash bargaining solution. Recall that the latter is characterized by the equation $n_A\varphi'(n_A) + n_E = 0$.

Since the actors are specimens of *homo economicus*, they optimize given their beliefs. Thus, if he is cast in the role of Adam, Larry needs to check that he wants to play Adam's part as the playwright has written it. That is to say, he has to check that $a = \alpha_t(\lambda)$ is optimal given his information. For this purpose, he requires the probability PROB ($\mu|\lambda$ is Adam) that he should attach to the event that, if he were to demand $x = \alpha_t(\mu)$ like Michael, then the result could be compatible with the demand made by Eve. Observe that

$$\text{PROB}\,(\mu|\lambda \text{ is Adam }) = 1 - G(\gamma_t(\mu)|\lambda),$$

[8]With such α_t and β_t, a player can deduce the label of an opponent *after* the opponent's demand has been observed. Recall that, in the story of a deal at time 0, when Larry encounters Michael, he acquires Michael's entire stock of experience.

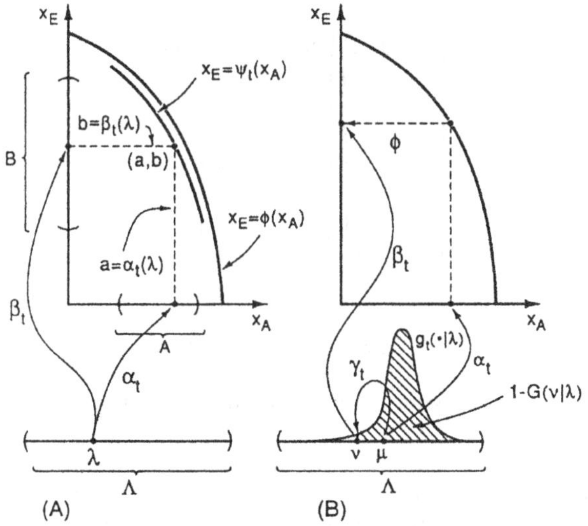

Figure 3: The functions α_t and β_t

where $\gamma_t : \Lambda \to \Lambda$ is defined by $\gamma_t = \beta_t^{-1} \circ \varphi \circ \alpha_t$ and

$$G(\nu|\lambda) = \int_{\theta \leq \nu} g_t(\theta|\lambda) d\theta \, .$$

Figure 3B illustrates these observations. Thus a necessary condition that Larry's part is optimal for him when he is cast in the role of Adam is that the expected payoff

$$\alpha_t(\mu)\{1 - G(\gamma_t(\mu)|\lambda)\} \tag{3}$$

be maximized when $\mu = \lambda$. A similar discussion is now required for Larry in the role of Eve. The only difference is that $\text{PROB}\,(\mu|\lambda \,\text{is Eve}) = G(\gamma_t^{-1}(\mu)|\lambda)$. The necessary condition for Larry's part to be optimal for him when he is cast in the role of Eve is that

$$\beta_t(\mu) G(\gamma_t^{-1}(\mu)|\lambda) \tag{4}$$

be maximized when $\mu = \lambda$.

The next step is to differentiate (3) and (4). The results are then evaluated respectively at points λ_1 and λ_2 with $\lambda_2 = \gamma_t(\lambda_1)$ (see Figure 4A). This yields

$$\begin{aligned} \alpha_t'(\lambda_1)\{1 - G(\lambda_2|\lambda_1)\} - \alpha_t(\lambda_1) g_t(\lambda_2|\lambda_1)\gamma_t'(\lambda_1) = 0 \\ \beta_t'(\lambda_2) G(\lambda_1|\lambda_2) + \beta_t(\lambda_2) g_t(\lambda_1|\lambda_2)(\gamma_t^{-1})'(\lambda_2) = 0 \end{aligned} \tag{5}$$

The terms $g_t(\lambda_2|\lambda_1)$ and $g_t(\lambda_1|\lambda_2)$ are now eliminated between these two equations using (2). After simplifying, the result is

$$-\frac{a_1}{b_2}\{\varphi'(a_1)\}^2 \frac{\alpha_t'(\lambda_1)}{\beta_t'(\lambda_2)} = \frac{h_t(\lambda_1)}{h_t(\lambda_2)} \frac{\text{PROB}\,(\lambda_2|\lambda_2 \text{ is Eve})}{\text{PROB}\,(\lambda_1|\lambda_1 \text{ is Adam})} \tag{6}$$

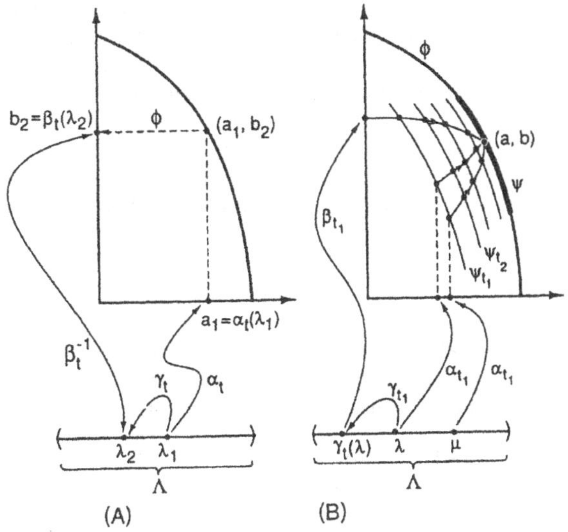

Figure 4: The function γ_t

where $a_1 = \alpha_t(\lambda_1)$ and $b_2 = \beta_t(\lambda_2)$.

To make something of this expression, a list of *substantive* assumptions are necessary. The first imposes the requirement that beliefs be sharp when t is large. It is that PROB $(\lambda|\lambda$ is Adam$) \to 1$ and PROB $(\lambda|\lambda$ is Eve$) \to 1$ as $t \to \infty$, for each λ. The remaining assumptions are about smoothness and convergence. It will be required that $h_t \to h$, $\alpha_t \to \alpha$, $\alpha_t' \to \alpha'$, $\beta_t \to \beta$, $\beta_t' \to \beta'$, $\gamma_t \to e$ and $\gamma_t' \to e' = 1$ as $t \to \infty$, where e is the identity function, and convergence is uniform on compact subsets of Λ.

Take a fixed value of $\lambda \in \Lambda$, and write $a = \alpha(\lambda)$ and $b = \beta(\lambda)$. Suppose that $\lambda_1 \to \lambda$ as $t \to \infty$ in (6). Then $\lambda_2 = \gamma(\lambda_1) \to \lambda$ as $t \to \infty$. Thus, $a_1 \to a$ and $b_2 \to b$ as $t \to \infty$. Moreover, $\gamma_t'(\lambda_1) = \alpha_t'(\lambda_1)\varphi_1(a_1)/\beta_t'(\lambda_2)$, and hence $\beta_t'(\lambda_2)/\alpha_t'(\lambda_1) \to \varphi'(a)$ as $t \to \infty$. In the limit as $t \to \infty$, (6) becomes

$$a\varphi'(a) + b = 0\,, \tag{7}$$

and hence (a, b) is necessarily the *Nash bargaining solution* (n_A, n_E). This is what the argument set out to prove.

It should be emphasized that the conclusion is by no means inevitable. Different conclusions will result if sufficient irregularities are permitted. One might, for example, make the space Λ discrete, or allow the graph of Ψ_t to wriggle as it approaches the graph of φ. (See Figure 4B.) The latter is ruled out in the argument by the insistence that $\gamma_t' \to 1$ as $t \to \infty$. However, all is not completely lost without this strong requirement. It remains the case that α and β are *constant*.[9] This is a result of some significance.

It implies that, in the long run, everybody will be planning to do much the same thing as each other when cast as Adam, and much the same thing as each other when cast as Eve. Thus Pareto inefficiencies due to coordination failures are ruled out.

[9] If $\alpha'(\lambda) > 0$ or $\beta'(\lambda) > 0$, then $\beta_t'(\lambda_2)\alpha_t'(\lambda_1) \to \beta'(\lambda)\alpha'(\lambda) = \Psi'(\lambda)$ as $t \to \infty$, where Ψ is defined locally by $\Psi = \beta \circ \alpha^{-1}$. Since $\gamma(\lambda) \to \lambda$ as $t \to \infty$, Ψ is simply a restriction of φ, and hence $\Psi'(a) = \varphi'(a)$ and the result follows.

More surprising perhaps is a second implication. To understand this second implication, return to the case in which what distinguishes the actors to begin with is whether they are dealt a joker or a knave from a deck of cards in which the fraction of jokers is p. After a long time interval t, nearly all actors will experience a history that leads them to estimate p fairly accurately. But some actors will have atypical histories that lead them to estimate p inaccurately as being close to q. Nevertheless, both typical and atypical actors will *do* much the same thing. Thus the actual realization of the chance move that selects p is not relevant to the final result. Chance therefore plays no role in determining what the actors coordinate on. What matters is only their *a priori* common understandings.

3 Conclusions

Some readers may think that the results in this paper and its predecessor, Binmore [4], are too flimsy to admit the generalizations proposed in this brief conclusion. In reply, I can do no better than to quote the following immortal lines from Plato's *Timaeus*:

> ...we may venture to suggest that what has been said by us is probable and will be rendered more probable by investigation. Let us affirm this.

What I would like to affirm is the following:

- Insofar as it is meaningful to speak of a social contract, one should seek its origins in the biological and social history of the human species.
- Biological and social librations should not be expected to select the same equilibrium.
- Utilitarian outcomes are possible when man is modeled as *homo behavioralis* and evolution is biological.
- Utilitarian outcomes should not be anticipated when man is modeled as *homo economicus* and evolution is social.
- *Homo sapiens* is more like *homo economicus* than *homo behavioralis*.
- We should be pleased at not being specimens of *homo behavioralis* living in a utilitarian society. Such a society would be unacceptably brittle because of its inflexibility in the face of unanticipated change.

References

[1] R. Aumann. Correlated equilibrium as an expression of Bayesian rationality. *Econometrica*, 55:1–18, 1987.

[2] K. Binmore. Modeling rational players, I and II. *Economics and Philosophy*, 3 and 4:179–214 and 9–55, 1987.

[3] K. G. Binmore. Social contract I: Harsanyi and Rawls. *Economic Journal*, 99:84–102, 1989.

[4] K. G. Binmore. Social contract III: evolution and utilitarianism. *Constitutional Political Economy*, 1:1–26, 1990.

[5] J. Harsanyi. Cardinal utility in welfare economics and in the theory of risk-taking. *Journal of Political Economy*, 61:434–435, 1953.

[6] J. Harsanyi. Cardinal welfare, individualistic ethics, and the interpersonal comparison of utility. *Journal of Political Economy*, 63:309–321, 1955.

[7] J. Harsanyi. Ethics in terms of hypothetical imperatives. *Mind*, 47:305–316, 1958.

[8] J. Harsanyi. Games with incomplete information played by 'Bayesian' players, Parts I - III. *Management Science*, 14:159–182, 1967.

[9] J. Harsanyi. *Rational Behavior and Bargaining Equilibrium in Games and Social Situations*. Cambridge University Press, Cambridge, 1977.

[10] J. Nash. The bargaining problem. *Econometrica*, 18:155–162, 1950.

[11] P. Young. *The Evolution of Bargaining Conventions*. Technical Report, University of Maryland, 1989.

Sex Differentiation in Annuities: Reflections on Utilitarianism and Inequality

Kenneth J. Arrow
Joan Kenney Professor of Economics
Stanford University
Stanford, California

No contemporary has done more to provide a foundation for utilitarianism or to argue more forcefully for its value in the determination of social morality than John Harsanyi. This little study is gratefully dedicated to him.

I consider here a simple problem in applied ethics and ask the implications of a utilitarian analysis. Simple as the issue is, it turns out to create some interesting questions as to how utilitarianism should be interpreted in application when the moral agents are not all alike and as to the meaning of equality and its linkages with utilitarianism. I am more interested in provoking questions than answering them.

The principle that men and women should, as such, be treated identically will be taken by all of us as axiomatic. Gender, like race, is a morally irrelevant category. In the utilitarian formulation, the sum of utilities runs over both genders equally.

However, there is a moral problem created when morally irrelevant categories happen to be correlated with categories that are less clearly irrelevant. Let us take the recurring discussion of sex differentiations in pensions and annuities. It happens to be statistically true, in most though not all societies that women have greater life expectancy than men. Consider then a man and a woman who at the age of retirement have accumulated equal claims to pensions. They have, we may stipulate, worked equally hard all their lives, have received equal compensation, and have identical needs by any reasonable definition. On an actuarial basis, the woman will receive a smaller annual payment, though of course for a longer period of time. Is this just?

Put this way, there seem to be conflicting arguments. The woman is being forced to lead a more constricted life. On the other hand, she already has an advantage; I assume that living is preferable to dying. Then an egalitarian might argue that the man should be compensated for the natural inequality.

There is one relevant argument that I want to abstract from for the point of this discussion. Sex is not the only reason for differences in longevity. Personal habits, such as smoking, exercise, and choice of foods, and hereditary factors such observed longevity of parents and grandparents also can be used to predict the survival distribution. It can be argued that use of sex alone to differentiate annuities is discriminatory.

I therefore take the case where the survival distribution of each individual is known. The actual age of death is unknown, a random variable whose distribution differs from one individual to another, but there is common knowledge of all the distributions.

Before setting forth the model, there is one more remark. An economist might object to the absence of differentiated annuities for given premium payments as interfering with efficiency. However, this objection is not properly an objection to the equal-annuity solution as such, if it is indeed the just solution. It is rather a question about the achievement of the just solution by decentralized decision-making. We will see that, properly speaking, there is no contradiction between efficiency and justice.

To take the simplest situation, we assume that each individual lives for either one or two periods. Let p_i be the probability individual i lives for two periods. Let c_{1i} be the consumption of individual i in period 1. Let c_{2i} be individual i's consumption in period 2 if he or she is living then. Let U(c) be the utility derived from consumption by any individual in any period during his or her life. It is the same in either period, so that there is no time preference. Utility in a period when the individual is not alive is taken to be 0; U(c) is assumed to be positive otherwise. Then the expected utility of individual i is,

$$U(c_{1i}) + p_i\, U(c_{2i}). \tag{1}$$

On the supply side, I assume a fixed volume of resources which can be used in either period. That is, resources can be carried forward from one period to the next with neither gain nor loss. Thus, the productivity of investment is zero. The consumption of individual i in period 2 is a random variable, being c_{2i} with probability p_i and 0 with probability 1-p_i, and therefore has an expected value, $p_i\, c_{2i}$. The expected value of total consumption in the two periods is thus,

$$c_{1i} + \Sigma\, p_i\, c_{2i}.$$

I assume a large economy, so that by the law of large numbers, expected consumption equals actual consumption. If R is the total volume of resources, then the resource constraint is,

$$\Sigma\, c_{1i} + \Sigma\, p_i\, c_{2i} = R. \tag{2}$$

Let us find the utilitarian solution to this moral dilemma. It would take as maximand the sum of the expected utilities (1).

$$\text{Max } \Sigma\, U(c_{1i}) + \Sigma\, p_i\, U(c_{2i}) \text{ subject to (2).} \qquad (3)$$

Straightforward Lagrangean analysis shows us that,

$$c_{1i} = c_{2i} = c_{1j} = c_{2j} \text{ for all i and j,} \qquad (4)$$

in total agreement with the view that annuities should not depend on longevity.

This is clearly inegalitarian in terms of individuals conceived of as living an entire life. Expected utility as given by (1) is an increasing function of the probability of survival. What we have seen, then, is that utilitarian and egalitarian arguments lead to apparently different conclusions.

Introducing the economist's intuition that individuals should be charged their respective actuarial values for annuities does not change the result but only sharpens the measure of inequality. Obviously the utilitarian optimum is necessarily efficient and therefore can be decentralized with a price system. If each individual is given an income and told to purchase an annuity (here symbolized by second-period consumption), then the price of the annuity will (in a large economy) be the probability of survival of that individual. Let c be the common value of all individual's consumption in both periods, as described in (4). Let individual i be given income $y_i = (1+p_i)\, c$. Then individual i will in fact consume $c_{1i} = c$ in period 1 and purchase annuity $c_{2i} = c$, as can easily be seen by maximizing (1) subject to the budget constraint,

$$c_{1i} + p_i\, c_{2i} = (1+p_i)\, c.$$

Thus in a market system, the utilitarian optimum leads to inequality measured by money incomes y_i.

Yet there persists an element of egalitarianism in the conclusion. One way of looking at it is to say that (4) can be stated as equality of person-years, though not of persons.

Another way to consider the allocation problem is as one of dynamic programming. Suppose that, however arrived at, there is a total allocation of C_1 to consumption in period 1, so that $R - C_1$ is available for period 2. In period 2, there are some survivors. Clearly either egalitarianism or utilitarianism would imply that the survivors share equally. Clearly then the annuity is the same for all survivors. This approach however leaves open the possibility that period 1 consumptions might differ by way of compensation, with those more likely to survive receiving a lower period 1 consumption, in effect, paying a higher premium. The dynamic programming approach can be put in utilitarian terms, in which case it generalizes the one already taken by avoiding reliance on the law of large numbers.

Specifically, the number of period 2 survivors, n, is a random variable. Let P(n) be the probability that there are n survivors. Then by the argument of the last paragraph, if there are n survivors, each one receives a consumption of $(R - C_1)/n$ in the second period. Hence, the total utility of second-period consumption by the survivors is,

$$n\, U[(R-C_1)/n],$$

and therefore the sum of expected utilities of all individuals in both periods is,

$$\Sigma\, U(c_{1i}) + \Sigma\, P(n)\, n\, U[(R - C_1)/n]. \qquad (5)$$

For given C_1, we choose clearly $c_{1i} = C_1/N$, where N is the total initial size of the population. We then choose C_1 to maximize (5), which is equivalent to the statement that the marginal utility of period 1 consumption equals the expected marginal utility of of the period 2 consumption of the survivors. This clearly generalizes (4) to the case where the actuarial expectations of the number of survivors are not very accurate.

Is there a paradox here? The conclusion simply restates the conclusion of utilitarians such as Edgeworth that more should be given to those with higher ability to transform consumption into utility. In this case superior longevity causes more efficiency as a "pleasure-machine." Although the avoidance of sex differentiation in annuities was seen by its proponents as based on egalitarian considerations, I think the real lesson is that, in this case at least, utilitarianism is a better guide than egalitarianism.

Utilitarianism, the Difference Principle, and the Veil of Ignorance: An Application of the Theory of Social Situations

John E. Roemer
Department of Economics, University of California
Davis California 95616 USA

1. Introduction

Both Harsanyi and Rawls have been credited with introducing the idea of the veil of ignorance as a thought experiment to motivate deliberation about the correct standard of justice for a society.[1] Harsanyi argues that behind the veil of ignorance, 'agents' would (or should) choose the utilitarian social choice rule, and Rawls argues they would choose the maximin rule. In this paper, I use Joseph Greenberg's (1990) innovative theory of social situations to study the problem of choice of income distribution behind a veil of ignorance.

The model is adapted from Howe and Roemer (1981). In that paper, we claimed to show that the maximin income distribution was the unique point in the core of a game that agents could be conceived of as playing behind the veil of ignorance. The core was taken to model the notion of a social contract, reflecting Rawls's contractarian views. In fact, however, we really showed that the maximin allocation was a 'core-like' outcome in a 'game-like' situation. With hindsight, it now appears that Greenberg's 'social situation' is what we really needed.

I will pose the problem more precisely after reviewing Greenberg's model in the next section. The reader will have to judge for himself whether the 'theory of social situations' sheds new light on the issues.

2. The theory of social situations[2]

For convenience, this section presents the key definitions of Greenberg's theory. In Greenberg's analysis, it is assumed that coalitions have a finite number of agents, and the environment is deterministic. In our application, we

---

[1] In fact, Vickrey (1945) proposed the basic idea somewhat earlier.

[2] This extremely abbreviated presentation will not suffice to teach the theory to those who have not learned it from Greenberg's papers and book.

will relax both of these assumptions; the necessary modifications of Greenberg's concepts will be introduced at that time.

Definition 1. A position, G, is a triple $G=(N(G), X(G), \{v^t(G)\}_{t\varepsilon N(G)})$, where N(G) is the set of players, X(G) is the set of all potentially possible outcomes, and $v^t(G):X(G)\to R$ is the utility function of player t in position G over the outcomes. (Thus, for all $x,y\varepsilon X(G)$, and for all $t\varepsilon N(G)$, $v^t(G)(x)>v^t(G)(y)$ if and only if t prefers, in position G, the outcome x to the outcome y.)

Definition 2. A situation is a pair $(\mathscr{F},\Gamma)$ where Γ is a set of positions, and the mapping $\mathscr{F}$, called the inducement correspondence, satisfies: for all $G\varepsilon\Gamma$, $S\subset N(G)$, and $x\varepsilon X(G)$, $\mathscr{F}(S|G,x)\subset\Gamma$, and if $H\varepsilon\mathscr{F}(S|G,x)$ then $S\subset N(H)$.

Definition 3. Let Γ be a collection of positions. A mapping σ that assigns to each position $G\varepsilon\Gamma$ a subset $\sigma(G)$ of X(G) is called a standard of behavior (SB) for Γ. The set $\sigma(G)$ is called the solution for G (assigned by σ).

Definition 4. Let σ be a SB for the situation $(\mathscr{F},\Gamma)$. We say that σ is an optimistic stable standard of behavior (OSSB) for $(\mathscr{F},\Gamma)$ iff for all $G\varepsilon\Gamma$: $y\varepsilon X(G)\backslash\sigma(G)$ if and only if there exist $S\subset N(G)$, $H\varepsilon\mathscr{F}(S|G,y)$ and $z\varepsilon\sigma(H)$ such that $v^t(H)(z)>v^t(G)(y)$ for all $t\varepsilon S$.

Definition 5. Let σ be a SB for the situation $(\mathscr{F},\Gamma)$. We say that σ is a conservative stable standard of behavior (CSSB) for $(\mathscr{F},\Gamma)$ iff for all $G\varepsilon\Gamma$: $y\varepsilon X(G)\backslash\sigma(G)$ if and only if there exist $S\subset N(G)$ and $H\varepsilon\mathscr{F}(S|G,y)$ s.t. $\sigma(H)\neq\emptyset$ and for all $z\varepsilon\sigma(H)$, $v^t(H)(z)>v^t(G)(y)$ for all $t\varepsilon S$.

The inducement correspondence summarizes the socially acceptable or legal opportunities that a coalition S has available to it, when facing a particular outcome x in a position G. I will also call this the set of rules by which society lives. It is best to think of a stable standard of behavior as an equilibrium concept associated with a particular inducement correspondence. Informally, a standard of behavior σ, which recommends certain outcomes in any position, is stable relative to an inducement correspondence $\mathscr{F}$ if, when faced with an outcome that σ recommends, no coalition has the motivation to object, given the possibilities available to it as summarized by $\mathscr{F}$. 'Optimistic' and 'conservative' are two decision procedures that members of the coalition may use to evaluate whether it should register an objection. Below, I shall introduce an expected-utility decision procedure as well.

3. The economic environment

For the construction, we work with a continuum of agents. I assume a finite-type economy, where there is a continuum of agents of each of a finite number of types. Let the set of agents be $I=[0,1]$. There are r types; the Lebesgue measure of type i agents is p_i, $\sum p_i=1$. We view a set of agents of measure 1, of any type, as possessing 1 unit of labor. A type i is characterized by a utility function $u^i(Y,L)$ and a skill level s^i. One good (Y) is produced from labor (L) according to the individual production functions $Y=s^iL$.

A tax scheme is a function $\tau(i,Y)$ that assigns to a unit mass of agents of type i who have produced income Y a tax τ. The agent's optimization problem, facing tax scheme τ, is:

$$\max_L u^i(s^iL-\tau(i,s^iL), L).$$

Call the solution $L(i,\tau)$, and define $Y(i,\tau)\equiv s^iL(i,\tau)$. We interpret $(L(i,\tau),Y(i,\tau))$ as the labor expended and output produced by a set of agents of type i of unit measure.

A feasible tax scheme is a function $\tau(i,Y)$ such that

$$\sum_i p_i\tau(i,Y(i,\tau)) = 0 . \qquad (3.1)$$

Define the maximin utility for the economic environment $\xi=((p_1,u^1,s^1),\ldots,(p_r,u^r,s^r))$ as

$$\overline{u}_{min}=\sup_{\text{feasible } \tau} \min_i u^i(Y(i,\tau)-\tau(i,Y(i,\tau)), L(i,\tau)) . \qquad (3.2)$$

We assume that the supremum in (3.2) is achieved for some feasible τ. It follows that $\overline{u}_{min}$ is achieved for a measurable subset of I (of measure at least $\min p_i$). There may be more than one feasible tax scheme that achieves $\overline{u}_{min}$; call the set of feasible tax schemes that achieve $\overline{u}_{min}$, $\tilde{T}$. These are the maximin tax schemes.

Similarly, for any feasible tax scheme we may define the integral of utility as:

$$\sum_i p_iu^i(Y(i,\tau)-\tau(i,Y(i,\tau)),L(i,\tau)) \qquad (3.3)$$

and define u^* as the supremum over feasible τ of this integral. This supremum is achieved by a set of tax schemes T^*, which we call the utilitarian tax schemes.[3]

---

[3]For both 'maximin' and 'utilitarian' concepts to make sense, we must assume that the utility functions are both cardinal unit comparable and level comparable. We shall assume even more below. Note that the sum (3.3) does not require the utility functions to be von Neumann Morgenstern, as there are no expectations being taken.

We assume that following (extremely weak) non-degeneracy assumption:

A0. *Let τ be any tax scheme for the economic environment* (p,u,s). *Let* $i(\tau)$ *be a type whose utility is lowest at* τ, *call it* $\underline{u}$. *Then there exists a tax scheme* τ' *(perhaps equal to* τ*) at which the best-off type has utility greater than* $\underline{u}$.

Call $(p_1,\ldots,p_r)=f$ a distribution and $\Delta=((u^1,s^1),\ldots,(u^r,s^r))$ a set of characteristics. A coalition is a measurable non-null subset of I. Let B be a coalition of measure $\lambda(B)$, where λ is Lebesgue measure. Consider the class of measurable functions $\pi:B\to\{1,2,\ldots,r\}$ such that

$$\lambda(\pi^{-1}(i))=p_i\lambda(B) \quad \text{for all } i=1,r. \tag{3.4}$$

Call this class $\Pi(B)$. Call $\pi\epsilon\Pi(B)$ a draw of the distribution f for the coalition B. Any draw assigns the r characteristics to 'agents' in the coaltion B in the proportions dictated by the distribution f.

We define the realization of (f,Δ) for B by π as the society (f,Δ,B,π), whose agents are the set B, and agent $t\epsilon B$ is endowed with characteristic $(u^{\pi(t)},s^{\pi(t)})$, where $\pi\epsilon\Pi(B)$. For short, I will write (B,π) for a society, as (f,Δ) will be fixed throughout. Each realization of (f,Δ) for B by some π induces an allocation for B with respect to any feasible tax scheme τ: call the allocation $B(\pi,\tau)$. Note that feasibility for τ on B is defined by

$$\textstyle\sum_i p_i\lambda(B)\tau(i,Y(i,\tau))=0$$

which is equivalent to (3.1); and so all societies (B,π), as π varies, have the same set of feasible tax schemes. In particular, once (f,Δ) is given, then for any (B,π) with $\pi\epsilon\Pi(B)$, the maximin utility is $\overline{u}_{min}$ as defined in (3.2). It is in order to make the same set of feasible tax schemes available to all coalitions that we must work with a continuum of agents.

The full society, (f,Δ,I,π), is a realization of (f,Δ) for I by some $\pi\epsilon\Pi(I)$. The associated allocation for I is (f,Δ,I,π,τ), where τ is the feasible tax scheme that is chosen. At an allocation, the agents $t\epsilon\pi^{-1}(i)$, who are endowed with characteristic (u^i,s^i), are equally treated and expend in total $p_iL(i,\tau)$ labor and consume $p_i(Y(i,\tau)-\tau(i,Y(i,\tau))$ output. Behind the veil of ignorance agents in I know (f,Δ), but not π. Their problem is to agree upon a choice of τ, before knowing π. Then some $\pi\epsilon\Pi(I)$ is drawn by nature, and the allocation (f,Δ,I,π,τ) is realized.

4. The decision problem as a social situation

Because their decision problem depends upon the draw π that nature chooses (the 'birth lottery'), it will be necessary to define a probability measure on the class of draws that each coalition faces. We view the agents of the economy as points in the interval [0,1]. They live in the Borel probability space $([0,1], \mathcal{B}, \lambda)$, where $\mathcal{B}$ is the Borel σ-field and λ is Lebesgue measure. It is assumed that, for any $B \varepsilon \mathcal{B}$, there is a probability space $\xi_B = (\Pi(B), \mathcal{F}_B, P_B)$, where $\mathcal{F}_B$ is a σ-field of draws in $\Pi(B)$, with the property that 'the probability that nature chooses a draw that assigns a given agent $t \varepsilon B$ to type i is p_i.' That is, define:

$$\text{for any } t \varepsilon B, \qquad D_i(t) = \{ \pi \varepsilon \Pi(B) \mid \pi(t) = i \} ;$$

$D_i(t)$ is in $\mathcal{F}_B$, and $P_B(D_i(t)) = p_i$. We will assume that the birth lottery, which assigns souls behind the veil of ignorance to types, is governed by this probability distribution.[4]

We must now reformulate the definitions of CSSB and OSSB to make sense in this continuum environment. When we write "for a.a. agents in S," we mean "for all $t \varepsilon S$ except for a λ-null set." When we write "for a.a. $\pi \varepsilon \Pi(B)$," we mean "for all $\pi \varepsilon \Pi(B)$ except for a P_B-null set of draws." Now the natural reformulation of OSSB is to allow a coalition S to 'block' an allocation if it can choose a tax scheme according to the inducement correspondence such that there is a non-null P_B-measurable event in $\mathcal{F}_B$ under which almost all (λ) its members do better than at the given allocation. The reformulation of CSSB is to allow a coalition S to block an allocation only when it can choose a tax scheme under the inducement correspondence such that under almost all (P_B) draws in $\Pi(B)$, almost all (λ) its members do better than at the allocation. (Note, we never need a measure on the space of tax schemes, since the choice of a tax scheme is deterministic.).

Recall that we shall denote an allocation (f,Δ,B,π,τ) as $B(\pi,\tau)$, for any coalition $B \subset I$. Define, for each (B,τ), a position $G^{B,\tau}$ as follows: $N(G^{B,\tau})=B$, $X(G^{B,\tau})=\{B(\pi,\tau) \mid \pi \varepsilon \Pi(B)\}$, and for any $t \varepsilon B$:

$$v^t(G^{B,\tau})(B(\pi,\tau))=u^{\pi(t)}(Y(\pi(t),\tau)-\tau(\pi(t),Y),\ L(\pi(t),Y)) := u^{\pi(t)}(\tau).$$

---

[4] It might be casually assumed that, if the population distribution of types is $(p_1,\ldots,p_r)$, then it naturally follows that the probability of receiving type i in the birth lottery is p_i. This is not so, for the latter claim is a statement of a probability measure on a space of functions that assign agents to types; hence, the construction of this paragraph is required. It is indeed not easy to show that the probability spaces ξ_B exist. This construction is carried out in Green (1990).

We abbreviate the utility on the r.h.s. as $u^{\pi(t)}(\tau)$, for once π and τ are given, the utility of agent t is determined.

Let Γ be the set of all such positions:

$$\Gamma=\{G^{B,\tau} \mid B \text{ a coalition of } I,\ \tau \text{ feasible}\}.$$

The interpretation is that, for any coalition and any feasible tax scheme, outcomes are possible for any (measurable) draw π that assigns types Δ to agents in B according to the distribution f. An agent's utility at an outcome is just the utility he derives from his optimal choice of labor and after-tax income, given his realized characteristics in the draw π, and under the tax scheme τ.

We next propose a restriction on the nature of admissible standards of behavior for the veil of ignorance model. No agent, coalition, or social planner can ever control the draw that nature makes, which assigns types to agents. Therefore, we insist that a standard of behavior σ, which recommends some allocations at a position $G^{B,\tau}$ must either choose all of the possible outcomes or the empty set:

A1. *For any* σ, B, *and* τ, *if* $\sigma(G^{B,\tau}) \neq \emptyset$ *then* $\sigma(G^{B,\tau}) = \{B(\pi,\tau) \mid \pi\in\Pi(B)\}$.

We now give a precise definition of a CSSB in this continuum environment:

Definition 5'. Let σ be a standard of behavior for the situation $(\mathfrak{F},\Gamma)$. We say that σ is a conservative stable standard of behavior (CSSB) for $(\mathfrak{F},\Gamma)$ iff for all $G^{B,\tau}\in\Gamma$: $B(\pi,\tau)\in X(G^{B,\tau})\backslash\sigma(G^{B,\tau})$ if and only if there exists a coalition $S\subset B$ and a position $G^{A,\rho}\in\mathfrak{F}(S \mid B(\pi,\tau))$ s.t. $\sigma(G^{A,\rho})\neq\emptyset$ and for almost all π'(w.r.t. P_A), $A(\pi',\rho)\in\sigma(G^{A,\rho})$, $u^{\pi'(t)}(\rho) > u^{\pi(t)}(\tau)$ for almost all (w.r.t. λ) $t\in S$.

Notation. The symbol $\overline{\forall}$ will mean "for almost all."

We may think of a CSSB as recommending certain tax rules for certain coalitions. If it recommends τ for B, then no matter what draw π is realized, the allocation $B(\pi,\tau)$ must be immune to objections that coalitions are allowed to make against it. The form of a coalition's objection is: under some (other) society and tax rule that it is allowed to propose under $\mathfrak{F}$, under a.a. draws by nature, a.a. its members would be better off. It is best to interpret the proposal-objection scenario happening before the (first) draw has been made: the veil of ignorance is not lifted until the social contract has been made.

6. Maximin

We next define two versions of a natural inducement correspondence $\gamma(S\,|\,G,x)\subset\Gamma$ for any $G^{B,\hat{\tau}}\in\Gamma$, $S\subset B$, $B(\pi,\hat{\tau})\in X(G^{B,\hat{\tau}})$, where S is a coalition:

$$\gamma^{L1}(S\,|\,G^{B,\hat{\tau}},\,B(\pi,\hat{\tau}))=\{G^{S,\tau}\,|\,\tau \text{ is feasible}\};$$
$$\gamma^{L2}(S\,|\,G^{B,\hat{\tau}},\,B(\pi,\hat{\tau}))=\{G^{B,\tau}\,|\,\tau \text{ is feasible}\}.$$

Thus, under γ^{L1} at a position $G^{B,\hat{\tau}}$, which we interpret as the set of possible allocations realized for coalition B with tax scheme $\hat{\tau}$ as π varies over the class $\Pi(B)$, any coalition can object to a proposal by arguing that it could withdraw from society, choose any feasible tax scheme <u>for itself,</u> and subject itself to a draw π from $\Pi(S)$. Under γ^{L2}, a coalition can object by proposing any other tax scheme <u>for the whole society</u>, B, of which it is part. The 'L' stands for liberal, for under these inducement correspondences, a coalition can choose <u>any</u> tax scheme. These are the <u>liberal inducement correspondences.</u>

Define the standard of behavior $\tilde{\sigma}$ as follows:

$$v^t(G^{B,\tau})(B(\pi,\tau))=u^{\pi(t)}(Y(\pi(t),\tau)-\tau(\pi(t),Y),\ L(\pi(t),Y)) := u^{\pi(t)}(\tau).$$

$$\tilde{\sigma}(G^{B,\tau})=\begin{cases} X(G^{B,\tau}) \text{ if } \tau\in\tilde{T} \\ \emptyset \text{ if } \tau\notin\tilde{T} \end{cases}.$$

$\tilde{\sigma}$ recommends all possible realizations, over $\pi\in\Pi(B)$, of (f,Δ) for B with respect to the maximin tax schemes, and none otherwise.

<u>Proposition 1</u> *Assume A0 and A1.*

A. $\tilde{\sigma}$ *is the unique conservative stable standard of behavior for* (Γ,γ^{L1}), *and* $\tilde{\sigma}$ *is the unique optimistic stable standard of behavior for* (Γ,γ^{L1}).

B. $\tilde{\sigma}$ *is the unique conservative stable standard of behavior for* (Γ,γ^{L2}). *There is no optimistic stable standard of behavior for* (Γ,γ^{L2}).

<u>Proof:</u>

1. First, we show that $\tilde{\sigma}$ is a CSSB for γ^{L1}; that is, we show that for any $G^{B,\hat{\tau}}$:

$$y\in X(G^{B,\hat{\tau}})\backslash\tilde{\sigma}(G^{B,\hat{\tau}}) \Leftrightarrow (\exists S\subset B)$$

$$(\exists \text{feasible } \tau)(\forall z\in\tilde{\sigma}(G^{S,\tau}))(\forall t\in S)(v^t(G^{S,\tau})(z)>v^t(G^{B,\hat{\tau}})(y))$$

Case 1 $\hat{\tau} \notin \tilde{T}$.

Then $\tilde{\sigma}(G^{B},\hat{\tau})=\emptyset$, so y is any outcome $B(\hat{\pi},\hat{\tau})$. At this allocation for B, let S be the coalition of agents who receive utility less than $\bar{u}_{min}$. It immediately follows that for any $\bar{\tau}\in\tilde{T}$, for any $z\in\tilde{\sigma}(G^{S},\bar{\tau})$, i.e., $z=S(\pi,\bar{\tau})$, and any $\pi\in\Pi(S)$:

$$(\forall t\in S) \quad v^{t}(G^{S},\bar{\tau})(S(\pi,\bar{\tau}))\geq\bar{u}_{min}>v^{t}(G^{B},\hat{\tau})(B(\hat{\pi},\hat{\tau})) \tag{6.1}$$

The converse direction holds automatically, because $\tilde{\sigma}(G^{B},\hat{\tau})=\emptyset$.

Case 2 $\hat{\tau}\in\tilde{T}$.

In this case, $X(G^{B},\hat{\tau}) \setminus \tilde{\sigma}(G^{B},\hat{\tau})=\emptyset$. We need to show that for any $y\in X(G^{B},\hat{\tau})$, say $y=B(\hat{\pi},\hat{\tau})$, there exists no SCB and no τ such that, for almost all $\pi\in\Pi(S)$, for almost all $t\in S$:

$$v^{t}(G^{S},\tau)(S(\pi,\tau) > v^{t}(G^{B},\hat{\tau})(B(\hat{\pi},\hat{\tau}). \tag{6.2}$$

We know that $v^{t}(G^{B},\hat{\tau})(B(\hat{\pi},\hat{\tau})) \geq \bar{u}_{min}$ for almost all $t\in B$ because $\hat{\tau}\in\tilde{T}$. For any coalition S and any τ , for all draws $\tilde{\pi}\in\Pi(S)$, there is a coalition $R_{\tilde{\pi}}$ of S which receives utility less than or equal to $\bar{u}_{min}$: $\forall t\in R_{\tilde{\pi}}$ $v^{t}(G^{S},\tau)(S(\tilde{\pi},\tau)) \leq \bar{u}_{min}$. Therefore, the falsity of (6.2) is shown.

This demonstrates that $\tilde{\sigma}$ is a CSSB for γ^{L1}. Indeed, the reasoning establishes that $\tilde{\sigma}$ is the unique CSSB for γ^{L1}.

2. We next establish that $\tilde{\sigma}$ is the unique OSSB for γ^{L1}. σ is an OSSB for γ^{L1} if and only if for all $G^{B},\hat{\tau}$:

$$\begin{gathered} y\in X(G^{B},\hat{\tau}) \setminus \sigma\,(G^{B},\hat{\tau}) \Leftrightarrow \\ (\exists S\subset B)(\exists G^{S},\tau)(\forall \pi \text{ in a non-null set of } \Pi(S))(\exists S(\pi,\tau)\in\sigma(G^{S},\tau)) \\ (\forall t\in S)(v^{t}(G^{S},\tau)(S(\pi,\tau))>v^{t}(G^{B},\hat{\tau})(y)). \end{gathered} \tag{6.3}$$

This time, we will show that if σ is an OSSB, then $\sigma=\tilde{\sigma}$. To this end, let σ be an OSSB for γ^{L1}.

Case 1 $\hat{\tau}\in\tilde{T}$. Let B be any coalition, and $y\in X(G^{B},\hat{\tau})$, $y=B(\hat{\pi},\hat{\tau})$. We will show that the right-hand side of equivalence (6.3) is false; it will follow that y must also be in $\sigma(G^{B},\hat{\tau})$, and hence $\sigma(G^{B},\hat{\tau})=X(G^{B},\hat{\tau})$. Let S be any coalition in B. We have:

$$(\forall t\in S) \quad v^{t}(G^{B},\hat{\tau})(B(\hat{\pi},\hat{\tau}))\geq\bar{u}_{min} \tag{6.4}$$

because $\hat{\tau}\in\tilde{T}$. At <u>any</u> allocation $S(\pi,\tau)$, there must be a coalition R of S such that

$$(\forall t\in R) \qquad v^{t}(G^{S},\pi)(S(\pi,\tau))\leq \bar{u}_{min}$$

by definition of $\overline{u}_{min}$. Therefore, the right-hand side of (6.3) is false.

Case 2 $\hat{\tau}\notin\tilde{T}$. We show that for any $y\in X(G^{B},\hat{\tau})$, the right-hand side of equivalence (6.3) is true. This implies that $\sigma(G^{B},\hat{\tau})=\emptyset$. Let S be a coalition of B receiving utility less than $\overline{u}_{min}$ at $\hat{\tau}$. Let $\overline{\tau}\in\tilde{T}$, and let $S(\pi,\overline{\tau})$ be any element of $\sigma(G^{S},\overline{\tau})$, which is equal to $X(G^{S},\overline{\tau})$ by Case 1. Then:

$$(\forall t\in S)(v^{t}(G^{S},\overline{\tau})(S(\pi,\overline{\tau}))\geq \overline{u}_{min} > v^{t}(G^{B},\hat{\tau})(B(\hat{\pi},\hat{\tau}),$$

which verifies the claim; hence $\sigma(G^{B},\hat{\tau})=\emptyset$. Furthermore, this holds for any draw π.

From the two cases, we have shown that $\sigma=\tilde{\sigma}$. A similar argument shows that $\tilde{\sigma}$ is, indeed, an OSSB.

3. The argument that $\tilde{\sigma}$ is a CSSB for (Γ,γ^{L2}) follows almost verbatim argument 1. above. One need only replace allocations $S(\pi,\overline{\tau})$ with allocations $B(\pi,\overline{\tau})$.

4. Suppose σ were an OSSB for γ^{L2}. Then for some position $G^{B},\tilde{\tau}$, $\sigma(G^{B},\tilde{\tau})\neq\emptyset$, and let $B(\tilde{\pi},\tilde{\tau})\in\sigma(G^{B},\tilde{\tau})$. Let S be a sub-coalition of the worst-off coalition at $B(\tilde{\pi},\tilde{\tau})$ of measure $\min_i \lambda(B)p_i$. We conclude, from assumption A0, there is some tax scheme τ, and a set of draws $\Pi^{*}\subset\Pi(B)$ of P_B-measure at least $\min p_i$, under which a.a. members of S do better at $B(\pi,\tau)$, for $\pi\in\Pi^{*}$, than they do at $B(\tilde{\pi},\tilde{\tau})$. This contradicts the assumption that σ is an OSSB for γ^{L2}. ■

A comment is in order on why 'the difference principle' is a stable standard of behavior under both 'infinitely risk averse' decision behavior (CSSB) and 'infinitely risk taking' decision behavior (OSSB) for γ^{L1}. It is easy to understand why maximin emerges under CSSB -- this is the usual intuition for why the difference principle is arrived at by infinitely risk averse agents. If the maximin income distribution holds, then a coalition of infinitely risk averse agents will not propose an alternative income distribution, for under that alternative, every agent has a positive probability of being worse off. The same explanation motivates why $\tilde{\sigma}$ is a CSSB for γ^{L2}. That maximin is also stable for γ^{L1} for infinitely risk-loving agents (OSSB) is due to the restriction that coalitions can only object by <u>withdrawing</u> from society, so to speak. A coalition of agents is willing to gamble on an alternative income distribution to maximin as long as there is some positive probability that <u>almost all</u> its agents will gain. It is the italicized words that force maximin under γ^{L1}.

7. Utilitarianism

When a coalition decides to object to an allocation y at position G, it does so only if it can induce another position in which there exists a positive probability that nature will choose a draw at which all its members do better than at y (if it is optimistic), or only if it can induce another position in which under almost every draw, almost all its members do better than at y (if it is conservative). It is natural to introduce a third kind of coalitional evaluation, which I call Bayesian. In this formulation, a coalition will object to a proposed allocation y at position G if and only if almost every member can expect to do better at the new position, where the expectation is taken over the probability distribution of allocations at that position. I will not formulate the definition of a Bayesian SSB completely generally, but just for the veil of ignorance model, where it is clear which probability distribution to use. I propose this notion because it seems to be the natural one that von Neumann-Morgenstern agents would use behind the veil of ignorance. When considering whether to object to a particular allocation $B(\pi,\tilde{\tau})$, the members of a coalition S wish to evaluate whether their expected utility under another tax scheme will be greater than their utility is at $B(\pi,\tilde{\tau})$, where the expectation is taken over a probability distribution on nature's possible draws, $\Pi(A)$, where (A,τ) is the counterproposal that S makes.

Let $G^{B,\tau}$ be a position. The expected utility of an agent $t \varepsilon B$ at that position is naturally defined as follows:

$$\sum_{i=1}^{r} p_i u^i(\tau) = E(\tau). \qquad (7.1)$$

To make sense of (7.1) as expected utility, we must now require that, as well as being cardinal-unit comparable and level comparable, the utility functions are von Neumann-Morgenstern functions over the set of lotteries that the agents face with nature[5]. That is, each agent has preferences over the pairs (i,τ), where i is type, that satisfy the von Neumann-Morgenstern axioms, and over lotteries over $\{(i,\tau)\}$ -- in which τ is certain but i is distributed according to probabilities p_i

---

[5] It is indeed asking alot to require that the same utility functions that give the expected-utility representation of preferences over lotteries are also the ones that are meaningful to sum to get a meaningful total utility in the population, in the sense that utilitarianism recommends. Weymark (in press) convincingly argues that the utilitarianism forthcoming from Harsanyi's theorem on utilitarianism (Harsanyi(1977, Chapter 4) has little to do with the utilitarianism with which philosophers are concerned. But in the rest of this section, we assume that the von Neumann-Morgenstern utility functions also happen to be ones which give meaningful comparisons of 'utility' between people.

-- and the utility of a lottery can be represented by the expression on the l.h.s. of (7.1). We also use the fact that the probability that nature chooses a draw which assigns type i to a given $t \varepsilon B$ is p_i.

Definition 6. Let Γ be the set of positions defined in section 5, and γ an inducement correspondence. A standard of behavior σ is Bayesian stable (BSSB) for (Γ,γ) if and only if for any B and feasible τ:

$$B(\pi,\tau) \varepsilon X(G^{B,\tau}) \setminus \sigma(G^{B,\tau}) \iff$$
$$(\exists S \text{ a coalition in } B)(\exists G^{A,\rho} \ \varepsilon\gamma(S \mid G^{B,\tau},B(\pi,\tau))) \text{ such that}$$
$$E(\rho) > u^{\pi(t)}(x(\pi(t),\tau) \text{ for almost all } t \varepsilon S.$$

We next define the utilitarian[6] SSB as:

$$\sigma^{-\infty}(G^B,\tau)= \begin{cases} X(G^{B,\tau}) \text{ if } \tau \varepsilon T^* \\ \varnothing \qquad \text{if } \tau \notin T^* \end{cases}.$$

What must an inducement correspondence γ look like if $\sigma^{-\infty}$ is a BSSB for it? Let $\tau^* \varepsilon T^*$. Since $G^{B,\tau^*} \setminus \sigma^{-\infty}(G^{B,\tau^*}) = \varnothing$, it must be that no coalition S can block the proposal τ^* at an allocation $B(\pi, \tau^*)$. In particular, let $\underline{S}$ be the coalition of agents who achieve the lowest utility at $B(\pi,\tau^*)$. Then, from the definition of BSSB, it must be that

$$\forall G^{A,\rho} \varepsilon \gamma(\underline{S} \mid B(\pi,\tau^*)) \text{ (there is a coalition T of } \underline{S})$$
$$(E(\rho) \leq u^{\pi(t)}(x(\pi(t),\tau^*)) \text{ for a.a. } t \varepsilon T.$$

(Without loss of generality, we may take $T=\underline{S}$.)

That is, $\underline{S}$ is only allowed to induce, at the allocation $B(\pi,\tau^*)$, tax schemes whose expected utility is not greater than the actual utility members of $\underline{S}$ get in the utilitarian allocation. This is certainly a set of lousy tax schemes to choose from.

I summarize our observations thus far as follows. We interpret the inducement correspondence for a society as a summary of the opportunities that coalitions may put forth to object to a proposed income distribution. The more liberal the inducement correspondence (i.e., the more tax schemes coalitions are allowed to propose as objections to a proposal), the less cause citizens have to be dissatisfied with the equilibrium income distribution (the allocation recommended by the standard of behavior). Because $\tilde{\sigma}$ is a conservative stable

[6]The choice of notation will become clear in section 8.

standard of behavior for γ^{L2} under the most liberal possible inducement correspondence, we conclude that maximin is a very stable social arrangement: badly off citizens have little basis to revolt against the rule γ^{L}. (A similar statement holds for γ^{L1}.)

However, the appeal of the conclusion concerning maximin is weakened by the premise that agents are not expected utility maximizers under CSSB -- they are 'infinitely risk averse.' We now suppose that they are expected utility maximizers, to which corresponds the concept of SSB called Bayesian. We have observed that, with Bayesian decision makers, the utilitarian tax scheme can only be supported as a stable standard of behavior if the opportunities society allows to badly off coalitions are decidedly restricted. Hence, we might conjecture that with Bayesian decision makers any society with a set of rules (γ) for which utilitarianism is a stable standard of behavior is prone to revolt by those who end up badly off after nature's draw: they can argue that the consensus for utilitarianism was purchased only by depriving them of meaningful opportunities to object.

If not utilitarianism, then what standards of behavior are Bayesian-stable for less restrictive (more liberal) inducement correspondences? The above argument shows that a standard of behavior σ is only Bayesian-stable for an inducement correspondence γ if γ restricts the proposals that the worst-off coalition can make to tax schemes whose expected utility is not greater than the utility its members receive at the recommended allocation. It immediately follows that <u>the inducement correspondence that is most liberal for the worst-off coaltion is associated with the standard of behavior</u> $\tilde{\sigma}$. Thus, if one of a Bayesian society's goals is to live by rules (γ) that are most immune to attack from those who are worst-off, then it must adopt a rule γ whose BSSB is maximin.

We can, however, cast utilitarianism and the difference principle in a somewhat different light, which we do next.

8. <u>Maximizing expected utility with a floor</u>

We have thus far studied stable standards of behavior when decision makers are infinitely risk averse (CSSB: a coalition blocks a proposal if and only if a.a. its members receive an improvement with probability one in the birth lottery), infinitely risk loving (OSSB: a coalition blocks a proposal with another one if and only if there is some positive probability in the birth lottery that a.a. its members will gain), and Bayesian (BSSB: block a proposal if and only if a.a. one's members can expect to do better). We now propose a more general class of decision procedures. Coalitions will block a proposal if and only if their members can expect to do better under another tax scheme subject to a floor that no coalition receive utility less than M. The number M will be taken as given for the entire σ-field of coalitions, and can be any number in the interval $[-\infty, \overline{u}_{min}]$. We will say that decision makers are Bayesian^M to describe this kind of behavior[7]. The associated definition of stable standard of behavior is:

Definition 7. A standard of behavior σ is Bayesian^M stable (B^MSSB) for (Γ, γ) if and only if for any coalition B and feasible τ:

$$B(\pi,\tau) \in X(G^{B,\tau}) \setminus \sigma(G^{B,\tau}) \Leftrightarrow$$

$$(\exists S \text{ a coalition of } B)(\exists G^{A,\rho} \in \gamma(S \mid G^{B,\tau}, B(\pi,\tau))) \text{ such that}$$

$$E(\rho) > u^{\pi(t)}(x(\pi(t),\tau)) \quad \text{for a. a. } t \in S$$

$$\text{and } u^{\pi'(t)}(x(\pi'(t),\rho) \geqq M \text{ for a.a. } t\in A, \text{ for a.a. } \pi' \in \Pi(A).$$

Consider the inducement correspondence:

$$\gamma^E(S \mid B(\pi,\tau)) = \{G^{B,\rho} \mid E(\rho) > E(\tau)\}.$$

γ^E seems to be an attractive rule: it allows a coalition to object to an allocation if it can propose another tax scheme for the society that is subjected to the draw under which the generic agent could expect to do better than he does at τ.

Let T^M be the set of tax schemes that maximize expected utility subject to a utility floor of M. Define:

$$\sigma^M(G^{B,\tau}) = \begin{cases} X(G^{B,\tau}) & \text{if } \tau \in T^M \\ \varnothing & \text{if } \tau \notin T^M \end{cases}$$

Proposition 2. *For all* M, σ^M *is a* B^MSSB *for* γ^E.

Proof:
Case 1. $\tau \in T^M$. A coalition S will block the allocation $B(\pi,\tau)$ only if it can propose, according to γ^E, a tax scheme for B whose expected utility is higher

[7] I do not propose Bayesian^M behavior as rational. One justification might be that, with respect to their own welfare, agents are 'rational' in the sense of expected-utility maximization, but they do not wish to live in a society where there are coalitions that are extremely badly off. With this justification, we must view utility as that welfare that derives from one's own consumption bundle; it does not include the welfare effect of the consumption of others.

than $E(\tau)$, and guarantees a floor of M. But there are no such tax schemes. Hence any allocation $B(\pi,\tau)$ is stable.

Case 2. $\tau \notin T^M$. In this case, σ^M recommends no allocation. And it is the case that, for any allocation $B(\pi, \tau)$ the worst-off coalition has available, under γ^E, tax schemes with a floor of M under which almost all its members can expect to do better: any $\tau^* \varepsilon T^M$ will do. ■

If $M = -\infty$, then Bayesian^M behavior is just Bayesian behavior: maximize expected utility, and $T^M = T^*$. In this case, the inducement correspondence in fact offers each coalition no proposals with which to counter the utilitarian tax scheme. When $M = \overline{u}_{min}$, at the maximin tax scheme which σ^M recommends, the inducement correspondence offers coalitions a large set of tax schemes with which to counter: any scheme whose expected utility is higher than expected utility under the maximin scheme. This class need not consist of all feasible tax schemes, although it will in the case of certain well-behaved economic environments. Proposition 2 suggests that apparent lack of opportunities which coalitions are offered to object to proposals, in the case where $M = -\infty$, is not due to the inducement correspondence, but to the decision procedure agents use. The same inducement correspondence generates utilitarianism and the difference principle as stable standards of behavior when 'rationality' of agents is defined by $M=-\infty$ and $M=\overline{u}_{min}$, respectively.

9. Conclusion

Imagine two economic environments which, in terms of the distribution of preferences and skills of their agents, are identical: but we observe that as societies they have as equilibria very different income distributions. Greenberg's theory of social situations encourages us to try and explain this by asking if the rules delineating how coalitions may legally, or politically, object to proposed social arrangments differ in the two societies. The rules are the inducement correspondences, and the equilibria are stable standards of behavior. The exercise here has been to ask: What kinds of rules would give rise, behind the veil of ignorance, to the two main contenders for the just income distribution, utilitarianism and the difference principle? The strategy has been to evaluate the different rules to which the different income distributions correspond as equilibria.

Unfortunately, there is another key variable. Utilitarianism and the difference principle are associated not only with different social rules (inducement correspondences), but with different views on what constitutes rational decision making. If decision makers are infinitely risk averse, then we showed that the difference principle is the natural social equilibrium for a very attractive set of rules, namely, the liberal inducement correspondence. This would seem to give rise to a very stable society, as the opportunities inherent in γ^{L2} are broad.

It would seem, however, that under conditions of decision making under uncertainty, which characterize the veil of ignorance situation, a more appropriate form of rationality is Bayesian. We first showed that utilitarianism can only arise as a social equilibrium in a veil of ignorance situation with Bayesian decision makers if society's rules are very restrictive. This seems to be a mark against utilitarianism. Indeed, if we take maximal freedom for the worst-off in society to object to the income distribution as a desirable feature of social rules, then, even with Bayesian decision makers, maximin is the recommended social equilibrium. (Nevertheless, even an inducement correspondence for which maximin is a stable standard of behavior, when agents are Bayesian, restricts opportunities for the worst-off coalition quite a bit.)

Finally, we proposed a continuum of possible decision behaviors (maximize expected utility subject to a floor of M) that might characterize societies, and showed that there exists a single inducement correspondence, γ^{E}, with respect to which a natural continuum of income distributions (that maximize expected utility subject to a floor of M) are the social equilibria in the respective societies. In this light, utilitarianism and maximin are each stable standards of behavior with respect to the same inducement correspondence, but for different definitions of rationality. Our verdict on which income distribution is justified by the kind of contractual agreement that the veil of ignorance models seems to depend, then, entirely on our view concerning rational decision behavior.

References

E. Green, "Individual-Level Randomness in a Nonatomic Population," Dept. of Economics, University of Minnesota, 1990 (xeroxed)

J. Greenberg, 1990. *Social Situations: A Game-Theoretic Approach*, Cambridge University Press

J. Harsanyi, 1977. *Rational Behavior and Bargaining Equilibrium in Games and Social Situations*, New York: Cambridge University Press

R.E. Howe and J.E. Roemer, 1981. "Rawlsian Justice as the Core of a Game," *American Economic Review* 71, no. 5, 880-95

W. Vickery, 1945. "Measuring marginal utility by reactions to risk," *Econometrica* 13, 319-33.

J. Weymark, in press. "A reconsideration of the Harsanyi-Sen debate on utilitarianism," in J. Elster and J. Roemer (eds.), *Interpersonal Comparisons of Well-Being*, Cambridge University Press

Bernoulli, Harsanyi, and the Principle of Temporal Good

John Broome
Department of Economics, University of Bristol
40 Berkeley Square, Bristol BS8 1HY, England

Section 1

Take a person who has preferences between uncertain prospects. Then between the prospects, there is a *preference relation*:

the person prefers __ to __ or is indifferent between them,

where the blanks are to be filled in with prospects. Expected utility theory lays down a number of axioms for this relation. Let us call the preferences *coherent* if they conform to these axioms. The theory shows that, provided the preferences are coherent, then probabilities and a utility function can be defined for them, having the following two properties. First, the utility of a prospect is the expectation of the utilities of its possible outcomes, evaluated according to the probabilities. And second, for any prospects X and Y, the utility of X is at least as great as the utility of Y if and only if the person prefers X to Y or is indifferent between them. If a utility function has the first property I call it *expectational*. If it has the second I say it *represents* the preference relation. In effect, utility is defined as that which the person maximizes the expectation of, and the axioms ensure that she maximizes the expectation of something.

An expectational utility function is defined uniquely up to increasing linear transformations.[1] The special property of increasing linear transformations is that they preserve the ratio of utility differences. Suppose A, B, C and D are four prospects with utilities $U(A)$, $U(B)$, $U(C)$ and $U(D)$, according to a utility function U. Suppose $V(A)$, $V(B)$, $V(C)$ and $V(D)$ are the utilities of these prospects according to another utility function V that is an increasing linear transform of U. Then the ratio of the utility differences $\{U(A) - U(B)\}$ and $\{U(C) - U(D)\}$ is the same as the ratio of $\{V(A) - V(B)\}$ and $\{V(C) - V(D)\}$. So the ratio of utility differences is not an artifact of the particular utility function that is chosen to represent the preferences. A utility function that is defined up to increasing linear transformations is often called *cardinal*.

---

[1] In most theories, that is. Jeffrey's (1983) is an exception, which I shall have to ignore in this paper.

Amongst the prospects that a person has preferences over, will be special ones whose results are not uncertain, because they lead to a particular outcome for sure. Let us call these *certain prospects*. The outcome of a certain prospect will be good or bad for the person to some degree: it will bring her some amount of wellbeing, harm or benefit. If a particular outcome is good for a person to particular degree, then presumably the certain prospect that certainly leads to this outcome is good for her to the same degree. So we may identify the goodness of a certain prospect with the goodness of its certain result.

Each certain prospect, like any prospect, will have a utility. Let *A*, *B*, *C* and *D* be certain prospects. Suppose *A* has a greater utility than *B*, and *C* than *D*. And suppose the difference in utility between *A* and *B* is greater than the difference in utility between *C* and *D*. Does this necessarily imply that a change from *B* to *A* would bring more benefit to the person than a change from *D* to *C* would? The person's utility function is defined to represent her preferences. I am asking whether it also represents her good to this extent, so that one difference in good is greater than another if and only if the corresponding difference in utility is greater. More generally, is the ratio of differences in good the same as the ratio of utility differences? If so, utility is an increasing linear transformation of good. In that case, we may say that utility represents good *cardinally*. The question of whether a person's utility necessarily represents her good cardinally is the subject of this paper.

It is a vital question for welfare economics, and also for utilitarian ethics in general. (At the end of Section 4 below, I shall mention one of the important issues that depends on the answer to this question.) Consequently, it has attracted a great deal of debate. It has interested John Harsanyi throughout his life. His opinion is that a person's utility does, indeed, necessarily represent her good cardinally. Harsanyi has returned to the argument several times in his life,[2] and his ideas have largely determined the course of the debate.

Though this is an old argument, I hope to shed a new and perhaps unexpected light on it. I shall set out from a formal theorem that Harsanyi published in 1955. Harsanyi used this theorem at the time to support his claim that utility represents good cardinally. I shall extend the theorem to a new application, in order to make a connection between the claim that utility represents good cardinally and a particular view

---

[2] For instance, Harsanyi (1953, 1955, 1975, 1977a, 1977b, 1982, and 1987).

about the make-up of a person's lifetime good. I shall connect the claim with a view I call the 'Principle of Temporal Good': roughly, that the goodness of a person's life as a whole is simply an aggregate of the good that comes to her at particular times in her life. At the end of the paper, I shall briefly discuss the acceptability of the Principle of Temporal Good.

Section 2

My question is whether a person's utility necessarily represents her good cardinally. But as I have posed it so far, this question cannot possibly have the answer yes. For several reasons, it is obviously not true that a person's utility necessarily represents her good cardinally. In this section, I shall bypass the obvious objections. I shall refine the question step by step, to make it into the important and interesting question that really needs to be asked.

The first reason why the question, as I have posed it, cannot have the answer yes is this. A person's utility function, as I have defined it, represents her preference relation. And for most people, a utility function defined this way does not even exist. Most people's preferences do not conform to the axioms of expected utility theory.[3] Consequently, they cannot be represented by an expectational function in the manner of expected utility theory. Evidently, then, a person's good cannot be represented by utility defined this way.

However, though most people's preferences do not conform to the axioms, I believe that the preferences of a fully rational person will do so. I have argued for this claim elsewhere,[4] and in this paper I shall take it for granted. Therefore, the preferences that a person *would* have if she was rational are coherent. So they can be represented by an expectational utility function. Let us, then, take an expectational function that represents the relation

> the person would prefer __ to __ or be indifferent between them
> if she was rational,

and ask whether *it* necessarily represents the person's good cardinally.

The answer to this question is no. Suppose prospect *X* has a greater utility than *Y*. Then we know that the person would prefer *X* to *Y* if she

---

[3] I think the evidence of Tversky and Kahneman (e.g. 1986) is sufficient to make this point. But some authors disagree; see Jeffrey (1987), who quotes Davidson (1980, 273).

[4] Broome (1991, Chapter 5).

was rational. But it does not follow that *X* is better for her than *Y*. The person might rationally prefer the alternative that is worse for her. This means that her utility function may not even represent her good ordinally, let alone cardinally.

One reason the person might rationally prefer the alternative that is worse for her is *altruism*. She might prefer one alternative to another because it is better for someone else, even though it is worse for her herself. Evidently we need to impose another condition. Let us consider the relation

> the person would prefer __ to __ or be indifferent between them, if she was rational and self-interested.

I mean 'self-interested' in a strong sense. I call a person self-interested if and only if she always prefers, of two alternatives, the one she believes to be better for herself. Many people who are self-interested in a common or garden sense are not self-interested in this strong sense. They are not altruistic but they are *imprudent*. They sometimes prefer a smaller amount of their good coming in the nearer future to a greater amount coming in the further future. A person who is self-interested in my sense is not imprudent. (It is a common opinion that imprudence is irrational. A.C. Pigou held this opinion, for instance.[5] If it is correct, then imprudence does not need to be ruled out by the self-interestedness condition, because it is already ruled out by the rationality condition.)

The relation I have just defined may also be represented by an expectational utility function. Will *this* function necessarily represent the person's good cardinally?

Again the answer is no. Even a rational and self-interested person might prefer an alternative that is worse for her to one that is better for her. She might have a mistaken belief. She might prefer the worse alternative because she believes it is better. This belief, although mistaken, may be rational. It might be rationally derived from the information the person has available. We shall have to apply the further condition that the person has beliefs that are not mistaken. So now consider the relation:

> the person would prefer __ to __ or be indifferent between them, if she was rational and self-interested, and if she had no mistaken beliefs.

This relation, too, can be represented by an expectational utility

---

[5] Pigou (1962, 24-26).

function. Will *it* necessarily represent the person's good cardinally?

Yet again the answer is no. If the person was rational and self-interested, and if she had no mistaken beliefs, she would very probably be a quite different sort of person from the sort she actually is. Therefore, different things would probably be good for her, if she attained this remarkable state, than are actually good for in her present condition. I have been progressively refining the person's preference relation by adding counterfactual conditions. Several authors have tried to come at a person's good by refining preferences in this way.[6] But, at least if they use conditions like the ones I have applied, this strategy will be unsuccessful.[7]

Nevertheless, we are very close now to a relation that properly captures the person's good. Rather than adding more and more conditions, let us instead move directly to the person's good itself. Let us work with the relation:

__ is at least as good as __ for the person.

I call this the person's *betterness relation.*

The blanks in this formula are, as usual, to be filled in with prospects, whose results may be uncertain. But one might doubt that the notions of goodness and betterness can legitimately be applied to uncertain prospects. One might think that only events that actually occur can be good or bad, better or worse. I have argued elsewhere, however, that this doubt is misplaced, and the notions are properly applicable to uncertain prospects. Furthermore, I have argued in the same place that a person's betterness relation, applied to prospects, will satisfy the axioms of expected utility theory: betterness is coherent, we may say.[8] Therefore, there will be an expectational utility function that represents it. From now on, when I mention a utility function, I mean an expectational function that represents a person's betterness relation. In effect, this utility is defined as that which it is best for the person to maximize the expectation of.

Such a utility function is defined to represent the person's good *ordinally.* That is to say, one prospect or outcome has at least as great a utility as another if and only if it is at least as good. The subject of this paper is the question: does this utility function also necessarily represent the person's good *cardinally*? Because the function is expectational, it is in itself a cardinal utility function: it is unique up to

[6] For instance, Brandt (1979).
[7] See Griffin (1986, 11).
[8] Broome (1991, Chapter 6).

increasing linear transformations. Expected utility theory tells us this. But that does not mean it necessarily represents good cardinally. It need not be an increasing linear transform of the person's good. Whether it is or not is something we shall have to discover by argument.

Section 3

Consider:

> *Bernoulli's Hypothesis*. Of two alternative prospects *X* and *Y*, *X* is at least as good for a person as *Y* if and only if the expectation of the person's good is at least as great under *X* as it is under *Y*.

Daniel Bernoulli believed that it was best for a person to maximize the expectation of her '*emolumentum*'[9] (he wrote in Latin), and the lexicon translates *emolumentum* as benefit or advantage - good, in my terms. So I feel justified in attributing this hypothesis to him.

It is equivalent to the claim that a person's utility function necessarily represents her good cardinally. To see this, notice first that Bernoulli's Hypothesis says it is best for a person to maximize the expectation of her good. But the person's utility is defined as that which it is best for her to maximize the expectation of. According to Bernoulli's Hypothesis, therefore, the person's good is, in fact, a utility function for her. Consequently, any utility function is an increasing linear transform of her good. So her utility represents her good cardinally.

Conversely, suppose a person's utility represents her good cardinally. Then her good is an increasing linear transform of her utility. But any increasing linear transform of an expectational utility function is another expectational utility function that represents the same betterness relation. So the person's good is an expectational utility function for her. Therefore, by the definition of expectational utility, the better of two prospects is the one with the greater expectation of good. And this is Bernoulli's Hypothesis.

Consequently, the question addressed in this paper is, put differently: is Bernoulli's Hypothesis true?

Let us have an example. Compare the two prospects shown in Table 1.

---

[9] Bernoulli (1738). The translator of Bernoulli's paper in *Econometrica* translates *emolumentum* as 'utility', thereby thoroughly obscuring the question that is the subject of this paper: whether utility, as expected utility theory defines it, measures good cardinally.

Table 1

States of nature		States of nature	
H	*T*	*H*	*T*
£1,000	£1,000	£2,100	£100

Prospect *A* Prospect *B*

The two states of nature are equally probable. Suppose the prospects *A* and *B* happen to be equally good for the person. Then, for an expectational utility function *U* representing her betterness relation, they must have equal expected utility:

$$U(1000) = \{U(2100) + U(100)\}/2.$$

That is, $$U(2100) - U(1000) = U(1000) - U(100).$$

So these utility differences are equal. Bernoulli's Hypothesis implies that the differences in good are also equal:

$$G(2100) - G(1000) = G(1000) - G(100).$$

Bernoulli's Hypothesis can very plausibly be denied. One very plausible objection is that a person's good is not an arithmetic quantity in the first place. It is a vaguer notion, too vague to allow the precise comparison of expectations of good required by the hypothesis. And even if these comparisons are possible, a second objection is that it is very plausible to think that the better of two alternatives for a person is not necessarily the one with the greater expectation of her good. That implies risk neutrality about good, which may seem dubious. It is plausible that, say, prospects that are less risky in terms of good are better than more risky ones. If this is so, then utility will be a strictly concave function of good. In the example, $\{G(2100) - G(1000)\}$ will be greater than $\{G(1000) - G(100)\}$.

Because of these doubts, Bernoulli's Hypothesis fell into disrepute in the first half of the twentieth century. Expected utility theory lost popularity as a result.[10] When the axiomatic version of the theory came into being around 1950,[11] there was a good deal of confusion, largely caused by a muddle over the word 'utility'.[12] Axiomatic expected utility

[10] Tintner (1942), for example, was careful to avoid it in analysing choices under uncertainty.

[11] Particularly with the work of Neumann and Morgenstern (1947) and Savage (1954).

[12] The confusion is documented in Ellsberg (1954) and Broome (forthcoming).

theory (applied to the betterness relation) defines a person's utility in such a way that the better of two prospects for her necessarily has the greater expectation of utility. But economists also tend to use 'utility' to stand for a person's good. So it seemed to many of them that expected utility theory implied that the better of two prospects for a person necessarily has the greater expectation of her good. This is Bernoulli's Hypothesis. So these economists felt they had either to accept Bernoulli's Hypothesis or reject expected utility theory. For instance, Maurice Allais' dislike of Bernoulli's Hypothesis seems to have been his main motivation in attacking expected utility theory.[13]

Actually, however, axiomatic expected utility theory does not imply Bernoulli's Hypothesis.[14] One may consistently accept the theory and reject the hypothesis. And the consensus reached amongst economists after a few years, very well represented in an article by Daniel Ellsberg published in 1954, was that that was the right thing to do. Expected utility theory became widely accepted. But the view - equivalent to Bernoulli's Hypothesis - that utility is a cardinal measure of good was widely rejected.[15]

There is, nevertheless, an argument to be made for Bernoulli's Hypothesis. Go back to the example of Table 1. Determining which of *A* or *B* is better for the person is a matter of weighing two considerations against each other. In favour of *B* is a one-in-two chance of getting £2,100 rather than £1,000. In favour of *A* is a one-in-two chance of getting £1,000 rather than £100. To speak generally, assessing alternative prospects in the face of uncertainty is a matter of weighing possible gains and possible losses against each other. The argument for Bernoulli's Hypothesis is that it is precisely this sort of weighing up that gives the notion of a person's good a quantitative meaning. In the example, we assumed that the two alternatives were equally good for the person. That is to say, the consideration in favour of *B* - £2,100 instead of £1,000 - exactly balances the consideration in favour of *A* - £1,000 instead of £100. It would be very natural to express this fact by saying that the difference in good for the person between £2,100 and £1,000 is the same as the difference in good between £1,000 and £100. To say that the differences in good are equal may simply *mean* that they balance when weighed against each other in evaluating prospects.

In responding to Bernoulli's Hypothesis earlier, I said that the former difference in good might actually be greater. But - it is

---

[13] Allais (1979) - a translation of a paper written in 1953.
[14] Despite Allais's attempts to prove otherwise; see Broome (1985).
[15] Arrow (1963, 10) stated the consensus view clearly.

reasonable to ask - what could this mean if it is not that it counts more in judging which of the alternative prospects is better? And actually we know that it counts exactly the same: the two prospects are equally good, so the two considerations are exactly balanced. Evidently I was trying to maintain a distinction between quantities of good and how much these quantities *count* in judgements under uncertainty. But this distinction may be empty. It is hard to see what use we can have for the notion of quantities of good except when we are weighing up differences in good in evaluating alternatives.

I raised two plausible objections to Bernoulli's Hypothesis. One was that good may not be a sufficiently quantitative notion. To this, the argument responds that judgements under uncertainty may actually be what makes good quantitative to the required extent. The other objection was that neutrality does not seem to be the only appropriate stance towards risk to good. The response to this is that talk about risk neutrality, or any other stance, already presupposes a metric of good. And it may be that the only metric available derives from judgements under uncertainty. It may derive from them in such a way that risk neutrality is automatic.

I think this is a good defence of Bernoulli's Hypothesis. But it is not conclusive. Certainly, the truth of the Hypothesis depends on the metric of good. To settle its truth we shall have to decide where the metric is determined: where the notion of quantities of good gets its meaning. And I am sure it is determined in some context where differences in good are weighed against each other. But uncertainty is not the only context where differences in good are weighed. Another is where differences in one person's good are weighed against differences in another person's, in evaluating distributions of good between people. So let us turn to that context.

An example is shown in Table 1, where distributions of wealth for two people are to be compared. Which is better? This is again a matter of

Table 2

People		People	
P	*Q*	*P*	*Q*
£1,000	£1,000	£2,100	£100

Distribution *A* Distribution *B*

weighing one difference in good against another. In favour of Distribution *B* is that it gives one person £2,100 instead of £1,000. In favour of *A* is that it gives the other person £1,000 instead of £100. Suppose we decide these considerations are exactly balanced, so the distributions are equally good. We might well think that, since the two differences in good are exactly balanced in judging the distributions, they are actually equal differences. They *count* equally in overall good, so that gives us a reason for thinking they *are* actually equal.

We have, then, two contexts where differences in good are weighed: when judging which of two uncertain prospects is better, and when judging which of two distributions between people is better. (Later we shall come across a third.) Either is a candidate for the context that gives a meaning to quantities of good.[16] And our question of whether utility represents good cardinally has now come down to this: which is the meaning-giving context? If it turns out to be uncertainty, then Bernoulli's Hypothesis will be true, and utility will measure good cardinally. Otherwise it will not.

Section 4

I now want to bring a theorem to bear on the question. It will require some preliminary work.

We have been dealing with a single person's betterness relation. Let us now consider the *general* betterness relation:

__ is at least as good as __.

This, like an individual's betterness relation, is a two-place relation between prospects. It is sometimes called a *social preference* relation. I have argued elsewhere that it, like an individual betterness relation, conforms to the axioms of expected utility theory: it is coherent.[17] So it can be represented by an expectational utility function.

Suppose there are *h* people, each with betterness relations defined over the same set of prospects. Suppose a general betterness relation is defined over the same set. The following principle connects these relations together:

---

[16] In saying that the notion of quantities of good acquires its meaning in a context where differences in good are weighed against each other, I am not adopting the operationalist view that the meaning of a quantity is determined by the operations that measure it empirically. In neither of the contexts I have mentioned is there any question of empirical measurement. In the context of uncertainty, the question is not which of the prospects the person chooses (which is empirically verifiable) but which would be better for her (which is not). Similarly, in the context of distribution between people, the question is which distribution is better.

[17] Broome (1991, Chapter 6).

> *Principle of Personal Good.* If two prospects are equally good for everybody, then they are equally good. And if one prospect is at least as good as another for everybody, and definitely better for somebody, then it is better.

Briefly: all good is people's good.

At first sight, the Principle of Personal Good may seem innocuous. It is only the Pareto Principle reformulated in terms of good. Harsanyi calls the Pareto Principle 'a very weak and hardly objectionable moral postulate'.[18] But it is actually by no means innocuous, and it conflicts with some natural egalitarian intuitions. What makes it contentious is applying it, not just to outcomes, but to uncertain prospects.

Consider the example shown in Table 3. The two states of nature *H* and *T* are equally likely. The numbers in the cells show how well off the two

Table 3

		States				States	
		H	*T*			*H*	*T*
People	*P*	1	2	People	*P*	1	2
	Q	1	2		*Q*	2	1

Prospect *C* — Prospect *D*

people will be in each state of nature, under the two alternative prospects *C* and *D*. (It is not essential to the example that the good of the people can be measured numerically, but it is a convenient simplification.)

Plainly, alternative *C* is exactly as good for person *P* as *D*, since it brings her exactly the same prospect of benefit. The same is true for *Q*, since *H* and *T* are equally likely. The Principle of Personal Good therefore says that *C* and *D* are equally good. But an egalitarian might easily doubt this conclusion. For sure, *C* will lead to equality between the people, whereas for sure *D* will lead to inequality. So an egalitarian may well think *C* better.

This is a serious objection to the Principle of Personal Good. But I think the Principle is sound nevertheless. There is a good reply to the objection. Let us ask what, if anything, is wrong with inequality. What does an egalitarian have to complain about? I think her complaint has to be that in an unequal society some people suffer unfairness: they are

[18] Harsanyi (1982, 48).

receiving less than they should, compared with how well other people are doing. This is a harm that is done them as individuals. The badness of inequality, that is to say, is a bad that comes to individuals. It is suffered by people individually, rather than by the society as a whole. Now, the figures in Table 3 show the whole good of the people: how well off they are, taking everything into account. Therefore, if either person is suffering unfairness under *D*, that harm she is suffering has already been taken into account. If inequality is bad, its badness is already incorporated in the figures. There is therefore no further reason why *C* should be better than *D*. So the egalitarian is wrong to think *C* better.

I conclude that, although the Principle of Personal Good is not as innocent as it seems at first sight, it is nevertheless correct. I have defended it in much greater detail elsewhere.[19] From now on I shall take it for granted.

This theorem can be proved, given some technical assumptions I need not go into:

> *Theorem 1.* Suppose the general betterness relation is coherent, and so is each person's. And suppose the Principle of Personal Good is true. Then the general betterness relation can be represented by a utility function U that is the sum of utility functions $U_1 \ . \ . \ . \ U_h$ representing the people's betterness relations:
>
> $$U(X) = U_1(X) + . \ . \ . + U_h(X) \text{ for all prospects } X.$$

Remember there are many utility functions representing a person's betterness relation, each a positive linear transform of the others. Theorem 1 says you can pick one function for each person, out of all her functions, in such a way that the total of all the functions you pick represents the general betterness relation.

The premises of Theorem 1 - coherence and the Principle of Personal Good - can be supported by argument. So the conclusion needs to be taken seriously. The theorem is a version of one that was first proved in 1955 by John Harsanyi.[20] One of Harsanyi's purposes was to support Bernoulli's Hypothesis, or equivalently the claim that utility represents good cardinally. And indeed, the theorem does bolster the case for Bernoulli's Hypothesis. I shall explain how.

---

[19] Broome (1991, Chapters 8 and 9). The defence includes a fuller account of 'individualistic egalitarianism', which I have described only briefly here.

[20] Harsanyi (1955). There is a rigorous version of Harsanyi's proof in Fishburn (1984). Proofs within different versions of decision theory include Broome (1990) and Hammond (1983).

Go back to the example of Table 2. Each of the distributions *A* and *B* can be treated as a prospect: the certain prospect of this particular distribution. So Theorem 1 tells us that the goodness of the distributions can be represented by utilities that are the sum of utilities representing the people's betterness relations. So

$$U(A) = U_1(A) + U_2(A)$$

and

$$U(B) = U_1(B) + U_2(B),$$

where U_1 and U_2 are utility functions for the people and U is a general utility function. Therefore

$$U(A) - U(B) = \{U_2(A) - U_2(B)\} - \{U_1(B) - U_1(A)\},$$

Suppose for simplicity (this is not essential to the argument) that the goodness of the distributions for the individuals depends only on their own wealth. Then we may write $U_1(A)$ simply as $U_1(1000)$, $U_2(B)$ as $U_2(100)$, and so on. So

$$U(A) - U(B) = \{U_2(1000) - U_2(100)\} - \{U_1(2100) - U_1(1000)\}.$$

This means that distribution *A* is better than *B* if and only if

$$\{U_2(1000) - U_2(100)\} > \{U_1(2100) - U_1(1000)\}.$$

The judgement between *A* and *B*, then, depends on comparing utility differences for the two people. These utility differences determine which is the better distribution. They determine whether a gain to the first person, from £1,000 to £2,100, counts for more or less than a gain to the second person, from £100 to £1,000, in evaluating the overall distribution. And, as I said before, since these utility differences determine which gain counts more, there is a case for thinking that they also determine which actually *is* the greater gain in good. So there is a case for thinking that the people's utilities represent quantities of their good.

But the point of Theorem 1 is that these utilities are derived from expected utility theory. In expected utility theory they tell us the value of uncertain prospects. They determine how differences are weighed against each other in the context of uncertainty. What we have now discovered is that these very same utilities *also* determine how differences are weighed between people.[21] So they determine the weights when making comparisons

[21] For each person there are many utility functions, each an increasing linear transformation of the others. Each can be used to determine how differences are weighed under uncertainty. Theorem 1, though, picks out just one function for each person. Only *this* particular function also serves the second purpose of determining how

in two distinct contexts. This makes it very much more plausible that they measure good cardinally. It is very plausible that these two contexts together determine the metric of good by giving meaning to quantities of good. This is how Theorem 1 supports Bernoulli's Hypothesis. This, if I understand him, is Harsanyi's own argument, in defending Bernoulli's Hypothesis against Amartya Sen.[22]

But that is not the end of the argument. If someone disagrees with Bernoulli's Hypothesis, what she needs to do now is argue that quantities of good get their meaning in some other way. What she needs is a *another* context in which differences of good are weighed against each other. The one to turn to, I think, is the weighing up of the goods that come to a person at different times in her life. The good in a life may be distributed across time in different ways: it may be spread evenly or unevenly, or it may come early or late. In assessing the overall goodness of different possible lives, good at one time has to be weighed against good at another. Perhaps this is the context where the metric of good is determined. So in Section 5 I shall turn to these intertemporal comparisons.

Before I do that, though, I want to draw one other conclusion from Theorem 1. This theorem says that the general goodness of a prospect is given by the *sum* of the utilities of individuals. So, if we manage to establish Bernoulli's Hypothesis that a person's utility represents her good cardinally, we shall be very close to concluding that general good is the sum of individuals' good. Some extra argument will be required to reach this conclusion, but I think it will be inescapable in the end.[23] And this conclusion is a central component of utilitarianism; I call it the 'Utilitarian Principle of Distribution'. Bernoulli's Hypothesis, then, together with Theorem 1, leads to the Utilitarian Principle of Distribution. This very much increases the significance of Bernoulli's Hypothesis. It is not just a claim about how to evaluate prospects in the face of uncertainty. It has a major implication about how to evaluate the distribution of good between people. Harsanyi himself uses Theorem 1 to support the Utilitarian Principle of Distribution.[24]

Section 5

Suppose there are a number of 'times', say T altogether. For some person

---

differences are weighed between people.

[22] Harsanyi (1975, 1977a); Sen (1976, 1977).

[23] See Broome (1991, Chapter 10).

[24] Harsanyi (1982).

j and time *t* consider the 'dated' betterness relation

__ is at least as good as __ for *j* at *t*,

where, as before, the blanks are to be filled in by prospects from a given set.

It is natural to be dubious about the existence of dated betterness relations like this. But we do often speak of how good things are for a person at one time or another. For instance, we might say that vaccinating a child against measles is bad for her at the time it is done (because of the risks) but it gives her a better prospect for the future. To be sure, we might be a little reluctant to say it is bad for her at the time it is done. If we think it is actually good *for her*, that may make us reluctant to say it is bad for her at any time. But I think this reluctance is easily overcome. There is really no difficulty in recognizing that the vaccination may be bad for the child at one time, even though the badness may be outweighed by a greater benefit later. So I think there is no real difficulty in recognizing the existence of dated betterness relations.

My expression 'at least as good for *j* at *t*' may be misleading. The betterness relation relates prospects that have a particular degree of uncertainty. The comparison between them is conditional on that uncertainty. It is done, if you like, from the viewpoint of a time when that uncertainty exists. As time progresses, and each time t is reached, the uncertainty will be progressively resolved. But the dated betterness relation, being a relation between the prospects, is not concerned with that. The prospect created by vaccinating a child is better for her five years from now than the prospect created by not vaccinating her, because she is less likely to catch measles then. Of course, if the vaccination damages her brain, it may *actually* turn out worse for her five years from now. But the *prospect* is nevertheless better for her five years from now.

The arguments that show the coherence of the individual and general betterness relations apply to dated betterness relations too. So I shall take it for granted that each dated betterness relation is coherent. It can therefore be represented by a utility function U_{jt}.

The real worry over dated betterness relations is not, I think, whether they exist, but whether they capture all that is good or bad for a person. The claim that they do is expressed in this

> *Principle of Temporal Good.* If two prospects are equally good for a person at every time, then they are equally good for her. And if one prospect is at least as good as another for the person at every time and definitely better for her at some time, then it is better for her.

This principle says that all goods are dated; there are no goods that cannot be pinned down to a particular time; lifetime good is an aggregate of dated goods. I shall discuss the truth of this principle in Section 6. But first I want to explain its importance.

A simple reinterpretation of Theorem 1 gives us

Theorem 2. Suppose a person's betterness relation is coherent, and so is each of her dated betterness relations. And suppose the Principle of Temporal Good is true. Then the person's betterness relation can be represented by a utility function U_j that is the sum of utility functions $U_{j1} \ldots U_{jT}$ representing her dated betterness relations:

$$U_j(X) = U_{j1}(X) + \ldots + U_{jT}(X) \text{ for all prospects } X.$$

Putting Theorems 1 and 2 together gives us

Theorem 3. Suppose the general betterness relation is coherent, and so is each person's betterness relation, and so are all of each person's dated betterness relations. And suppose that both the Principle of Personal Good and the Principle of Temporal Good are true. Then the general betterness relation can be represented by a utility function U that is the sum of utility functions U_j representing each person's betterness relation. And each of these in turn is the sum of utility functions U_{jt} representing the person's dated betterness relations:

$$\begin{aligned} U(X) &= U_1(X) + \ldots + U_h(X) \\ &= U_{11}(X) + \ldots + U_{1T}(X) \\ &\quad + \ldots + \\ &\quad U_{h1}(X) + \ldots + U_{hT}(X) \qquad \text{for all prospects } X. \end{aligned}$$

I have claimed that the various betterness relations are coherent, and that the Principle of Personal Good is true. So the truth of the conclusions of these theorems turns on the Principle of Temporal Good.

Suppose the conclusions are true; what would that tell us? It would tell us that the same utility functions determine how differences in good weigh against each other in *three* different contexts: across time within a life, across people, and under uncertainty. That would follow from a simple extension of the arguments in Section 4. And it would make it very hard to resist the conclusion that these functions represent quantities of good. I can think of no other context where the notion of quantities of good could get its meaning. Whenever there is weighing up to be done, according to the theorems, the total of utility always determines what is best. The same utilities always determine how much good counts in every context. In these circumstance, it would be very hard to maintain the distinction between good and how much good counts.

All this is a consequence of the Principle of Temporal Good. If you believe this principle, you ought to believe in Bernoulli's Hypothesis. You ought to believe that utility, which is defined within expected utility theory to represent the betterness relation ordinally, also represents quantities of good cardinally. Furthermore, since (as I explained at the end of Section 4) Bernoulli's Hypothesis comes close to implying the Utilitarian Principle of Distribution, you may well find yourself committed to this Principle. Conversely, if you do not think that utility represents good cardinally, you will need to reject the Principle of Temporal Good.

Remember, though, what the style of my argument has been. It has been about meaning. I argued that, if the Principle of Temporal Good is true, then the notion of a quantity of good gets its meaning in such a way that Bernoulli's Hypothesis is automatically true. At first Bernoulli's Hypothesis looks like a substantive claim about what is good for people in the face of uncertainty. But actually its truth or falsity depends simply on how a metric of good is constructed. Furthermore, suppose Bernoulli's Hypothesis really does lead to the Utilitarian Principle of Distribution, so that the Principle of Temporal Good indirectly implies the Utilitarian Principle. Then the conclusion will be that the metric of good is constructed in such a way that general good is the sum of the good of individuals. The Utilitarian Principle seems like a substantive and important claim about how good should be distributed between people. But this argument suggests it may simply reflect the construction of a metric of good. It may be less important than it seems.

Section 6

My main aim in this paper was to make the connection between Bernoulli's Hypothesis and the Principle of Temporal Good. That I have now done. But I shall conclude with some brief remarks about the truth of the Principle itself. I shall try, at least, to indicate what is at issue.

The Principle of Temporal Good is formally similar to the Principle of Personal Good. One is about intertemporal comparisons of good, the other about interpersonal comparisons. But otherwise they are analogous. I think the Principle of Personal Good is true. So if the analogy between the principles is close enough, it may support the Principle of Temporal Good too. The critical question is how close the analogy is.

On the face of it, it is not close at all. Suppose for simplicity

Table 4

		States	
		H	*T*
Times	*P*	1	2
	Q	1	2

Prospect *C*

		States	
		H	*T*
Times	*P*	1	2
	Q	2	1

Prospect *D*

there are only two times in a person's life, and consider the example shown in Table 4. The numbers show how well off the person is at the different times, under two alternative prospects. This example is exactly parallel to the example of Table 3, which I used to illustrate a difficulty for the Principle of Personal Good. Table 4 illustrates a difficulty for the Principle of Temporal Good, too. But it is actually not an analogous difficulty at all.

Alternatives *C* and *D* are equally good for the person at both times, since at both times they each give her an equal chance of one or two units of good. The Principle of Temporal Good, then, implies that prospects *C* and *D* are equally good. But it is plausible that actually *D* is better. This is because *D* gives the person, for sure, three units of good during her life, whereas *C* gives her either two units or four. *C* is risky from the viewpoint of the person's life as a whole, and *D* is not. So if there is any value in avoiding risk to the person's total good, *D* is better. Its extra value appears from the standpoint of the person as a whole, and does not show up in either of the times taken separately. The Principle of Temporal Good assumes that all good occurs at some particular time or other, and this example suggests that may be incorrect.

In the example of Table 3, I gave a reason for thinking, not that prospect *D* might be better than *C*, but that *C* might be better than *D*. In that example, *C* leads to equality between people, and *D* does not. A similar thought is plausible with our new example too. Prospect *C*, for sure, gives our subject a life of even tenor, whereas *D* leads to ups and downs. Perhaps evenness might be valuable in a life, just as equality might be valuable between people. This is a reason for thinking *C* better than *D*. But the point I want to make is that in Table 4 the standpoint of a life as whole gives grounds for the opposite thought that *D* may be better than *C*. Nothing analogous is plausible in Table 3. There is no standpoint analogous to the standpoint of a life as a whole. One might sensibly be concerned about risk to a person's total good through her lifetime, but it does not seem sensible to be concerned with risk to the total good of people taken together. In Table 3, both people individually suffer the same risk under both alternatives, and therefore avoiding risk gives no reason to prefer one alternative to the other.

On the face of it, then, the analogy between the Principles of Temporal and Personal Good is weak. There is a reason for rejecting the Principle of Temporal Good that is not also a reason for rejecting the Principle of Personal Good. This reason arises from the standpoint of a life as a whole. This is a standpoint that links together the different times in a person's life, and there is no analogous standpoint that links together different people.

Nevertheless, there is a strong tradition in economics that makes the analogy closer than it seems. This tradition treats a person as a succession of different 'selves', rather than one continuing being. For instance, Robert Strotz says: 'The individual over time is an infinity of individuals, and the familiar problems of interpersonal utility comparisons are there to plague us.'[25] Similar talk of successive 'selves' occupying one body, is to be found in J.A. Mirrlees's 'The economic uses of utilitarianism',[26] and in many other places in the literature of economics. In speaking of successive selves, Thomas Schelling confesses: 'It is only in talking with economists that I feel at all secure in using the terminology of "selves".'[27]

I shall not here try to assess the merits of this tradition.[28] But I do want to point out its consequences. It assimilates the relations between different times in a person's life to the relations between different people. So it will assimilate the Principle of Temporal Good to the Principle of Personal Good, at least in this respect. This by no means provides a complete argument for the Principle of Temporal Good.[29] But nevertheless someone within this tradition might very plausibly repudiate the standpoint of a life as a whole. Because it depends on the standpoint of a life as a whole, she would not find my example convincing. So she might well accept the Principle of Temporal Good. If she takes this step, the argument in this paper shows she ought as a consequence to accept Bernoulli's Hypothesis. She ought to believe that a person's utility, defined to represent her betterness relation ordinally, will necessarily represent her good cardinally.

---

[25] Strotz (1955-6).

[26] Mirrlees (1982).

[27] Schelling (1984).

[28] Criticisms appear in Haslanger (1989), Thompson (1983), Wiggins (1980), and in many other places.

[29] See Broome (1991, Chapter 11).

REFERENCES

Allais, Maurice, (1979), 'The foundations of a positive theory of choice involving risk and a criticism of the postulates and axioms of the American School', in Allais and Hagen (1979).

Allais, Maurice, and Hagen, Ole, (eds) (1979), *Expected Utility Hypothesis and the Allais Paradox*, Reidel.

Arrow, K. J., (1963), *Social Choice and Individual Values*, Second Edition, Yale University Press.

Bernoulli, Daniel, (1738), 'Exposition of a new theory on the measurement of risk', *Commentarii Academiae Scientiarum Imperialis Petropolitanae*, 5, translated by Louise Sommer in *Econometrica*, 22 (1954), 23-36.

Brandt, Richard, (1979), *A Theory of the Good and the Right*, Oxford University Press.

Broome, John, (1985), 'A mistaken argument against the expected utility theory of rationality', *Theory and Decision*, 17, 313-318.

Broome, John, (1990), 'Bolker-Jeffrey expected utility theory and axiomatic utilitarianism', *Review of Economic Studies*, 57, 477-502.

Broome, John, (1991), *Weighing Goods*, Blackwell.

Broome, John, (forthcoming), '"Utility"', *Economics and Philosophy*.

Butts, R., and Hintikka, J., (1977), *Foundational Problems in the Special Sciences*, Reidel.

Davidson, Donald, (1980), *Essays on Actions and Events*, Oxford University Press.

Ellsberg, Daniel, (1954), 'Classic and current notions of "measurable utility"', *Economic Journal*, 64, 528-556.

Feiwel, George R., (ed.) (1987), *Arrow and the Ascent of Modern Economic Theory*, New York University Press.

Fishburn, Peter C., (1984), 'On Harsanyi's utilitarian cardinal welfare theorem', *Theory and Decision*, 17, 21-28.

Griffin, James, (1986), *Well-Being: Its Meaning, Measurement and Moral Importance*, Oxford University Press.

Hammond, Peter J., (1983), 'Ex-post optimality as a dynamically consistent objective for collective choice under uncertainty', in Pattanaik and Salles (1983, 175-205).

Harsanyi, John C., (1953), 'Cardinal utility in welfare economics and in the theory of risk-taking', *Journal of Political Economy*, 61, 434-435, reprinted in Harsanyi (1976, 3-5).

Harsanyi, John C., (1955), 'Cardinal welfare, individualistic ethics, and interpersonal comparisons of utility', *Journal of Political Economy*, 63, 309-321, reprinted in Harsanyi (1976, 6-23).

Harsanyi, John C., (1975), 'Nonlinear social welfare functions: do welfare economists have a special exemption from Bayesian rationality?', *Theory and Decision*, 6, 311-332.

Harsanyi, John C., (1976), *Essays on Ethics, Social Behavior, and Scientific Explanation*, Reidel.

Harsanyi, John C., (1977a), 'Non-linear social welfare functions: a rejoinder to Professor Sen', in Butts and Hintikka (1977).

Harsanyi, John C., (1977b), *Rational Behavior and Bargaining Equilibrium in Games and Social Situations*, Cambridge University Press.

Harsanyi, John C., (1982), 'Morality and the theory of rational behavior', in Sen and Williams (1982, 39-62).

Harsanyi, John C., (1987), 'Von Neumann-Morgenstern utilities, risk taking, and welfare', in Feiwel (1987, 545-558).

Haslanger, Sally, (1989), 'Persistence, change and explanation', *Philosophical Studies*, 56, 1-28.

Jeffrey, Richard C., (1983), *The Logic of Decision*, Second Edition, University of Chicago Press.

Jeffrey, Richard C., (1987), 'Risk and human rationality', *The Monist*, 70 223-236.

Mirrlees, J. A., (1982), 'The economic uses of utilitarianism', in Sen and Williams (1982, 219-238).

Neumann, John von, and Morgenstern, Oskar, (1947), *Theory of Games and Economic Behavior*, Second Edition, Princeton University Press.

Pigou, A. C., (1962), *The Economics of Welfare*, Macmillan.

Savage, Leonard J., (1954), *The Foundations of Statistics*, Wiley.

Schelling, Thomas C., (1984), 'Self-command in practice, in policy, and in a theory of rational choice', *American Economic Review Papers and Proceedings*, 74, 1-11.

Sen, Amartya, (1976), 'Welfare inequalities and Rawlsian axiomatics', *Theory and Decision*, 7, 243-262, reprinted in Butts and Hintikka (1977).

Sen, Amartya, (1977), 'Non-linear social welfare functions: a reply to Professor Harsanyi', in Butts and Hintikka (1977).

Sen, Amartya, and Williams, Bernard, (eds) (1982), *Utilitarianism and Beyond*, Cambridge University Press.

Strotz, R. H., (1955-6), 'Myopia and inconsistency in dynamic utility maximization', *Review of Economic Studies*, 23, 165-180.

Thomson, Judith Jarvis, (1983), 'Parthood and identity across time', *Journal of Philosophy*, 80, 201-220.

Tintner, Gerhard, (1942), 'A contribution to the non-static theory of choice', *Quarterly Journal of Economics*, 56.

Tversky, Amos, and Kahneman, Daniel, (1986), 'Rational choice and the framing of decisions', *Journal of Business*, 59, 250-278.

Wiggins, David, (1980), *Sameness and Substance*, Blackwell.

VI. Reflections

Economics and Thermodynamics: von Neumann's Problematic Conjecture

Paul A. Samuelson
Massachusetts Institute of Technology, E52-383C
Cambridge, Massachusetts 02139, U.S.A.

Of the attempts to find analogies between thermodynamics and economies there is alas no end. But the mathematical genius John von Neumann has earned the right to command our investigation when he suggests that his growth model (1945; earlier 1932 and 1937) defines a function

$$\text{Value of Inputs}_t/\text{Value of Outputs}_{t+1}$$

$$= \sum_i^n\sum_j^m p_i A_{ij} x_j / \sum_i^n\sum_j^m p_i B_{ij} x_j$$

$$= \phi(P,X)$$

whose "role seems to be similar to that of thermodynamic potentials in phenomenological thermodynamics." (1945, p.1)

Recently Andrew Brody (1989) has interestingly addressed the suggested relation and I shall attempt here to make further explorations. The goal is not to have economics shine brighter in the reflected light of physics's most perfected axiomatic system, but rather to guard against unwarranted pretensions.

Motivation

It is a fair bet that von Neumann was influenced by the following aspects of his $\phi(P,X)$ function.

1. Just as an economic system has a pair of conjugate vectors--in his case the <u>intensive</u> <u>prices</u> of n goods, $P = (p_1 \ldots p_n)$, and the <u>extensive quantities</u> of the m different activity levels, $X=(x_1 \ldots x_m)$--so does a Gibbs thermodynamic system have an <u>extensive vector of quantity magnitudes</u>, (specific volume, entropy,...) or (V,S,...) or more abstractly $(x_1 \ldots x_n)$, and a <u>conjugate vector of corresponding intensive forces</u>, (-pressure, absolute temperature,...) or (-P,T,...) or more abstractly $(y_1 \ldots y_n)$.

2. von Neumann (1928) had already worked out 2-person zero-sum game theory that involved each player's vector of conjugate mixed strategies: $(x_1 \ldots x_m)$ and $(y_1 \ldots y_n)$, $n \gtrless m$. The 1928 game theory involved saddlepoint or Min-Max properties and so did his growth model's ϕ function: namely, for $(\overline{P},\overline{X})$ <u>equilibrium</u> vectors of the growth system and (P,X) arbitrary vectors, von Neumann could proclaim

$$\phi(P,\overline{X}) \leqq \phi(\overline{P},\overline{X}) \leqq \phi(\overline{P},X) \tag{2}$$

Classical mathematical economics had primarily concerned itself with Min-Min, or what is the same thing with Max-Max, rather than with Min-Max. No doubt this was one reason why von Neumann, with and without his economist collaborator Oskar Morgenstern, used to proclaim that the social sciences would need a new kind of mathematics rather than rely on the post-Newtonian calculus that had been devised primarily to serve the needs of physics.

3. Brody [1989] spells out a possible third reason to motivate a connection between thermodynamics and the 1945 economics. If I interpret Brody [1989, pp. 145-6] properly, (2)'s first inequality corresponds to "..the first criterion of equilibrium: maximal entropy." The second inequality of (2) is "analogous in the Gibbs system to the dual criterion of equilibrium; minimal energy."

Doubts

On reflection some possible disquiet develops on these three suggestive points.

A. In thermodynamics, the intensive and extensive conjugate vectors have the <u>same</u> number of elements: n=n rather than $n \gtrless m$, as in game theory, the growth model, Danzig linear programming, and Gale-Kuhn-Tucker concave programming. [Caveat: the number of <u>non-zero</u> p's and x's will not exceed min(n,m).]

B. In Gibbs thermodynamics, as in classical mathematical economics, there is a basic role for Min-Min rather than initially for Min-Max. Gibbs [1873], his first thermo paper, builds on an (internal) Energy function that is convex (from below) in its arguments--in both of them individually and collectively:

$$\text{Energy} = U(V,S) \tag{3a}$$

$$-P = \partial U(V,S)/\partial V \text{ or } U_v(V,S) \tag{3b}$$

$$T = U_s(V,S) \tag{3c}$$

$$U_{vv}>0,\ U_{ss}>0,\ U_{vv}U_{ss} - U_{vs}U_{sv}>0 \tag{3d}$$

At once Gibbs's function G(P,T) is definable by a Min-Min property

$$\underset{(V,S)}{\text{Min-Min}}\{U(V,S) + PV - TS\} = G(P,T) \tag{4a}$$

We can verify that -G(P,T) has the same properties in its arguments as U(V,S) had in its; actually, U and -G are "dual to each other." Corresponding to (4a) goes

$$\underset{P,T,}{\text{Min}}\{-G(P,T) + PV - TS\} = U(V,S) \tag{4b}$$

Corresponding to (3) goes

$$V = G_p(P,T),\ S = -G_T(P,T) \tag{4c}$$

$$-G_{PP}>0,\ -G_{TT}>0,\ G_{PP}G_{TT}-G_{PT}G_{TP}>0 \tag{4d}$$

Corresponding to von Neumann's $\phi(P,X)$, for the case of n=2=m, we can now write the 4-variable thermodynamic function

$$U(V,S) + PV - TS - G(P,T)$$
$$= \Phi(-P,T,V,S)$$
$$= \Phi(y_1,y_2,x_1,x_2) = \Phi(Y,X) \quad (5)$$

Unlike (2)'s Min-Max property, we translate from (4) the following symmetric Min-Min properties that any equilibrium $(\overline{Y},\overline{X})$ must possess

$$\Phi(Y,\overline{X}) \geqq \Phi(\overline{Y},\overline{X}) = 0 \leqq \Phi(\overline{Y},X) \quad (6)$$

Note the contrast between (6)'s inequality signs that point inward instead of showing the parallel Min-Max pattern of von Neumann's (2).

C. The analogy achieved for von Neumann's model with thermodynamics is evidently thin at spots. To dramatize this deficiency let me demonstrate how much closer is the thermodynamics isomorphism with classical mathematical economics that scholars like Harold Hotelling [1932] were already working out just when von Neumann was first unveiling his growth model before Princeton's 1932 mathematicians. Hotelling had begun the duality developments later associated with R. Roy (1942), L. Court (1941), Samuelson (1938, 1947, 1953, 1960, 1962, 1965, 1972), R. Shephard (1953), D. Champernowne (1953); and Hotelling had also chanced already to use the terminology of a "potential function." I shall be cruelly brief in my exposition.

Classical Math Economics and Thermo

Define a firm's gross revenue as a concave function of its n inputs, X = (x_j), which it buys at prescribed competitive prices, Y = (y_j). Then

$$\underset{x_j}{\text{MAX}}\{R(x_1,\ldots,x_n) - \sum_1^n y_j x_j\} = -R^*(y_1,\ldots,y_n) \quad (7a)$$

$$y_j = \partial R(X)/\partial x_j,\ x_j = \partial R^*(Y)/\partial y_j, j=1,\ldots n \quad (7b)$$

Exactly dual to each other are R(X) and R*(Y).

Now, in addition to von Neumann's ϕ and Gibb's Φ, let's define Hotelling's

$$\Psi(Y,X) = R(x_1,\ldots,x_n) + R^*(y_1,\ldots,y_n) - \sum_1^n y_j x_j \quad (8)$$

Then, in <u>exact</u> analogy with thermodynamics, pre-Neumann mathematical economics does escape (2)'s parallel-arrowed Min-Max in favor of (6)'s Max-Max: namely

$$\Psi(Y,\overline{X}) \leqq \Psi(\overline{Y},\overline{X}) \geqq \Psi(\overline{Y},X) \quad (9)$$
$$\Psi(\overline{Y},\overline{X}) \geqq \Psi(Y,X)$$

D. We are still left with the task of doing full justice to Brody's "maximal entropy" and "minimal energy" interpretations for (2). I shall return to this after we notice that other Legendre- transformation potential functions--other than Gibbs's U(V,S) and G(P,T)--do permit of some Min-Max interpretations. Perforce I must discuss Gibbsian "Enthalpy", H(P,S), and Gibbsian "Free Energy à la Helmholz", F(V,T).

E. Economists go whole afternoons not thinking about the intricacies of thermodynamics, and they prefer that way of life. However, if we are to do serious justice to the proposed topic we cannot just talk about thermodynamics; we must actually do some. Otherwise someone might get snowed by mere cocktail chatter.

Mixed Thermo Functions and Minimax

The Enthalpy and Free Energy potentials are defined respectively as follows:

$$\begin{aligned}\text{Enthalpy} &= H(\bar{P},\bar{S}) \\ &= \operatorname*{Min}_{V}\{U(V,\bar{S}) + \bar{P}V\}\end{aligned} \tag{10a}$$

$$H_P(P,S) = V,\ H_S(P,S) = T \tag{10b}$$

$$\begin{aligned}\text{Free Energy (Helmholz)} &= F(\bar{V},\bar{T}) \\ &= \operatorname*{Min}_{S}\{U(\bar{V},S) - \bar{T}S\}\end{aligned} \tag{11a}$$

$$F_V(V,T) = -P,\ F_T(V,T) = -S \tag{11b}$$

As yet still no thermodynamic Min-Max. So far only two separate single Mins. But now when we proceed to examine the convexity or concavity of H and F as functions, we find that they differ from the <u>uniform</u> convexity of U and -G assured in (3d) and (4d). We do find convexity in one variable and concavity in the other, which does smack of some kind of Min-Max.

Specifically, double differentiation does deduce from (3) and (10) and from (4) and (11)

$$H_{PP}(P,S) < 0 < H_{SS}(P,S),\ \partial(V,T)/\partial(P,S) < 0 \tag{12a}$$

$$F_{TT}(V,T) < 0 < F_{VV}(V,T),\ \partial(P,S)/\partial(V,T) < 0 \tag{13a}$$

From these opposite-signed curvatures, we can verify the respective saddlepoints

$$H(P,\bar{S})-P\bar{V}-\overline{TS} \leqq H(\bar{P},\bar{S})-\overline{PV}-\overline{TS} \leqq H(\bar{P},S)-\overline{PV}-\bar{T}S \tag{12b}$$

$$F(\bar{V},T)+\overline{PV}+T\bar{S} \leqq F(\bar{V},\bar{T})+\overline{PV}+\overline{TS} \leqq F(V,\bar{T})+\overline{PV}+\overline{TS} \tag{13b}$$

At last we have deduced in (12b) and (13b) a saddlepoint property in some thermodynamic functions. Does this confirm von Neumann's conjecture? Only minimally--in the following sense. Von Neumann's saddlepoint in (2) is a different species from the saddlepoints of (12b) or (13b), even though they belong to the same genus. In (2) the left-hand-most and right-hand-most expressions allow half the 2n variables to be freely variable; in (12b) and (13b), only half-of-half the 2n variables are permitted to be freely variable in the exterior expressions.

Actually, we are now alerted to a more fundamental mismatch between the von Neumann and Gibbs models. For von Neumann, after conventional normalization, there is generically a unique $(\overline{P},\overline{X})$ equilibrium. Almost always, technology determines a mode of maximal growth that is a unique (normalized) $\overline{X}$ vector; almost always it determines a vector of relative prices that is a unique (normalized) $\overline{P}$ vector. For standard Gibbs or Hotelling we have parametric equilibria: for each of the n-fold infinity of possible $\overline{X}$ specified, there is a unique $\overline{Y}$ equilibrium; for each of the infinity of specified $\overline{Y}$, there is a unique $\overline{X}$ equilibrium; indeed, for each of the 2^n arbitrary ways of specifying a vector whose elements can be each either one of the pair (y_j,x_j), there is a unique equilibrium vector whose elements are the equilibrium values of the other ones of the various (y_j,x_j) pairs, $j=1, 2,\ldots, n$.

In sum, after a tedious post mortem, we find that the purported von Neumann analogy between his growth model and classical thermodynamics is a limping analogy--a kind of a logical pun rather than a congruent isomorphism.

Meaning of Entropy Maximization

C. P. Snow in a moment of enthusiasm proposed as a test of whether one belonged to "the two cultures" the ability to understand and appreciate both Shakespeare and the Second Law of Thermodynamics. I fear the last is an over-stringent test for it would leave heaven underpopulated. Rarely do two scientists agree on how to resolve the paradoxes of time-symmetric microdynamic equations leading to time-asymmetric evolution in entropy. Fortunately, if we stick resolutely to phenomenological, macroscopic, classical thermodynamics, we sidestep deep issues.

Gibbs wisely took as his credo the words of Clausius:

> The Energy of the World is Constant.
> Its Entropy Ever Grows.

Time's arrow does not treat energy and entropy symmetrically: one is not maximized while the other is minimized; one is conserved as a constant while the other is being maximized, which is not quite the same thing as a Min-Max property.

Dr. Brody is not in error when speaking of minimal energy as a criterion of thermo<u>static</u> equilibrium. Thus, economists recognize the interchangeability of "An efficient demand allocation <u>maximizes</u> (ordinal) utility for fixed total income." and "An efficient allocation <u>minimizes</u> total expenditure for fixed utility attainment." In Samuelson [1960] I followed Gibbs's thermostatics and spoke of "minimizing total energy for fixed entropy." But neither Gibbs nor I were addressing time's asymmetric arrow when doing this. To elucidate the time-directional point, the present section is needed. (Interestingly, consumers-surplus economists after Dupuit [1849] anticipated the equal-area criterion that Maxwell admired so much in Gibbs's treatment of super-cooled liquids near freezing point regions where van der Waal's kind of relation obtains.)

Figure 1 uses a Jevons-like and Edgeworth-like box diagram of economics to handle the following case. A unit mass of a vapor fills the inside of a rigid and insulated container: its temperature, if measured, would be found to be about 110° centigrade--above the boiling point. Its internal energy, if measured, would be found to be initially $\overline{U}_1$. A second similar unit mass is initially at 120° centigrade with a higher initial energy of $\overline{U}_2$.

Now let the two subsystems be connected by a conducting inner-wall, a piston. Their temperature difference disturbs the initial equilibrium: their temperatures finally settle to a common level, say at somewhere near 115°C; likewise they achieve a common pressure. By symmetry, they end up at equal energies and specific volumes. Symbolically, we have observed the spontaneous ("irreversible") transformation

$$(\overline{T}_1,\overline{U}_1,\overline{V}_1,\overline{S}_1;\overline{T}_2,\overline{U}_2,\overline{V}_2,\overline{S}_2) \rightarrow (T^*,U^*,V^*,S^*;T^*,U^*,V^*,S^*) \qquad (14)$$

Because the external walls of the whole system are kept rigidly at the initial total volume, $\overline{V}_1+\overline{V}_2$, and because the Law of Conservation of Energy says the initial total of energy is conserved at $\overline{U}_1+\overline{U}_2$, we can write

$$\overline{V}_1+\overline{V}_2 = V_1^*+V_2^* \qquad (15a)$$

$$\overline{U}_1+\overline{U}_2 = U(V_1^*,S_1^*)+U(V_2^*,S_2^*) \qquad (15b)$$

Subject to the constraints of (15a) and (15b), Nature moves to maximize the new total of entropy:

$$\underset{V_j^*,S_j^*}{\text{Max}} \{S_1+S_2\} = S_1^*+S_2^* \qquad (15c)$$

So long as U is a strictly convex function, the symmetry of the (15) system does assure a unique symmetric equilibrium. Since (15a) and (15b) give us only two relations for four variables, we pick up the needed two further relations from the vanishing of Newtonian derivatives. After elimination of

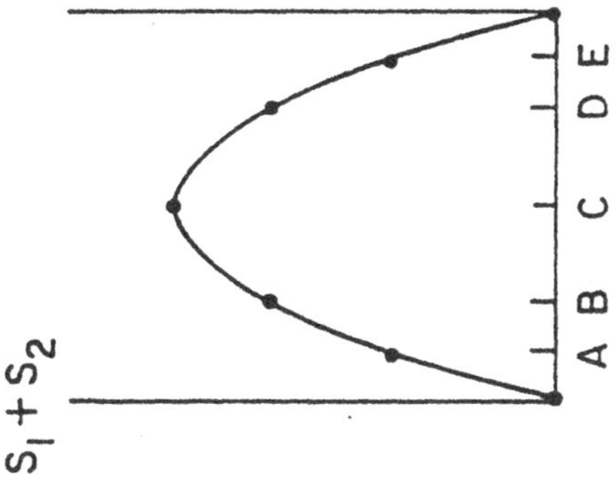

Fig. 1b

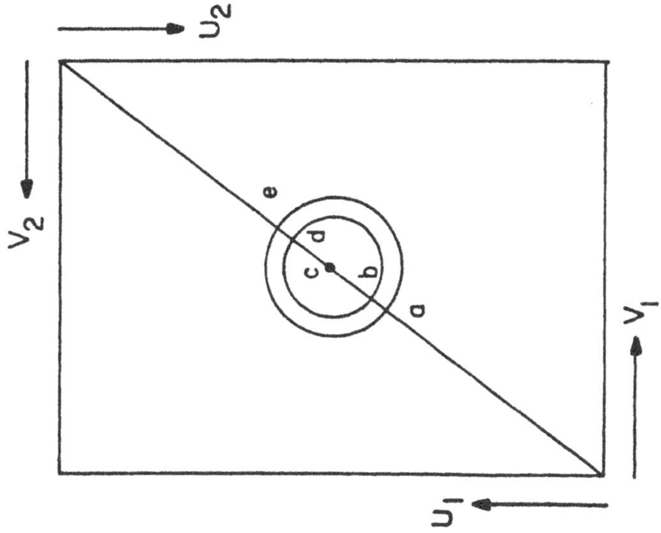

Fig. 1a

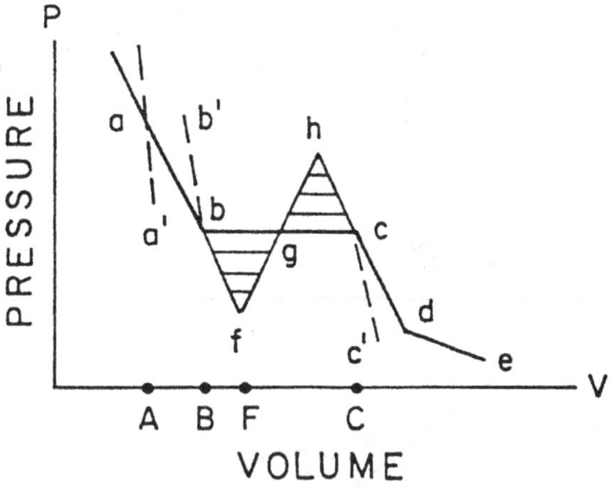

Fig. 2

Lagrange multipliers, these can be verified to be equivalent to the conditions of equal pressures and temperatures:

$$-U_V(V_1^*,S_1^*) = -U_V(V_2^*,S_2^*) = P^* \tag{15d}$$

$$U_S(V_1^*,S_1^*) = U_S(V_2^*,S_2^*) = T^* \tag{15e}$$

In Figure 1a the contours represent levels of equal total entropy. Figure 1b records how the symmetry point in the box diagram, whose sides respectively measure initial total volume and total energy ($\Sigma\overline{V}_j$ and $\Sigma\overline{U}_j$), does achieve a maximum of final entropy (ΣS_j*).

How Min-Energy Works Only for Thermostatics

The Max-Entropy conditions of (15) are seen to define a different final equilibrium from the following Min-Energy algorithm:

Subject to

$$\overline{V}_1+\overline{V}_2 = \hat{V}_1+\hat{V}_2 \tag{16a}$$

$$\overline{S}_1+\overline{S}_2 = \hat{S}_1+\hat{S}_2 \tag{16b}$$

$$\underset{V_j,S_j}{\text{Min}}\{U(V_1,S_1)+U(V_2,S_2)\} \tag{16c}$$

$$= \hat{U}_1+\hat{U}_2$$

This turns out to entail similar vanishings of Newtonian derivatives as before; after we eliminate the new Lagrange multipliers of this new problem, we do get (15d) and (15e), namely

$$-\hat{P} = U_V(\hat{V}_1,\hat{S}_1) = U_V(\hat{V}_2,\hat{S}_2) \tag{16d}$$

$$\hat{T} = U_V(\hat{V}_2,\hat{S}_2) = U_V(\hat{V}_2,\hat{S}_2) \tag{16e}$$

However, with the disagreement of the (15b) and (16b) relations, the systems (15a)-(15b)-(15d)-(15e) and (16a)-(16b)-(16d)-(16e) do differ and necessarily the deduced equilibria differ. Specifically,

$$S_1^*+S_2^* > \overline{S}_1+\overline{S}_2 = \hat{S}_1+\hat{S}_2 \tag{17}$$

$$\hat{U}_1+\hat{U}_2 < \overline{U}_1+\overline{U}_2 = U_1^*+U_2^*$$

Dr. Brody's readers will realize that the alternative approaches cannot both correctly match Nature's true before-and-after scenario. But they and Gibbs's readers will also realize that where thermo<u>statics</u> is concerned, both approaches do lead to the identical (15d)-(15e) and (16d)-(16e). These last equilibrium relations are the only ones that are free of the initial-condition's barred variables, and therefore they do correctly identify all of the $(V_1,S_1,U_1,-P_1,T_1;V_2,S_2,U_2,-P_2,T_2)$ states that would be self-perpetuating (which is the very notion of a thermostatic equilibrium!).

Non-convexity: Phase Transitions and Non-Local Maxima

Figure 1's simple story holds when isotherms in the indicator diagram of the (V,P) space are well-behaved monotonically falling. Gibbs was never one wantonly to break his silence: his 1873 paper was motivated to explain phase transitions 'twixt solid and liquid or liquid and gas. The U(V,S) surface Maxwell fabricated from Gibbs's recipe violated convexity so as to explain coexistence of ice and water at melting-point transitions.

Figure 2 plots on the (V,P) indicator diagram an isotherm that violates convexity of the U(V,S) surface. (It pictures what obtains for, say, H_2O near standard atmosphere temperature and in the range of a couple of degrees Celsius and below 0°C. To aid the reader's eye my curvilinear isotherm is drawn to consist of line-segments.)

The isotherm abcdfghcde has the anomalous rising branch of fgh, so that latitudes between f and g intersect it three times. Mother Nature, tolerating no local instabilities, in fact maximizes its total Entropy by replacing the anomalous isotherm with the "stable" abcde locus. Gibbs taught the delighted Maxwell that the bc freezing point, at which both the solid and liquid would coexist, had to come at the precise P level where the shaded triangular areas would be equal. Beyond C all is ice; before B all is water; 'twixt B and C both ice and water obtain, with just enough B-like molecules and C-like molecules to precisely fill the specified $\overline{V}_1+\overline{V}_2$ total volume. (A skilled experimenter can actually realize a "supercooled" liquid along the bf branch; but shaking or otherwise perturbing the equipment will destroy the globally unstable equilibrium in favor of the stable bc branch. The dashed loci--aa', bb', cc', ... represent the adiabats, steeper than the isotherms they touch in accordance with the LeChatelier principle of convexity.)

Hotelling economics varies the Gibbs story. For it, label the vertical axis p_1, and the horizontal axis x_1. Holding $(p_2,...)$ constant, the demand for x_1 will not be abcde. Rather will it be ab for p_1 above the bc level, and cd for p_1 below that level; at the critical level for competitive p_1 of bc, the end-points b and c will be indifferently tied, but the points between them will be avoided; the level at which the bifurcation takes place is determined by the same Gibbs-Maxwell rule of equal areas!

Qualifications

Dr. Brody is right to call attention to the fact that the techniques in mathematics useful to economists in the Debreu age involve inequalities like those pioneered by von Neumann [1928, 1945]. It is appropriate to emphasize that Willard Gibbs himself peculiarly utilized inequalities. (I confine my remarks to his classical thermodynamics. His formulation of Newton-Lagrange dynamics in a manner reminiscent of Gauss's principle of least restraint is a related matter.)

Not for nothing am I the intellectual grandson of Gibbs (via his last protegé, E. B. Wilson, my Harvard master.) Before the epoch of linear and concave programming and prior to the introduction of fixed-point theorems, my prewar explorations of revealed preference (Weak Axioms and such) dealt with $\Sigma\Delta p_j\Delta x_j$ and $\Sigma p_j\Delta x_j$ expressions that were in the Gibbsian mold.

My strategy was to take from thermodynamics not its dubious analogies of content but rather what it had to offer by way of common useful techniques of maximization and implied comparative statics. I have had to stress, and I take this occasion to reinforce the point, that there is no Energy whose magnitude is conserved in the universe of economics. There is no one Entropy total that gets maximized in the econmic world. For profit-seeking firms and pleasure-pursuing consumers, a dollar profit function or a (ordinal) utility function has some of the properties that an Entropy function has for Clausius's domain.

Nicholas Georgescu-Roegen [1971] has sagely reminded us economists that the image of an hour glass (whose sands run out) may be a more appropriate logo than the image of a frictionless pendulum (whose equilibrium and conserved-energy is eternal). But the formalisms of Gibbs-Tisza thermodynamic potentials (and their Min-Min or Min-Max aspects) add naught to the Georgescu insight. As I wrote for the Yale 150th Birthday Convocation for Gibbs [Samuelson, 1990, p.266], eschewing the formalism of Equation (16) above, we capture in a simple way the fact that an economic equilibrium must run down if it depends indispensably on (say) a supply of copper, and if concentrated deposits of copper ore are finite and become exhausted in finite time. As soon as equilibrium displays hysteresis dependence on the integral of all copper ever mined, no steady state for Q* is possible at a positive level. Enough said.[1]

[1]For bibliographical reference, the formulation in (12b) and (13b) seems to be only implicit in the vast thermodynamic literature. What I have called Hotelling duality in (8) is actually borrowed from my 1960 homage to his Festschrift, and from my 1965 paper on "full duality." The new mathematics envisaged for economics by von Neumann involved, among other things, the topology of fixed points and all that. However, Gibbs's fruitful use of finite inequalities, which resembled my use of expressions like $\Sigma\Delta y_j\Delta x_j=0$, (applicable to real variables or to integral variables), would have been justly deemed by him part of physics' "old mathematics."

References

Brody, A. 1989. "Economics and Thermodynamics," in M. Dore, S. Chakravarty, and R. Goodwin, eds., John von Neumann and Modern Economics. Oxford: Clarendon Press.

Champernowne, D. 1953. "Commentaries on J. Robinson on Capital," Review of Economic Studies 21, 107-35.

Court, L. 1941. "Invariable Classical Stability of Entrepreneurial Demand and Supply Functions," Quarterly Journal of Economics 55, 134-44.

Georgescu-Roegen, N. 1971. The Entropy Law and the Economic Process. Cambridge, Massachusetts: Harvard University Press.

Gibbs, J.W. 1873, 1873, 1876-78. Thermodynamics papers in The Collected Works of J. Willard Gibbs, Volume I. New Haven, Conn.: Yale University Press.

Hotelling, H. 1932. "Edgeworth's Taxation Pardox and the Nature of Demand and Supply Functions," Journal of Political Economy 40, 577-616.

von Neumann, J. 1928. "Zur Theorie der Gesellschaftsspiele", Math. Annalen, 100, 295-320.

____________. 1945. "A Model of General Equilibrium," Review of Economic Studies 13, 1-9.

Roy, R. 1942. "De l'Utilité, Contribution à la Theorie des Choix." Paris: Hermann.

Samuelson, P. 1947 and 1983. Foundations of Economic Analysis. Cambridge, Mass.: Harvard University Press.

____________. 1938. "A Note on the Pure Theory of Consumer's Behavior," Economica, N.S. 5, 61-71.

____________. 1953. "Prices of Factors and Goods in General Equilibrium," Review of Economic Studies 21, 1-20.

____________. 1960. "Structure of a Minimum System," in R. Pfouts, ed., Essays in Honor of Harold Hotelling. Chapel Hill: University of North Carolina Press.

____________. 1962. "Parable and Realism in Capital Theory," Review of Economic Studies 29, 193-206, particularly Appendix.

____________. 1965. "Using Full Duality to Show That Simultaneously Additive Direct and Indirect Utilities Implies Unitary Price Elasticity of Demand", Econometrica 33, 781-96.

____________. 1972. "Unification Theorem for the Two Basic Dualities of Homothetic Demand Theory," Proceedings of the National Academy of Science, USA 69, 2673-74.

____________. 1990. "Gibbs in Economics" in D. G. Caldi and G. D. Mostow eds., Proceedings of the Gibbs Symposium, Yale University 1989. Providence, Rhode Island: American Mathematical Society and American Institute of Physics, 255-68.

Shephard, R. 1953. Cost and Production Functions. Princeton, N.J.: Princeton University Press.

Game Theory, Symmetry, and Scientific Discovery

Horace W. Brock
Strategic Economic Decisions, Inc.
Menlo Park, California

Formal theories in the natural and social sciences usually take the form of mathematical relationships specifying how one subset of variables depends upon another. Given this truism, it is not a little ironic that many of the most innovative and successful theories of twentieth century science have resulted from an *a priori* specification of how given subsets of variables (or relationships) do *not* depend upon other subsets. Indeed, a hallmark of successful theory construction has been an identification of apposite invariances, independence assumptions, and other essentially algebraic conditions prior to constructing a theory.

Conditions of this kind are conceptually distinct from other mathematically pregnant types of assumptions, e.g. monotonicty axioms (preference orderings in decision theories, variational principles in physics), and topological axioms (continuity in decision theories, analyticity in physics). The algebraic dimension of theory construction is arguably the most subtle and powerful one, and we shall focus on this alone in this essay. Part I suggests that John Harsanyi's success in advancing many frontiers within game theory and ethics has stemmed in good measure from his ability to identify and to exploit myriad symmetries which lie at the very heart of these disciplines. Bringing attention to this accomplishment will hopefully increase awareness of the fact game theory is to social science what relativity theory is to physics: essentially a theory of symmetry.

Part II extends the discussion to the philosophy of science, a field to which Harsanyi himself [16] has contributed. First, we abstract the fundamental reason why symmetry arguments should be expected to play a pivotal role in normative epistemology and efficient theory construction. The reason lies in the desirability and efficacy of a deep principle of epistemological consistency. Examples from six disciplines are given to support this view, and to suggest that the logic of scientific discovery could benefit from algebraic foundations that do not currently exist.

Next, in the spirit of this observation, we offer an algebraic reexamination of the venerable norm of simplicity in scientific theories. In the context of theory construction, it is argued that the desideratum of simplicity *per se* should be replaced by that of *optimal simplicity* -- an optimum reflecting the ineluctable tradeoff between simplicity and complexity that characterizes most successful scientific theories.

Algebraic considerations suggest two dimensions of the concept of optimal simplicity in a yet-to-be-constructed theory **T**. The first concerns the "correct" number of independent variables (and relationships) to aim for. We argue that this number should be equal to the number of essential parameters characterizing the full symmetry group $\mathbf{G_s^T}$ of **T**. An algorithm is sketched for constructing this group and assessing its size. But this algorithm does more. For it is embedded in a hypothetical Expert-System designed to encode important aspects of the "creativity" of original scientific thinkers. This Expert-System thus advances the twin objectives of modeling *retroduction* in the philosophy of science, and of creating a *normative epistemic engine* in the philosophy of mind.

The second dimension of optimal simplicity concerns whether the relationships of a yet-to-be-constructed theory should be linear or nonlinear. We argue for nonlinearity, partly because of a Machian prejudice that everything is coupled to everything else, and partly because of a pragmatic prejudice based on historical experience: linear probabilistic theories are typically superseded by nonlinear deterministic theories with better predictive and explanatory power. Examples supporting these twin prejudices are cited from game theory, physics, and mathematical aesthetics. All in all, our simplicity argument mirrors Whitehead's wise admonition: "Seek simplicity, and distrust it."

I. Symmetry Arguments in Harsanyi's Work

Were the author to give an introductory course in analytical social science, his first assigned reading would be Harsanyi's important yet rarely cited 1961 paper, "On the Rationality Postulates Underlying the Theory of Cooperative Games" [11]. The ostensible purpose of this paper was to rederive in a more satisfactory manner the Nash-Zeuthen-Harsanyi bargaining solution to two-person games. But much more is achieved, including a particularly lucid explication of the inescapable role of symmetry arguments in game theory -- and in other scientific theories as well. Harsanyi starts off by demonstrating that increasingly strong postulates are required as one ascends from single-person decision problems under uncertainty, to zero-sum games, and then to positive-sum games. He identifies these, and explicates the proper role of each axiom.

In particular, he demonstrates that in shifting from zero-sum to positive-sum theory, two new postulates are needed over and above those of expected utility maximization and symmetrical expectations of rationality: (i) a symmetry postulate whereby a solution function (e.g. Zeuthen's "concession rule") must be a symmetric function of all strategically relevant variables; and (ii) a restriction-of-irrelevant-*variables* axiom whereby the only variables that should enter into a theory of rational behavior are strategically relevant variables.

Regarding the symmetry axiom, Harsanyi shows how fundamental symmetry is to the very concept of a theory of rational behavior in game situations, regardless of whether one introduces strategically irrelevant variables (e.g. the different telephone numbers of the players). Indeed, in a demonstration replete with implications for theory construction in general, he establishes that any determinate solution theory *must* be a symmetric function of its variables as a matter of sheer logical necessity.

As for his postulate excluding irrelevant variables, Harsanyi successfully argues that for a theory of rational behavior to be a *bona fide* theory of rationality, strategically irrelevant variables must be excluded -- again as a matter of logical necessity. Failure to exclude them deprives us of one of the major achievements of game theory: clarification of what it *means* to be rational as opposed to irrational. Only with such a norm in hand can one then proceed to identify irrational behavior as such, and to explain how it is irrational. Given this perspective, one can immediately appreciate how various non-game theoretical approaches to bargaining (e.g. Hicks', Schelling's, and one of Raiffa's) necessarily entail elements of irrational behavior.

Harsanyi's arguments in this paper are essentially a rebuttal to Thomas Schelling's [34] attack on the symmetry postulate of game theory, and to his call for the introduction of "situational" variables that we would now deem strategically irrelevant, even if descriptively significant. As a former undergraduate student of both Raiffa and Schelling in the late 1960s, the present author can attest to the conceptual muddle that resulted from failing to introduce the distinctions that Harsanyi first introduced.

***Classical n-Person Cooperative Theory*:** Turning next to n-person bargaining theory, we come to Harsanyi's [13] 1963 generalization and synthesis of both the Nash two-person variable threat bargaining model and the classical Shapley value. This model is primarily known for its introduction of intrinsic "utility weights" which subsequently made it possible for Shapley to generalize his value to the non-transferable utility case [38]. This is somewhat unfortunate, for this particular bargaining model is highly important in its own right. It can be viewed in some respects as the culmination of "classical" n-person solution theory -- indeed as its *symmetry limit* in the sense we are about to describe.

To understand the critical role of symmetry assumptions in Harsanyi's 1963 theory, let's sketch how it is possible to rederive the solution by postulating a family of *a priori* symmetries on a one-by-one basis, and by witnessing how these constraints shape the resulting solution. Naturally, symmetry assumptions alone do not buy us Harsanyi's solution, but they go a surprisingly long way. Because his model is constructive rather than deductive in nature, many of the symmetries we introduce necessarily pertain to the *structural* properties of the bargaining game being formulated -- not to the formal properties of the outcome itself.

Assume that our goal is to construct an n-person bargaining solution for a non-transferable utility game given in normal form. As a first step, let's decide on which of two generic types of solution we want: one that is chronologically symmetric in the sense that every player *simultaneously* takes all other parties' actions as given when selecting his strategy? Or else one that is chronologically asymmetric and exploits the dynamics of the game? Let's opt for the first approach. This will restrict us to a fixed point solution theory of some kind as in classical (Arrow-Debreu) theories of competitive equilibrium. It is interesting to note that these classical "static" theories played a role in the evolution of economics analogous to that played by Harsanyi's bargaining model in the evolution of game theory. Today, both fields proffer truly dynamic models.

Next, let us ask who are to be the "actors" (as distinct from the players) in the n-person game? Symmetry considerations suggest that *all* possible actors should participate, barring good reasons to the contrary. But what in turn does this mean? Here we are forced to confront one of the deepest aspects of cooperative game theory, one which Harsanyi was the first to recognize. The actors will be the players viewed simultaneously as members of all possible $\mathbf{n(n-1)/2}$ bilateral bargaining games $\mathbf{G_{ij}}$ between all possible pairs of individuals, *and also* as members of all $\mathbf{2^n - 2}$ coalitions $\mathbf{S}$ that will be allowed to form and oppose complementary coalitions $\mathbf{S^*}$. But recognizing this, we must decide whether pairwise bargaining or coalitional decision-making comes first. Given the symmetrical spirit of our fixed point paradigm, we again opt for simultaneity and answer "neither".

There is, however, one elementary sense in which the bilateral bargaining games G_{ij} take logical precedence over coalitional behavior: coalitions consist of individuals and thus of pairs of individuals, whereas pairs do not consist of coalitions. Mathematically, this implies that the maximization problem to be solved by a given coalition **S** must be *constrained* by the particular sharing agreement (the bargain) reached by each pair of players within **S**.

Next, what should we assume about the structure of coalition formation in the game? Which of all possible coalitions will form, and in what order? Symmetry considerations suggest that unless there is an *a priori* case to the contrary, all possible coalitions will form. (This of course is Harsanyi's "pluralistic" axiom that individuals in life will be members of different, overlapping coalitions reflecting the diversity of their interests -- not members of some disjoint structure.) What about the order of coalition formation? Again, symmetry considerations imply that there should be no discrimination here, and that all coalitions **S** and their complements S^* form simultaneously.

Finally, what shall we assume about the behavioral rationales of the players both in their bilateral games and in their coalitional games? (Here our assumptions concern the nature of rational behavior *per se*, and not such "structural" n-person details as whether all coalitions are allowed to organize.) Consider first behavior in the bilateral bargaining games. Elementary Nash-Harsanyi bargaining theory provides a straightforward answer, namely symmetrical "split-the-difference" strategies awarding each player an *equal* (utility-weight weighted) utility gain above the pertinent disagreement payoff. This set of $n(n-1)/2$ sharing agreements will then constrain the optimization problems solved by the coalitions.

Note that the case for these symmetrical Nash-Harsanyi payoffs rests largely on familiar symmetry postulates, in particular invariance under relabeling of the players, invariance under cardinal transformations of the utilities, and symmetry of the solution function itself (already discussed above). At an even deeper level, Shapley [38, Theorem I] has used a group theoretical invariance argument to justify the use of cardinal rather than ordinal utilities in *any* genuine bargaining model.

Turning to the question of coalitional behavior, what should each coalition be assumed to optimize in selecting its (threat) strategy? Symmetry considerations suggest that any coalition **S** except for the grand coalition **N** will wish to choose a strategy maximizing the difference between the damage caused to its antagonist

S* and that caused to itself. But by symmetry, **S*** will pick *its* strategy with the aim of minimizing this same entity. The MAX MIN strategies selected on these grounds of symmetry correspond of course to the "modified" characteristic function introduced by Harsanyi in his original 1959 bargaining model [10], and axiomatized in the transferable utility case from extensive form axioms by Selten [35].

In their seminal "Power and Taxes" papers, Aumann and Kurz [2] have noted how much more appropriate this solution is for political contexts than are alternative solutions, in particular those resting on the concept of "blocking" used in the core and other classical cooperative theories. Aumann and Kurz recognize that blocking is *insufficiently symmetrical* as regards questions of reciprocal opportunity costs and incentives, whereas the modified characteristic function is not.

After articulating all these structural and behavioral symmetries, where do we end up? We possess the raw material with which to construct the symmetry group characterizing Harsanyi's 1963 solution concept, a group of order $[\mathbf{n} + 2(\mathbf{n} \cdot 2^{\mathbf{n}-1})]$, where **n** denotes the number of players. This number just happens to correspond to the number of independent equations defining a Harsanyi equilibrium. We shall discuss the significance of this observation in Part II below where we formally define symmetry groups, and discuss the concept of the "right" amount of symmetry (and hence simplicity) to build into a theory.

The important point here is that virtually all of the symmetries we have adduced are common sensical and *a priori* in nature, subject to our proviso of a classical game theoretical world of little or no friction. In principle, they could have been utilized systematically to expedite the construction of Harsanyi's theory. Indeed, many of them were involved in his reasoning process, either explicitly or implicitly. But one subtle structural symmetry was not, namely the simultaneity of "pairwise" and "coalitional" bargaining. The absence of this particular symmetry is what differentiates Harsanyi's 1963 solution from his original 1959 solution and makes it more compelling, as Harsanyi himself has pointed out at the beginning of [13].

By relaxing still other symmetries in turn, we obtain the symmetry groups of other solution concepts such as Harsanyi's [18, Chapter 13] "discriminatory solutions" (allow the order of coalition formation to matter), and various solutions lying off of the main diagonal of the hypercube generated by Owen's multilinear extensions of games. A much more dramatic modification of these symmetries retrieves the symmetry groups of classical concepts like the core (remove the "pluralism" symmetry, Shapley's "unwavering" invariances implying an essential role for cardinal utility, etc.). In short, Harsanyi's 1963 equilibrium incorporates about as many reasonable symmetries as are conceivable in the context of a classical, static, normal form bargaining game with complete information. It is in this sense that his solution can be viewed as a symmetry limit.

Of course, this theory is "constructive" as opposed to "deductive", and this raises the question of whether his solution can be viewed as a symmetry limit from the more formal standpoint of an axiomatic approach. Assuming that we forego a normal form representation and take a non-transferable utility (NTU) characteristic function as given, then recent results of Hart [20] suggest that it can be. His axiomatization of the Harsanyi NTU-value parallels Aumann's [1] characterization of the Shapley NTU-value and reveals the full symmetry of the Harsanyi solution.

In particular, the axiom set which Aumann originally introduced suffices for the Harsanyi case just as it does for the Shapley case provided only that the domain of the axioms becomes the set of *all* possible coalitions **S** and not just the grand coalition **N**. From a behavioral standpoint, this situation is consistent with the normal form spirit of Harsanyi's constructive model in which the role of coalitions and their threat strategies are real, not simply notional, as Aumann has pointed out to us. From a formal standpoint, the situation is clearly more symmetrical.

In bringing this brief review of classical cooperative theory to a close, it is worth noting that Harsanyi's subsequent applications of his bargaining equilibrium to the measurement of power and of social status [12] permitted a dramatic generalization of traditional quantitative measures, e.g. Simon's and Dahl's power measures, and the Shapely-Shubik power index. The increased explanatory and predictive power of Harsanyi's measures stems almost wholly from the symmetries incorporated in his underlying bargaining model. For example, his measure of a given player's true power reflects the symmetries implicit in such concepts as relative "opportunity cost" and "incentive power".

Post-Classical Theory: Turning very briefly to Harsanyi's contribution to post-classical solution theory, we find symmetry considerations continuing to play a central role in theory creation. In the theory of games with incomplete information [14], the "Harsanyi transformation" as it is now called in undergraduate texts [26] amounts to symmetrizing the information structure of the game by introducing a "basic probability distribution" that governs the lottery generating the "types" of possible players in a game. Symmetry at an even deeper level is achieved via Selten's "posterior lottery" model [14, Part III] which can be used to overcome the problem of mutually inconsistent subjective probability distributions.

Harsanyi's "tracing procedure" [15] for its part can be viewed as a dynamical model of convergent expectations whereby rational individuals symmetrize initially inconsistent expectations about each other's behavior. Of greater interest is the general non-cooperative solution theory of Harsanyi and Selten [19] of which the tracing procedure is a component part. While this monumental theory of equilibrium selection can be appraised from many different standpoints, we should like to suggest that it offers a classic example of theory construction via the process of "symmetry-breaking" familiar from physics (e.g. Rosen [27], Chapter 5).

The principle here is that as the domain of a theory is broadened to include increasingly complex cases, symmetries must be weakened if not shed outright. Consider the limiting case of "exact" charge symmetry in particle physics, a symmetry postulating equivalence of pairs of states differing only in that one or more protons in one are replaced by neutrons in the other, or vice versa. In reality, this symmetry is broken and only holds approximately due to the Pauli exclusion principle and electromagnetic interaction.

For its part, the Harsanyi-Selten theory starts off with classical non-dynamic games of complete information where straightforward dominance relationships and invariances such as "isomorphism" suffice to yield classical solution concepts like the Nash bargaining solution. Then when dynamics and incomplete information are incorporated, tempting new invariances present themselves. Yet some of these must be given up as the theory is crafted, e.g. invariance with respect to sequential agent splitting (cf. Impossibility Theorem 3.12.1).

Others must be severely modified. For example, invariance with respect to the elimination of inferior choices, duplicates, and semiduplicates must be replaced with *partial* invariances of the same kind which only apply "locally" to those indecomposable games derived in computing the Harsanyi-Selten solution. And so on. In a similar fashion but in a different context, Kohlberg and Mertens [23] discover the mutual inconsistency of various invariances under elementary game tree transformations in their analysis of forward versus backward induction.

How are we to react to a solution theory that asks us to strike compromises amongst competing *desiderata* regarding not only invariances, but also dominance relationships and continuity? The tendency is to regret the situation. But if physics is any guide here, we should welcome the process as the essence of serious theory construction. The ultimate contribution of Harsanyi and Selten has been to teach us how truly complex the concept of rational behavior really is when the domain of competitive decision-making is fully general. From this perspective, we should perhaps welcome the process of discarding desiderata such as agent splitting invariance which, *upon careful reflection*, may never have really been appropriate in general game situations.

Ethics: Finally, there is Harsanyi's well-known work in utilitarian moral philosophy. Once again, symmetry considerations played a central role in his ability to push back the frontiers of this field. In the case of his constructive model of "impersonal" social choice [18], symmetry asserts itself in the crucial assumption that each of n individuals has an *equal* probability **1/n** of being in the shoes of any of the players once the veil of ignorance is lifted. In his axiomatic derivation [18] of a utilitarian social welfare function, symmetry enters explicitly in the form of his axiom 4°° that the (additive) social welfare function **W** should be a symmetric function of the individual utility functions. At a more indirect level, algebraic considerations enter via the familiar axioms of expected utility maximization which underlie his first derivation of a linear welfare function.

More recently, Harsanyi [17] has refined his welfare function and argued for the superiority of "rule" rather than "act" utilitarianism. Here the scholar who for years urged us not to confuse the strategic considerations of game theory with the moral concerns of ethics somewhat reverses his position: he utilizes the Nash equilibrium point concept to clarify how rule-utilitarianism outperforms act-utilitarianism when coordination and expectations effects are properly taken into account. Interestingly, it is the "symmetrical-best-reply" perspective of Nash's non-cooperative theory that buys Harsanyi his principal results. His analysis of the "voter's paradox" offers an excellent case in point.

Regrettably, lack of space has required us to neglect many of Harsanyi's contributions both to game theory and ethics. Yet our objective in Part I will have been achieved if we have succeeded in demonstrating that Harsanyi's successes stem in good measure from his recognition of the importance of symmetry considerations in theory construction, and from his ability to exploit these both constructively and deductively as the case may be. We now turn to a more abstract perspective on the role of algebraic (in particular symmetry) assumptions in the logic of scientific discovery.

II. Algebra, Theory Construction, and "Simplicity"

Popper [25] amongst others has expressed the view that the process of scientific discovery is essentially psychological and lies outside the scope of logic. Herbert Simon and others [24, 43] have argued against this viewpoint in recent years. Indeed, the Carnegie-Mellon school urge that greatly increased attention should be given to the logical and replicable dimension of theory construction, and the associated problem of the "retroduction" of scientific theories. As support for this view, and the related argument that the logic of scientific discovery is not merely psychological, they cite the advent of artificially intelligent systems such as **BACON**. This rather elementary "search" system can derive Kepler's laws on its own if given sufficient data.

The suggestion we wish to make is that abstract algebraic considerations should play an important role in this endeavor, especially if the experience of twentieth century science is any guide. We start off with a brief analysis of *why* one might expect successful theory creation to have an essentially algebraic dimension. Once an answer is given, and a general principle established, a host of concrete examples are cited to substantiate our viewpoint. Thereafter, we turn to the time-honored desideratum of "simplicity" and suggest two ways in which this problematic concept can be clarified via algebraic reasoning.

The General Principle – Epistemological Consistency: Twentieth century science suggests that the search for truth is really a search for that subset of variables and relationships that are "fundamental" in the sense that they remain valid when all irrelevances have been swept away. At a formal level, this perspective suggests a strategy of theory construction in which algebraic considerations rise to the fore. For algebra alone gives us the precise tools with which to articulate *and* to sweep away irrelevances. The method, of course, is first to postulate all apposite additivity, linearity, independence, and related invariance assumptions; and then to utilize these as constraints in theory construction. Within algebra, group theory is of particular importance in this endeavor since it quite literally helps us discern "reality" once all irrelevances have been *transformed away* via admissible transformations.

More formally, suppose we start off with a normative meta-model **M** of efficient theory construction, that is, a rational procedure for arriving at a theory **T** generating a solution $\mathbf{S^T}$ (e.g. the laws of motion) to the problems posed within a given discipline $\mathbf{D}$ (e.g. classical mechanics). Suppose moreover that we have agreed upon some set $\mathbf{I^T}$ of irrelevances within $\mathbf{D}$, i.e. variables and/or relationships deemed irrelevant and to which $\mathbf{S^T}$ should thus be blind. Define the dual of this set $\mathbf{I^T}$ in $\mathbf{M}$ to be the set $\mathbf{E^T}$ of *equivalent problem reformulations*, i.e. all formulations of a given problem that differ only with regard to elements of $\mathbf{I^T}$. For example, the Nash-Harsanyi bargaining theory deems as (strategically) equivalent any two bargaining games $\mathbf{G_1}$ and $\mathbf{G_2}$ that differ only with respect to the choice of the utility unit and the telephone numbers of the players. Both games would then be elements of an appropriate set $\mathbf{E^T}$.

At this stage we can invoke a deceptively simple yet deep consistency axiom that has proven of great importance in inductive logic and science in recent decades: *meaningful theories will award one and the same solution* $\mathbf{S^T}$ *to all elements of* $\mathbf{E^T}$. In inductive logic, scholars such as Cox [7], Jaynes [22], and Rosenkrantz [28] have stressed the epistemological priority of this principle of "consistency." For motivation, consider the special case of probability theory where this principle requires that in problems where we have the same information, we should assign the same probabilities. The following example illustrates how this principle can be used to help characterize an "objective" prior distribution [28, p. xiv].

Assume we are sampling a normal distribution of known (unit) variance with unknown mean $\mathbf{m}$ and that we seek an "ignorance prior" $\mathbf{f(m)}$ on $\mathbf{m}$. Then the density is a function $\mathbf{h(x-m)}$ of $\mathbf{x - m}$, so that $\mathbf{m}$ can be viewed as a location parameter. A translation of coordinates $\mathbf{x' = x + b}$, $\mathbf{m' = m + b}$ leaves the normal form and spread of the density unaltered, and so "we have the same information" in two versions of the problem that differ solely by a translation of coordinates. If $\mathbf{f}$ and $\mathbf{g}$ are the prior densities of $\mathbf{m}$ and $\mathbf{m'}$ in two equivalent versions of the problem, then the consistency principle forces $\mathbf{f = g}$.

This leads straightaway to the functional equation $\mathbf{f(m) = f(m+b)}$ expressing the translation invariance of $\mathbf{f}$. The unique solution to the problem is the uniform prior originally advocated by Jeffreys in 1939 on somewhat *ad hoc* grounds. Note that all of the essential power in this approach stems from algebraic considerations, in particular **(i)** the articulation of the appropriate invariance, and **(ii)** the expression of our consistency requirement in the form of a functional equation whose solution is the parameter of the "objective" prior that is sought.

But, in fact, the role of algebra can run much deeper than this argument suggests. To see this, consider the problem of determining the *general* conditions under which "initial ignorance" about a set of statistical parameters can be translated into an objective prior distribution. Suppose as is usually the case that in asserting initial ignorance, we are not quite as ignorant as we pretend insofar as we *agree* that a certain set of operations transform our initial problem into equivalent problems. (These operations might be those of parameter rescaling and translation.) Suppose additionally that these operations constitute a valid transformation group the number of whose independent parameters equals the number of statistical parameters in the original problem.

Then in this happy case, the fundamental domain of the underlying group reduces to a point, and the form of the prior will be uniquely determined (e.g. Jaynes [22] and Wigner [42]); and the specification of the transformation group represents an *exhaustive description* of a state of knowledge. Regrettably, many problems are degenerate in this regard, and these are the cases where Jaynes recommends the use of maximum entropy (or minimum cross-entropy) distributions.

Demonstrable Success of the Principle: The epistemological consistency principle introduced above has played a critical if implicit role in several of the most significant scientific advances of the twentieth century. Both relativity theory and game theory are fundamentally theories of symmetry that fully exploit consistency. In their respective domains, they were arguably the first theories ever to be based upon laws about laws -- laws of symmetry, as it happens. General relativity theory postulated that if a set of laws can be deemed to possess objective truth, then these laws will hold true ("work") regardless of the space-time transformations involved.

Game theory for its part introduced the perspective of symmetrical rationality in contexts of strategic interdependence. Specifically, for a general behavioral theory to be deemed meaningful, it must assume that *all* players possess the rule book of the theory, and all must expect each other to play by it. When enriched by the symmetrical-best-reply concept of Nash's equilibrium point solution, this perspective lanced the Gordian knot of an infinite regress of expectations -- one of the great advances in the history of social science.

Arrovian social choice theory also relies heavily upon invariance conditions. Yet their true role was obscured until Sen [36] offered an ingenious new proof of Arrow's impossibility theorem based largely on invariance arguments. Sen's reformulation significantly enhances our understanding of the very meaning of Arrow's theorem.

In inductive logic, we have already cited the role of symmetry considerations in characterizing "objective" prior distributions under conditions of uncertainty as opposed to risk -- one of the most difficult problems of all. An equally challenging set of problems of a non-decision theoretical nature were also solved by the introduction of algebraic reasoning and invariance (independence) assumptions. We have in mind the interrelated problems of model identifiability, causality, and spurious correlation addressed by Herbert Simon and various Cowles Foundation scholars in the early 1950s. The issue of spurious correlation provides an excellent example of the reasoning process we have in mind, and we can do no better than to quote Simon himself:

> We shall show that the decision that a partial correlation is or is not spurious (does not or does indicate a causal ordering) can in general only be reached if *a priori* assumptions are made that certain *other* causal relations do *not* hold among the variables. ... The need for such assumptions follows from considerations of elementary algebra [40, Chapter 2.3].

Simon goes on to show that the required *a priori* assumptions are "zero constraints" of two generic kinds: the coefficients of certain variables deemed irrelevant must be zero; and certain error terms $\mathbf{u_i}$ and $\mathbf{u_j}$ are uncorrelated so that $\mathbf{E(u_i \cdot u_j) = 0}$.

The rapid development of information theory offers yet another compelling example of the power algebraic reasoning in theory construction. Jaynes [21] has shown that Shannon's entropy measure is the *only* measure that satisfies an algebraic consistency condition represented by a simple "composition law" of probability theory. In more recent work of great interest, Shore and Johnson [39] have generalized the inferential approaches of Shannon, Jeffreys, and Jaynes. Utilizing three elementary consistency axioms, they axiomatize a unique inferential operator that combines a prior distribution $\mathbf{p}$ with mean value constraints $\mathbf{D}$ to yield a posterior distribution $\mathbf{q = p \cdot D}$. The operator in question minimizes a generalization of entropy known as *cross-entropy*. Both Bayesian conditionalization and entropy maximization can be shown to be special cases of this procedure, and thereby receive novel support as modes of inference.

Quantum physics is not fundamentally a theory of symmetry in the way relativity theory is. Yet its linear structure permitted the full power of group representation theory to be unleashed in its behalf. As is well known, the results have been stunning. One classic example was the prediction of the existence of the Omega-minus baryon particle prior to its discovery -- a prediction of Gell-Mann and Ne'eman's group theoretical "eightfold way". Existence was ultimately confirmed at Brookhaven National Laboratory in 1964.

We have only been able to cite a few of the areas in which algebraic considerations of symmetry and consistency have played a key role. Other disciplines include ethics, economics, and aesthetics. In hindsight, all of this may seem quite natural -- a development to have been expected. But a reading of original sources starting with Einstein suggests that this is not the case. We find that *explicit* symmetry considerations often arise after the fact, when they assume a role that purges, clarifies, and unifies.

The *status quo* in the philosophy of science literature mirrors this reality: to the best of our knowledge, there has been little if any discussion of the generic role of algebraic reasoning in the disciplines of theory construction, much less in epistemology (but see [4; 30; 38]). In the final section of this essay, we reexamine the concept of simplicity from an algebraic perspective, and in doing so attempt to redress this methodological state of affairs in a small way.

"Simplicity" Revisited

Most analyses of simplicity pertain to the testing and ranking alternative theories that already exist. For example, the paucity-of-parameters criterion suggests that if two different theories fit the empirical facts equally well, then the one with fewer parameters is "better" in several different regards. Our own focus will be different, and will center on the creation of a theory in the first place. What might simplicity/complexity refer to in this case? How simple/complex should we try to make a theory on *a priori* grounds? Why? Note that the implicit question here is that of *optimal* simplicity. Our traditional prejudice on behalf of maximal simplicity amongst competing theories must be set aside prior to the existence of one (much less two) theories which end up fitting the facts.

When posed in this *a priori* context, the meaning of (optimal) simplicity factors quite naturally into two general questions: First, how many independent relationships (equations) should we attempt to postulate when creating the theory? The more of these, the less simple the theory in an elemental sense. Second, how simple should we expect these relationships to be -- in particular will they be linear or nonlinear equations? We shall now argue that algebraic considerations offer singular guidance in answering both questions.

Consider first the question of the number of independent relations to aim for. We are about to introduce a generalized 5-step algorithm which is partly intended to compute this number. Specifically, Step #4 identifies the "correct" number of independent relationships in **T** with the number of essential parameters characterizing the symmetry group of **T**. Yet our algorithm does more than this. To use a concept introduced above, it constitutes a *meta-model* of the logic of scientific discovery, so we shall designate it as **M**. Since three of the five steps of this procedure require interaction with human experts, so **M** is not technically an algorithm but rather an interactive *Expert-System* which calls up embedded algorithms as appropriate. If implemented, this Expert-System would be an order of magnitude more "creative" and powerful than existing systems surveyed in [24, 42].

The Expert System M and its Algorithms: Suppose we are trying to construct a theory **T** that will offer solutions to a class problems in some discipline **D** (e.g. physics). Then drawing only on *a priori* information, follow the steps specified in the following meta-model:

Step #1: Identify the Variables in T: Let **V** denote a universe of variables that could end up being interrelated by **T**. Identify a subset of truly relevant variables within **V**, and denote the set by **V***. These variables will serve as the domain of the theory being constructed. (This step is not trivial. Recall Harsanyi's [11] independence-of-irrelevant-*variables* postulate, and its crucial role in the proof of his own bargaining solution.) Note that at this stage, we cannot yet determine the number of variables within **V*** that will serve as the *independent* variables of **T**. But Step #4 below will help us do so. Note: Step #1 will clearly require interaction with "foundations thinkers" in **D**, and thus cannot be fully automated.

Step #2: Identify the Symmetries of T: Identify on a one-by-one basis the set $\mathbf{I}^{\mathbf{T}}$ of all appropriate "irrelevances" of the theory being constructed, and characterize each as a formal symmetry transformation of **T**. This identification can be carried out directly, or else indirectly by constructing the dual set $\mathbf{E}^{\mathbf{T}}$ of equivalent problem formulations within **T**. The caveat "appropriate" just above is important, since we want to end up with exactly the right amount of symmetry: not too much, and not too little. This point will be motivated below.

Denote the ith such symmetry transformation as $\mathbf{S}_i$, and denote the set of all these transformations as **S**. This set will always be non-empty, a fact that follows from the equivalence relations that *any* theory **T** will impose on the state space of the underlying system (cf. Rosen [27, Chapter 3] for a proof).

This second step is extremely important. Like Step #1 above and Step #5 below, it calls for considerable interaction with foundations thinkers, and thus cannot be fully automated. Nonetheless, if our Expert-System **M** is an effective one, this second component will play an important role in algorithmically eliciting, systematizing, and unifying the insights of experts who are interactively consulted.

Step #3: Construct the Symmetry Group of T: Assume that each symmetry transformation is *invertible*, as is almost always the case in applications. [This assumption can be relaxed, but then Step #4 would have to be altered.] Apply the various symmetry transformations $\mathbf{S_i}$ in a repeated *and* consecutive manner. By "repeated" we mean binary repetitions of a given transformation. By "consecutive" we mean binary combinations of operations across different transformation sets, e.g. a 90 degree rotation combined with a given translation. The result of any such binary combination will itself be a symmetry transformation possessing an inverse (cf. [27, Chapter 3] for a proof).

Now carry out these operations exhaustively until closure is achieved. Closure should occur in a finite number of steps provided that the repeated application of any discrete symmetry transformation in **S** closes in a finite number of steps, and that any continuous family of transformation in **S** has a finite number of essential parameters, as they almost always do in applications.

Define the set of all symmetry transformations that has resulted upon closure to be the symmetry group of **T**, denoted by $\mathbf{G_s^T}$. [Henceforth we shall delete the superscript **T**.] In the case where the various symmetry transformations generating it commute, $\mathbf{G_s}$ will take the form of a *direct product* of the symmetry subgroups associated with these transformations. Denote by $\mathbf{G_i}$ the symmetry subgroup associated with $\mathbf{S_i}$.

Finally, denote by $\mathbf{O(G_s)}$ the *order* of $\mathbf{G_s}$. Given our procedure for constructing this symmetry group, its order will be an appropriate measure of the *total amount of symmetry* which **T** must ultimately incorporate. In the event that $\mathbf{G_s}$ is a direct product of the symmetry subgroups $\mathbf{G_i}$, the computation of $\mathbf{O(G_s)}$ should be straightforward. Otherwise it may not be. Note: Step #3 could be automated algorithmically with little difficulty.

Step #4: Determine the Correct Number of Independent Relationships: Denote the number of *essential* (i.e. functionally independent) parameters of $\mathbf{G_s}$ by $\mathbf{O^*(G_s)}$. In certain cases, this number will simply equal the order of the symmetry group. More generally, it will be a non-trivial function of the number of generators in any discrete symmetry subgroups $\{\mathbf{G_i^D}\}$ that involve generators, and of the number of essential parameters in any continuous symmetry subgroups $\{\mathbf{G_j^C}\}$.

Next, let $\mathbf{r^*}$ denote the correct number of independent variables within the set $\mathbf{V^*}$ of relevant variables. Since **T** is by assumption to be a determinate theory, $\mathbf{r^*}$ will also denote the number of independent *relationships* involving the independent variables. Denote this set of relationships by $\mathbf{R^*}$, and henceforth think of $\mathbf{R^*}$ as the full mathematical expression of **T** that we are seeking.

The algorithmic instruction here in Step #4 is to set $\mathbf{r^* = O^*(G_s)}$. As we see in the "comment" just below, there is a compelling theoretical reason for this equality. Moreover, it holds in all cases we know of where symmetry analyses of the proposed kind have actually been carried out. Yet we know of no proof that equality should hold in general. Note: Step #4 could be automated with no difficulty, assuming that $\mathbf{r^* = O^*(G_s)}$. It is unclear whether automation would be possible in the absence of equality.

Comment: In crafting **R***, our fundamental algebraic requirement is that these relationships satisfy *exactly* the amount of symmetry captured by $\mathbf{G_s}$ -- no more, and no less. The principal reason for this goal is to avoid the costs associated with two types of errors in theory construction: that due to building in too much symmetry; and that due to incorporating too little symmetry. These errors are reviewed below where we see that too much symmetry necessarily translates into too few independent relations, with attendant costs; and too little symmetry yields too many relations, again with attendant costs. Thus there will always exist a "correct" number **r*** of independent relationships, lying in between these twin danger zones. And it is these **r*** relationships that will constitute the idealized formalism **R***. But what *is* the correct number?

It would seem to be $\mathbf{O^*(G_s)}$. The theoretical reason why this is true has been set forth by Sachs [30, p.138] in the context of relativistic field theories:

> When each of the parameters of a symmetry group is varied, the underlying field will change its values in a corresponding way. Each of the changes in the field that follows from the variation of each of the independent group parameters then gives rise to a different equation. When the parameters are independent (i.e. they are the essential parameters), then all the correlated equations that result from these parameter changes must also be independent. Thus, the number of equations (that is, relations between the field components) at each space-time point must be equal to the number of essential parameters of the underlying symmetry group.

At an empirical level, the equality $\mathbf{r^* = O^*(G_s)}$ holds true for the cases where symmetry analyses have actually been carried out. Specifically, it holds in the case of the conditions establishing the existence of a unique prior probability distribution in inductive logic [22]. Equality also holds for the algebraic generalization of general relativity theory [31] discussed below. And it holds for Harsanyi's 1963 n-person bargaining solution as we have indicated in Part I.

Step #5: Characterize the Relationships R* Themselves: Proceed now with the most difficult task of all, namely constructing the **r*** relations **R*** so as to give substantive content to **T**. At this stage, algebraic considerations based on symmetry insights will have to be supplemented by other mathematical strategems reflecting altogether different kinds of intuition. The use of monotonicities will play a particularly important role in this regard, e.g. Lagrangeans in physics; preference relations in economics and game theory. Of particular interest is the way in which symmetry considerations interact with monotonicities, e.g. the way in which a sequential specification of symmetries in physics constrains (and therefore helps to "create") the resulting Lagrangean of a theory. For an example in social science, consult Sen [36].

Note: Implementing Step #5 will require extensive interaction with foundations thinkers in **D**, so it will not be able to be automated. Even the creation of an expert-interactive subroutine here will be difficult given the magnitude of human creativity involved in this particular step. Such a subroutine would have to allow for the possibility of a feedback loop from Step #5 to previous steps.

The Twin Errors: Our fundamental point in Step #4 was that $\mathbf{R}^*$ must end up embodying exactly the right amount of symmetry -- not too much, not too little. Consider what goes wrong when this requirement is violated in either direction. First, a scientist could wittingly or unwittingly impose too many symmetries on some suboptimal formalism **R** -- i.e. any number *larger* than the symmetry group order $\mathbf{O^*(G_s)}$, the right amount of symmetry. Doing so will yield *too few* independent relationships compared to the optimum number $\mathbf{r}^*$. For each additional symmetry will necessarily reduce the number of independent variables/relationships, usually on a one-for-one basis. This follows from the essential role of symmetry conditions in theory construction.

As a result, **R** will be too constrained. It will not retain enough freedom to fit/explain the data. Viewed somewhat differently, a theory incorporating too much symmetry will not have the ability to generate as many (falsifiable) predictions as it ought to have. In his 1970 paper in *Nature*, Sachs [30] has demonstrated this point in the case of relativistic field theories. He claims that such theories lack "maximum generality" and necessarily mask information (predictions).

Second, the scientist could (wittingly or unwittingly) build too few symmetries into his theory -- any number *smaller* than $\mathbf{O^*(G_s)}$. Doing so will yield *too many* independent relationships compared to the optimum $\mathbf{r}^*$. A theory with this deficiency will be too loose, too unconstrained. It will be possible to fit the data in too many ways given the abundance of "free" parameters. Viewed somewhat differently, it will be sensitive to distinctions to which it should be blind. As a result, it may yield superfluous predictions that are false. As an example, consider those false predictions of classical physics stemming from the often implicit assumption that different space-time reference frames *do* have genuine physical significance, and thus matter.

"Simplicity" Generalized: Recall that we originally equated the optimal simplicity of a given theory **T** with the number $\mathbf{r}^*$. We then saw that $\mathbf{r}^*$ itself is typically equal to the amount of symmetry as proxied by $\mathbf{O^*(G_s)}$. Let us now set aside our concern with the right amount of simplicity for a given theory, and generalize our concept of simplicity so as to be able to rank different and possibly competing theories in terms of their *relative simplicity.* We will assume that all such theories are formally comparable in the sense that they address the same "reality" by interrelating the same set of relevant variables $\mathbf{V}^*$. In this comparative context, a theory that possesses fewer independent variables/relationships than another theory can be deemed the *simpler* of the two. [Note that we cannot state that simpler means better given our finding that there exists an optimum $\mathbf{r}^*$ consistent with all apposite symmetries embodied by the group $\mathbf{G_s}$.] Now we have seen that the number of independent relationships in any theory is inversely proportional to the number of symmetries that are imposed upon it. As a result, *simplicity is a linearly increasing function of symmetry across formally comparable theories.* This result gives formal expression to an aesthetically appealing intuition linking the concepts of symmetry and simplicity.

We now present some concrete examples from the history of science that support the overall viewpoint we have taken in constructing the epistemic engine **M**.

Examples from Physics: Pitfalls in the evolution of physics during the twentieth century offer a justification for the normative epistemological strategy embedded in the Expert-System **M**. Classical physical theories lacked the predictive and explanatory power of relativistic theories because they lacked the symmetries that Einstein added, namely special and general relativistic covariance with respect to transformations of the space-time parameters. In the case of general relativity, the underlying symmetry group is a 16-parameter Lie group representing covariance of the laws of nature under the group of continuous transformations of the space-time derivatives, i.e. covariance under the group of all possible "motions".

The formalism $\mathbf{R}^*$ consistent with this symmetry will be a 16-component metric tensor defining 16 independent relations at each space-time point. Note that $\mathbf{r}^* = \mathbf{O(G_s)}$ here. This has been proved by Sachs [30,31]. Einstein's tensor formulation **R** of 1916 did *not* have this property. For algebraic reasons following from the particular differential geometry chosen, **R** consisted of only 10 independent relations at each space-time point. This set of $\mathbf{r} = \mathbf{10}$ independent relations was sufficient to characterize gravitational phenomena, but not electro-magnetic phenomena as well.

We know from the foregoing analysis that if $\mathbf{r} < \mathbf{r}^*$ in two corresponding formalisms **R** and $\mathbf{R}^*$, then **R** must be accommodating too many symmetries. And indeed, Einstein's formalism was compatible not only with motion symmetry, *but* *with space-time reflection invariance as well.* The latter was an unwarranted symmetry having nothing to do with the *a priori* concept of relativity theory. As we have seen, the algebraic price paid for accommodating too many symmetries will be a smaller-than-optimal set of independent relations incapable of predicting a portion of the available experimental (in this case electro-magnetic) data.

This impasse kept open the hope that some generalization of general relativity might eventually make possible a field theory $\mathbf{R}^*$ with $\mathbf{r}^* = 16$ independent relationships unifying gravitational and electro-magnetic phenomena. It did, and algebraic considerations played the decisive role in this extension. The developments we have in mind originated in 1934 when Einstein and Mayer [8] pondered the spin representations of Dirac's wave equation of the electron, and found *quaternion algebra* emerging as a derived structure from the group representation theory. They did not utilize this fact to generalize general relativity. But even if they had wished to, the mathematics of quaternion calculus in a Riemannian space had not yet been developed.

Sachs [31] himself did pursue this route several decades later. Upon using field variables that obeyed the algebra of quaternions, *and* upon removing reflection invariance, he discovered the existence of a quaternion metrical field (a 16-component variable) and 16 independent relations at each space-time point that are solved by this field. The resulting formalism was shown to *factorize* into **(i)** a symmetric 10-component field that solves a generalization of Einstein's original tensor theory of gravitation, and **(ii)** an anti-symmetric 6-component field that solves a generalization of Maxwell's electro-magnetic theory. The fact that both formalisms are generalizations of their classical antecedents is important: both yield new, falsifiable predictions -- predictions masked by the earlier theories as must be the case whenever $\mathbf{r} < \mathbf{r}^*$.

We have drawn on physics to illustrate the concept of a theory whose formalism is maximally general and embodies "just enough" symmetry as gauged by the order of the irreducible symmetry group of the underlying theory. The crucial observation is that this group can often be deduced on *a priori* grounds *before* a theory has actually been constructed. It is thus available for guidance from the very start, provided only that we undertake the difficult fundamental thinking needed to construct it.

Examples from Social Science: Could a logic of scientific discovery of this kind find its place in the social sciences? In principle, yes, especially in the decision sciences where symmetry considerations have played a critical role ever since the publication of the von Neumann-Morgenstern treatise on game theory in 1944. Yet *a priori* reasoning filtered through group theoretical methods *per se* have not played a significant role to date. The recharacterization of Harsanyi's 1963 solution (and related solutions) sketched in Part I above offers one approach to such a research programme. Another interesting example can best be illustrated by the following thought experiment.

Suppose it is the year 1945 and we wish to fill in an important gap in the new von Neumann-Morgenstern theory, namely to determine a unique value for two-person pure bargaining games. Suppose, moreover, that we are struggling with the knotty problem of the right *informational basis* for such a value theory: ordinal versus cardinal utility? Interpersonal comparisons or not? In short, what are the appropriate informational invariances for a meaningful theory?

There are two routes to an answer. First, we could listen to the arguments of dogmatists of different persuasions, and then choose. Logical positivists (especially within economics proper) would insist on an ordinally invariant theory devoid of interpersonal comparisons. Philosophers concerned with "fairness" would insist on cardinally significant interpersonal comparisons. Few if any scholars would make the case for Nash-Harsanyi cardinal invariance, largely due to lack of any such tradition at that time. Choice here would indeed be difficult.

Second, we could pursue the group theoretical approach sketched above. Shapley [38, Theorem I] was to take this route many years later in 1969. He would show that elementary *a priori* reasoning about the essential nature of bargaining implies a symmetry group (his transitive, "unwavering" group G) calling for cardinal utilities but no interpersonal comparisons. Note where we come out here: Shapley's group has *just enough* symmetry for the bargaining problem at hand (as does Sachs' 16-parameter group for a general theory of physics-as-motion); the positivist-ordinalist group contains *too many* symmetries (as does Einstein's original formulation of general relativity), and the classical utilitarian "fairness" group possesses *too few* symmetries (as does the group of classical mechanics, or the Poincaré group of special relativity for that matter).

Simplicity and Nonlinearity

Thus far, symmetry arguments have helped us address the question of simplicity from the standpoint of the optimal number $\mathbf{r}^*$ of independent relationships a theory might contain on *a priori* grounds. Can we go one step further and ask how simple the equations themselves will be once found? In particular, should we aim to create a linear or nonlinear theory, and which is more simple? This is the final question we shall address, and we wish to make the case for nonlinearity.

From a purely mathematical standpoint, a set of nonlinear equations is clearly more complex than a corresponding set of linear ones. However, this is not necessarily the case from a *conceptual* standpoint. The history of science bears witness to a tendency for unsatisfactory linear theories of a piecemeal nature to be replaced by more unified nonlinear theories. These nonlinear theories typically strike us as "elegant" -- a concept often associated with simplicity. The emergence of general economic equilibrium theory, of NTU-bargaining theory, and of general relativity are cases in point.

On what grounds can these theories be deemed simple? Their nonlinear nature gives expression to the rather comforting adage that "everything *does* depend upon everything else", that everything is coupled in a generalized Machian sense. Such coupling between the components of a system is sometimes evidenced mathematically by the emergence of wholly new variables in nonlinear models, e.g. Harsanyi's intrinsic utility weights, and Einstein's metrical tensor indicating the (local) coupling of mass and the force of gravity.

Yet aesthetic considerations are not the only grounds for a prejudice in favor of nonlinearity. A more pragmatic reason is that those nonlinear models that supersede linear models are more satisfactory from a predictive and an explanatory standpoint. Of particular interest here is the fact that antecedent linear models often admit of -- and sometimes rely upon -- *probabilistic interpretations* that bespeak their incomplete and problematic evolutionary status. Three brief examples from game theory, quantum physics, and aesthetics support these assertions.

Game Theory: Shapley's 1953 axiomatic characterization [37] of a unique value for transferable-utility n-person games was highly significant. The formalism is fully linear and in this regard makes essential use of a strong additivity axiom, namely "when two independent games are combined, their values must be added player by player". From the start, Shapley's value suffered from two fundamental criticisms. First, it was only defined for transferable-utility games, a very restricted class. Second, being deductive rather than constructive in spirit, it lacked a compelling behavioral rationale. People could rightly ask: "Why would **n** players in a real-world bargaining game ever agree to a Shapley value outcome?"

Shapley himself addressed this second question at the conclusion of his original paper. As is well-known, he introduced the following probabilistic model of a constructive nature that yielded his value: each player is awarded his expected marginal contribution to the set of players that precede him, where the expectation is taken with respect to the uniform distribution over the set of all orders of the players. A criticism of this scheme is that a player is awarded his *entire* (utility) contribution to a coalition. This violates an intuitive symmetry whereby we would expect the player to receive only a portion of his contribution to the coalition (but see Rothblum [29] for an alternative interpretation).

Harsanyi's fully nonlinear 1963 bargaining solution for the n-person game took an altogether different approach to payoff determination. The model is constructive in nature, as we have seen above. Moreover, it is a normal form model that pays explicit attention to rational strategy determination both by pairs of individuals and by all possible coalitions. Probabilistic considerations play no role at all. Yet Harsanyi's solution turns out to be equivalent to an NTU-generalization of Shapley's value, and the classical TU-value emerges as a very special case of this generalization. In short, Harsanyi solved both problems cited above: he generalized the value to the NTU case -- something Shapley was to do in a slightly different manner six years later; and he rendered Shapley's combinatorial model an unnecessary source of support for the solution.

Quantum Physics: An interesting parallel to these developments in game theory took place in physics, albeit over a much longer time horizon. The advent of quantum mechanics in the 1920s was a milestone of twentieth century science. This theory is fully linear and intrinsically probabilistic. From the very start, however, debates raged as to the proper interpretation of quantum physics, especially as regards its probabilistic dimension and its lack of meaningful causality. Consider now Sachs' generalization of relativity theory discussed just above. It turns out that in the limiting "flat" space-time domain of low energy-momentum transfer, Sachs' matter field equations reduce to the linear eigenfunction form of the Dirac spinor-field equations of quantum physics -- equations which generalize the classical Shroedinger equation.

The standard quantum physical numerical results then follow from Sachs' theory [32], even the Lamb shift of quantum electro-dynamics. Yet these results are explicitly derived from a nonlinear, nonsingular, generally (not just specially) covariant, fully deterministic field theory *devoid of probabilistic considerations.* In commenting upon this theory and its ability to incorporate quantum phenomena, the quantum physicist and philosopher Clive Kilmister has remarked [31, p. xii]: "As one reads through this monograph it is possible to formulate a much deeper idea: that algebraic structures in general and quaternion algebra in particular arise because they are the key to a basic discreteness." Regrettably, lack of space prevents a further discussion of this matter, but we refer the reader to an excellent new book [33] which discusses these developments in the context of the great debates about physics (e.g. Einstein versus Bohr) during the past century.

Aesthetics: Surprisingly little has been done to explain at a fundamental and formal level what it is that makes things beautiful. Almost all of the classical literature dealt with applications and generalizations of the Pythagorean concept of the "golden mean" and (more recently) its relationship to Fibonacci series. Unfortunately, the domain of such theories is intrinsically limited. A significant advance resulted from an abstract attack on the problem by the mathematician G. D. Birkhoff [3] in 1930. Birkhoff sought a theory that could explain or at least predict "the quality of aesthetic experience" in a broad range of situations.

The theory he constructed was linear, and explained beauty in terms of density of order relations in an aesthetic object. Specifically, let **C** represent the *complexity* of an object. This measures the degree of "thoughtful perception" needed to understand an object. Let **O** represent the *order* of the object. Basically, this represents the extent of symmetry or harmony within the object. Then Birkhoff's measure of beauty **M** is given by $\mathbf{M} = \mathbf{O}/\mathbf{C}$.

This measure has not proven very successful. First, it postulates that the more complex, the less beautiful (for a given order **O**), and vice versa. Yet people react positively to "just enough" complexity, other things being equal, and they are bored by too much simplicity. Second the concept of order (symmetry) is not well defined. Some of Birkhoff's proposals for characterizing order (e.g. "undue repetition") reduplicate what is captured by complexity. As a result, the theory is internally redundant. Third, his measures of both order and complexity are subjective and can be given a problematic probabilistic interpretation. Nonetheless, Birkhoff was clearly on the right track, because questions of harmony, order, and complexity *are* intrinsic to aesthetics.

A nonlinear theory proposed independently by the author [6] overcomes these three problems, while retaining Birkhoff's fundamental insight. Briefly, most any object ranging from a Chippendale chair to a Bach fugue can be analyzed in terms of a theme, denoted by **t**, and its transformations ("echoes") $\mathbf{T_1(t),\dots,T_n(t)}$

elsewhere in the object. Indeed, any identifiable portion of the design can be designated as its theme (its *line element*, in effect), and all other portions can then be expressed as transformations of this theme. Note that this is as true for physical objects as it is for music possessing polyphonic structure.

Now define the complexity of the theme as the degree of the polynomial that best approximates its graph, and denote this by $\mathbf{C(t)}$. Next, define the complexity of any transformation $\mathbf{T_i(t)}$ by its degree, and denote this measure by $\mathbf{C[T_i(t)]}$. For simplicity, let us assume there is only one transformation so we can write $\mathbf{C[T(t)]}$. (More generally, one can define an "average complexity" $\mathbf{C_a[T_i(t)]}$ for the various transformations.)

Then the proposed measure of beauty is a quadratic form in $\mathbf{C(t)}$ and $\mathbf{C[T(t)]}$. Without going into details, this form expresses the idea that the more complex the theme is of an object we deem beautiful, then the more simple the transformations of that theme will be, *and vice versa*. In short, the new measure trades-off two different dimensions of simplicity/complexity, and it does so using objectively measurable criteria. "Boring" objects (simple themes given simple transformations) get a low score, but so will "overly complex" objects (complex themes given complex transformations).

Note the critical manner in which nonlinearity enters here. The measure does not contrast the complexity of one portion of the object (say the theme) with that of another portion *per se*, but rather with the complexity of the transformation mapping the theme into the other portion. What thus matters is how much aesthetic energy must be expended in shifting one's perception from one portion to another. Viewed relativistically, different starting points (themes) serve to *rescale* the complexity of the remaining portions of an otherwise identical object, and in doing so alter its beauty. Everything is coupled to everything else in this precise sense.

The examples we have given from three different fields support the idea that nonlinear theories can prove simpler than yet superior to their linear counterparts -- counterparts that are often antecedents. The claim for their simplicity lies in their "holism", itself a manifestation of the intrinsic coupling of all components of the system. The claim for their superiority stems from their success in overcoming the problems of their antecedents, and in offering a more profound *explanation* for the phenomena being analyzed. Recent chaos theory models not cited above further embarrass the linear stochastic tradition, and support our claim. All in all, both aesthetic and pragmatic considerations suggest an *a priori* prejudice for nonlinearity in efficient theory construction.

Conclusion

While the importance of symmetry as a concept and a phenomenon has long been appreciated (e.g. Weyl [41]), its role in the philosophy of science has not been. Yet when the history of twentieth science is written, the burgeoning efficacy of symmetry arguments in theory construction may itself emerge as one of the fundamental conceptual advances of our time. We have tried to substantiate this view in three ways. *First*, we have focused on the work of one particularly distinguished foundations thinker, and ascribed a good measure of his success to skill in identifying and exploiting apposite symmetries in game theory and ethics.

Second, we have attempted to abstract the fundamental reason why symmetry arguments should be expected to play a pivotal role in efficient theory construction, both in the social and natural sciences. This reason lies in the desirability and efficacy of a deep principle of epistemological consistency. Happily, this principle leads naturally to the utilization of group theory and other powerful modes of deduction. *Third*, drawing upon algebra, we have introduced the notion of the optimal simplicity/complexity of a yet-to-be-constructed theory, a notion based on symmetry considerations that turn out to be measurable. In doing so, we have sketched the nature of an abstract Expert-System designed to help automate a highly "creative" and efficient logic of scientific discovery.

Acknowledgment

Joseph Rosen and Mendel Sachs have been very helpful regarding the logic of the Expert-System. The author has also benefited from discussions with Kenneth Arrow, Robert Aumann, John Harsanyi, Clive Kilmister, Michael Redhead, and Patrick Suppes. Finally, Robert Wood kindly commented upon and edited the entire paper. All errors and opacities are, of course, the author's own responsibility.

BIBLIOGRAPHY

1. Aumann, R. J.: "An Axiomatization of the Non-Transferable Utility Value", *Econometrica*, 53 (1985), 599-612.

2. Aumann, R. J. and M. Kurz: "Power and Taxes in a Multi-Commodity Market", *Israel Journal of Mathematics*, 27 (1977), 145-159.

3. Birkhoff, G. D.: "Mathematics of Aesthetics", in J. R. Newman (ed.), *The World of Mathematics*, Volume IV, Simon and Schuster, New York, 1956.

4. Brock, H. W.: "The Role of Symmetry Groups in Alternative Theories of Fair Division", *Proceedings of the Naval Postgraduate School Conference on Mathematics Linkages*, Monterey, 1978.

5. Brock, H. W.: "A Game Theoretic Account of Social Justice", *Theory and Decision*, 11 (1979), 143-151.

6. Brock, H. W.: "Algebraic Foundations of Aesthetics", Unpublished private correspondence to Messrs. Blairman and Sons, London, 1982.

7. Cox, R. T.: "Probability, Frequency, and Reasonable Expectation", *American Journal of Physics*, 14 (1946), 1-13.

8. Einstein, A. and W. Mayer: "The Group Theoretical Structure of Relativistic Physics", *Preuss. Akad. Wiss. Phys. Math. Klass Sitz.*, 27 (1932), 522-534.

9. Harsanyi, J. C.: "Approaches to the Bargaining Problem before and after the Theory of Games: A Critical Discussion of Zeuthen's, Hicks', and Nash's Theories", *Econometrica*, 24 (1956), 144-157.

10. Harsanyi, J. C.: "A Bargaining Model for the Cooperative n-Person Game", in A. W. Tucker and R. D. Luce (eds.), *Contributions to the Theory of Games*, IV, Princeton University Press, Princeton, 1959, 325-355.

11. Harsanyi, J. C.: "On the Rationality Postulates Underlying the Theory of Cooperative Games", *Journal of Conflict Resolutions*, 5 (1961), 179-196.

12. Harsanyi, J. C.: "Measurement of Social Power in n-Person Reciprocal Power Situations", *Behavioral Science*, 7 (1962), 81-91.

13. Harsanyi, J. C.: "A Simplified Bargaining Model for the n-Person Cooperative Game", *International Economic Review*, 4 (1963), 194-220.

14. Harsanyi, J. C.: "Games with Incomplete Information Played by 'Bayesian' Players", Parts I-III, *Management Science*, 14 (1967), 159-182, and (1968), 320-334 and 486-502.

15. Harsanyi, J. C.: "The Tracing Procedure: A Bayesian Approach to Defining a Solution for n-Person Cooperative Games", *International Journal of Game Theory*, 4 (1975), 61-94.

16. Harsanyi, J. C.: *Essays on Ethics, Social Behavior, and Scientific Explanation*, Part C, D. Reidel Publishing Company, Dordrecht, Holland, 1976.

17. Harsanyi, J. C.: "Rule Utilitarianism and Decision Theory", *Erkenntnis*, 11 (1977), 25-53.

18. Harsanyi, J. C.: *Rational Behavior and Bargaining Equilibrium in Games and Social Situations*, Chapter 4, Cambridge University Press, Cambridge, 1977.

19. Harsanyi, J. C. and R. Selten: *A General Theory of Equilibrium Selection in Games*, M.I.T. Press, Cambridge, 1988.

20. Hart, S.: "An Axiomatization of Harsanyi's Non-Transferable Utility Solution", *Econometrica*, 53 (1985), 1295-1313.

21. Jaynes, E. T.: "Information Theory and Statistical Mechanics, I", *Physical Review*, 106 (1957), 620-630.

22. Jaynes, E. T.: "Prior Probabilities", *I.E.E.E. Transactions on Systems Science and Cybernetics*, SSC-4 (1968), 227-241.

23. Kohlberg, E. and J. Mertens: "On the Strategic Stability of Equilibria", *Econometrica*, 54 (1986), 1003-1037.

24. Langley, P. W. and H. A. Simon: *Scientific Discovery: Computer Explorations of the Creative Process*, M.I.T. Press, Cambridge, 1987.

25. Popper, K.: *The Logic of Scientific Discovery*, Basic Books, New York, 1959.

26. Rasmusen, E.: *Games and Information*, Basil Blackwell, Oxford, 1989.

27. Rosen, J.: *A Symmetry Primer for Scientists*, John Wiley and Sons, New York, 1983.

28. Rosenkrantz, R. D.: "Editor's Introduction", in R. D. Rosenkrantz (ed.), *E. T. Jaynes: Papers on Probability, Statistics, and Statistical Physics*, D. Reidel Publishing Company, Dordrecht, Holland, 1983.

29. Rothblum, U. G.: "Combinatorial Representations of the Shapley Value based on Average Relative Payoffs", in A. E. Roth (ed.), *The Shapley Value: Essays in Honor of Lloyd S. Shapley*, Cambridge University Press, Cambridge, 1988.

30. Sachs, M.: "On the Most General Form of a Field Theory from Symmetry Principles", *Nature*, 226 (1970), 138-139.

31. Sachs, M.: *General Relativity and Matter*, D. Reidel Publishing Company, Dordrecht, Holland, 1982.

32. Sachs, M.: *Quantum Mechanics from General Relativity*, D. Reidel Publishing Company, Dordrecht, Holland, 1986.

33. Sachs, M.: *Einstein versus Bohr: The Continuing Controversy in Physics*, Open Court Publishing Company, La Salle, Illinois, 1988.

34. Schelling, T.: "For the Abandonment of Symmetry in Game Theory", *Review of Economics and Statistics*, 41 (1959), 213-224.

35. Selten, R.: "Bewertung Strategischer Spiele", *Zeitschrift fur die Gesamte Staatswissenschaft*, 116 (1960), 221-282.

36. Sen, A. K.: "Information and Invariance in Normative Choice", in W. Heller, R. Starr, and D. Starrett (eds.), *Social Choice and Public Decision-Making: Essays in Honor of Kenneth J. Arrow*, Volume I, Cambridge University Press, Cambridge, 1986.

37. Shapley, L. S.: "A Value for n-Person Games", in H. W. Kuhn and A. W. Tucker (eds.), *Contributions to the Theory of Games*, II, Princeton University Press, Princeton, 1953, 307-317.

38. Shapley, L. S.: "Utility Comparisons and the Theory of Games", in *La Decision: Aggregation et Dynamique des Ordres de Preference*, Editions du Centre National de la Recherche Scientifique, Paris, 1969, 251-263.

39. Shore, J. E. and R. W. Johnson: "Axiomatic Derivation of the Principle of Maximum Entropy and the Principle of Minimum Cross-Entropy", *I.E.E.E. Transactions on Information Theory*, IT-26 (1980), 26-37.

40. Simon, H. A.: *Models of Discovery*, D. Reidel Publishing Company, Dordrecht, Holland, 1977.

41. Weyl, H.: *Symmetry*, Princeton University Press, Princeton, 1952.

42. Wigner, E. P.: *Group Theory*, Academic Press, New York, 1959.

43. Zytkow, J. M. and H. A. Simon: "Normative Systems of Discovery and Search", *Synthese*, 74 (1988), 65-90.

John C. Harsanyi, System Builder and Conceptual Innovator

Reinhard Selten,
University of Bonn, Dept. of Economics, Adenauerallee 24–42, D–5300 Bonn, Germany

1. Introduction

John C. Harsanyi is one of the few system builders of our time. The principles of Bayesian rationality are his point of departure. They are applied to practical, ethical, and epistemological decision making. This leads to three branches of Bayesian theory:

1) Decision and game theory, concerned with "practical" decision making

2) Ethical utilitarianism, the application of Bayesian principles to moral decision making

3) Epistemological Bayesianism, the extension of Bayesian rationality to the philosophy of science.

The contributions of John C. Harsanyi to these three areas form a coherent system: a unified general theory of rational behavior emerges from his writings. This system has reached a high degree of maturity, even if he is still working on it. He certainly has the potential to continue to surprise us by unexpected new developments.

The system of John C. Harsanyi enriches traditional elements of Bayesian and utilitarian thinking with deep conceptual innovations. He is an avid reader of scientific and philosophic literature. Over the years he has gained a tremendous knowledge of a wide range of subjects in various disciplines. Due to his background his work is a meeting place of a number of different intellectual traditions. He makes use of what he finds valuable, but he does not overlook difficulties and inconsistencies which prevent the construction of a coherent system by merely compiling traditional ideas. He overcomes such difficulties by surprising conceptual innovations.

At a time when the impossibility of interpersonal utility comparisons was an almost universally accepted dogma of economic theory, John C. Harsanyi found a convincing argument to the contrary. This argument is so simple that in retrospect one may wonder why it was not recognized long before. The same is true for Harsanyi's solution of the problem of incomplete information in game theory, which also seemed to be basically unsolvable before.

It would be an exaggeration to say that all of the work of John C. Harsanyi has been concentrated on the construction of his system. Some of his papers are only loosely connected to this goal. Among them are pioneering contributions to game theory and imaginative reconceptualizations of basic issues within social science.

This paper offers an appraisal of the work of John C. Harsanyi. It has been written by a close friend who hopes that he can do justice to his thinking even if there are differences of opinion. The disagreements mainly concern the problem of bounded rationality. The extremely critical views of the author of this paper on Bayesianism as a descriptive theory of human decision making have been expressed elsewhere (Selten, 1990). Harsanyi has a more optimistic attitude towards the assumption of at least approximate Bayesian rationality in descriptive models. Criticism based on this disagreement will be suppressed in this essay.

The organization of this paper will be guided by the distinction among practical, ethical, and epistemological rationality. Section 2 to 6 are on Harsanyi's work in game theory including bargaining. Section 7 reports on his investigations of basic issues in the social sciences. Section 8 is devoted to his work on utilitarian ethics and section 9 to his contributions to the philosophy of science. The final section 10 contains some biographical remarks.

2. A general solution for cooperative games

Harsanyi's first paper on game theory (1956) connected Nash's cooperative theory with Zeuthen's bargaining theory. This paper was the first step towards his general bargaining model for cooperative n–person games (1959, 1963). Harsanyi's bargaining model may be looked upon as a modification and generalization of two important solution concepts: Nash's cooperative 2–person theory with variable threats and the Shapley–value. A formula for the Shapley value determines the bargaining equilibrium on the basis of the equilibrium payoffs of threat games between opposing coalitions.

Most cooperative solution concepts are based on characteristic function games with or without side payments. Harsanyi's bargaining model was the first cooperative n–person theory for normal form games. With the help of an ingenious construction involving equilibrium utility weights and dividends payed by coalitions to players he succeded to define a bargaining solution which looks very similar to an equilibrium point of a non–cooperative game, even if his model does not have the structure of a non–cooperative game.

In the time in which John C. Harsanyi conceived his bargaining model, it was the agreed upon opinion that fundamentally different approaches are needed for cooperative and non–cooperative games. Moreover, cooperative theory was considered to be much more important than non–cooperative theory. Why should players fail to cooperate, if it is profitable to do so? Only later, Harsanyi was one of the first game theorists to recognize the necessity to model game situations with cooperative opportunities as non–cooperative games. In what might be called the first period of Harsanyi's thinking on game theory, he did not yet take this point of view.

Harsanyi's views at the end of this first period are systematically exposed in the book on "Rational Behavior and Bargaining Equilibrium in Games and Social Situations" (1977). The book begins with an excellent exposition of Bayesian decision theory followed by a chapter on utilitarian ethics. Most of the book is devoted to the development of his ideas on cooperative games which lead to his n–person bargaining model. In a relatively small part at the end he presents a theory of equilibrium selection in non–cooperative games which goes back to a paper in the "Advances in Game Theory" (1964). Even if this theory is now superceded by the work of Harsanyi and the author (1988) this early attempt shows that already at this time Harsanyi was fully aware of the equilibrium selection problem.

Interestingly, in his 1977 book Harsanyi still expresses the opinion that a player in a non–cooperative game should sometimes play a maximin strategy rather than an equilibrium strategy, e.g. in 2–person games with a uniquely determined equilibrium point which is completely mixed. In such games the equilibrium payoffs are equal to the maximin values. He could not avoid this deviation from a strict application of Bayesian principles before he had found a satisfactory interpretation of mixed strategies. We shall come to this ingenious discovery in section 4.

Harsanyi's n–person bargaining model is still one of the most important theories within cooperative game theory. It would be desirable to find a convincing non–cooperative game model which supports this theory. However, this does not seem to be an easy task.

The book of 1977 appeared after a long time of preparation. "In fact, the first version of the book was completed in 1963" he writes in the foreword. I remember that he brought a type–written manuscript to the Jerusalem workshop on game theory in 1965. When the book was finally published, he was far beyond his first period of thinking on game theory.

3. The incomplete information breakthrough

In the sixties John C. Harsanyi belonged to a group of game theorists working for the consulting firm Mathematica on research contracts with the Arms Control and Disarmament Agency. It was the task of this group to apply game theory to problems of arms control. Harsanyi was interested in the problem of arms control negotiations between the two superpowers. He felt that it was necessary to model the incomplete information of each of both superpowers on the strategic possibilities and the preferences of its opponent. This seemed to be impossible in view of the standard assumption of game theory that all players have full knowledge of the rules of the game, including strategic possibilities and preferences of their opponents.

Harsanyi was not the first one to recognize the problem of incomplete information. In a Rand Corporation Memorandum, J.W. Milnor (1951) had proposed an approach in which players make maximin assumptions on unknown features of their opponents. However, Milnor's approach suffered from the well known counterintuitive features of the maximin criterion and never became popular among game theorists. There seemed to be no satisfactory solution to the problem of incomplete information. For this reason game theory was often severely criticized. The lack of an adequate method to deal with incomplete information was a serious restriction of the scope of game theoretic modelling.

The author remembers that already in 1961, John C. Harsanyi gave a talk on the problem to a small group of people at Princeton. At that time he had not yet found his surprisingly simple solution; it was first presented in a type-written manuscript which was exstensively discussed at the game theory workshop at Jerusalem in 1965. He proposed to model games with incomplete information as starting with a lottery performed by nature. This lottery jointly determines features of the players known to themselves but not to the opponents. According to the features a player may have, he can be of several different "types". (Originally, Harsanyi did not use the word "type"; it was suggested by Maschler at the Jerusalem workshop.) The players do not know the types of their opponents but they know the joint distribution over the types. Therefore each player can form a conditional distribution over the types of the other players.

The whole game including the initial lottery on types is an ordinary extensive game. With this formalism Harsanyi succeeded in reducing incomplete information to imperfect information. This was an ingeniously simple solution of the problem. In the heated discussion among the 17 participants of the Jerusalem workshop – at that time the number of game theorists was still very small – the objection was raised that the conditional probabilities of types over the types of opponents may not be "consistent" in the sense that they are derivable from a joint distribution over the possible type combinations. However, as the author pointed out at the Jerusalem workshop, also this kind of "inconsistent" incomplete information can be reduced to imperfect information. For this purpose every type is modelled as a different player and payoffs are determined by appropriate lotteries at the end of the game.

Harsanyi presented his theory of games with incomplete information in three papers published in Management Science (1967–68). Today the economic literature abounds with models based on this work. For those who have grown up with his ingenious solution of the problem of incomplete information, it may be hard to understand why it was so difficult to find it. It is a mark of greatness of Harsanyi's discovery of the key to incomplete information that this conceptual innovation is now taken for granted.

4. A new interpretation of mixed strategies

A player who plays a mixed strategy makes use of a random mechanism in order to decide which pure strategy to play. This is the traditional interpretation of the notion of a mixed strategy. Why whould a player behave in this way? From the Bayesian point of view, a player must choose a best reply to what he expects the other players to do. If a mixed strategy is a best reply, then all pure strategies to which it assigns positive probabilities are best replies as well. Therefore it is not clear why a player should use a particular genuinely mixed best reply rather than a pure one or another mixed one. Consider a game with a unique equilibrium point which requires a player to use a mixed strategy. Even in this case a recommendation to use this strategy does not seem to have a firm basis in Bayesian rationality. The player maximizes his payoff just as well by any other best reply, if the other players use their equilibrium strategies.

Harsanyi found an ingenious solution to this problem (1973a). He argues that in every real game situation there are small randomly fluctuating influences like a player's present mood which may work more or less in favor of some strategic choices or others. In a normal form game these influences can be modelled as small independent continuous random variables, one for each strategy of a player. The realizations of these random variables are known to the concerning player, but not to the other players. In this sense the realizations are private information. The distributions are common knowledge of all players. We shall refer to normal forms enriched by such random influences on the players' payoffs as games with fluctuating payoffs.

It can be seen immediately that games with fluctuating payoffs fit into Harsanyi's theory of games with incomplete information. A type corresponds to a joint realization of all random variables influencing a player's payoff. A pure strategy in a game with fluctuating payoffs is a function which assigns a pure strategy of the normal form to every possible type of a player.

Under appropriate technical assumptions games with fluctuating payoffs always have equilibrium points in pure strategies. Moreover, for almost all types of a player the best reply to given strategies of the other players is uniquely determined; any deviation diminuishes the type's expected payoff.

From the point of view of an outside observer who does not see the realizations of the random variables, the behavior produced by a combination of pure strategies for a game with fluctuating payoffs is undistinguishable from a combination of mixed strategies for the underlying normal form game. Harsanyi has shown that in this sense pure strategy equilibrium points of games with fluctuating payoffs converge to mixed strategy equilibrium points of the underlying normal form game, if the random influences approach zero. Every mixed strategy equilibrium point which satisfies certain mild regularity conditions can be reached in this way.

Harsanyi's theory of games with fluctuating payoffs provides a convincing interpretation of equilibrium points in mixed strategies. The players only apparently play mixed strategies. Actually they play pure strategies in a slightly different game in which their choices depend on small privately experienced influences on their payoffs. Undoubtedly, this reinterpretation of the notion of a mixed strategy equilibrium point is a conceptual innovation of great importance. Without it a cornerstone of Harsanyi's Bayesian approach to game theory would be missing.

5. Non-cooperative modelling of cooperation

It is now a widely held opinion that, in the last analysis, a fully detailed model of a game situation must be a non-cooperative game. If binding agreements are possible, the offers and counteroffers which lead to them should be modelled as formal moves in an extensive game. From this point of view, cooperative theories can be looked upon as shortcuts which need to be justified by more detailed non-cooperative models.

Nash's work on bargaining published in Econometrica (1950, 1953) and Aumann's theory of supergames (1959) are early examples of non-cooperative modelling of cooperation. However, a general need for the justification of cooperative theories by non-cooperative models was not felt in the time when cooperative game theory was dominant within game theory. Cooperation was assumed rather than explained when profitable binding agreements were possible.

John C. Harsanyi and the author have proposed a theory of two-person bargaining under incomplete information (1972). When we worked on this problem we found it necessary to model the institutional background of bargaining as the rules of a non-cooperative game. It was a natural consequence of our common work that we became more and more convinced of the necessity to build non-cooperative models of cooperation.

In this spirit, Harsanyi (1974) created a pioneering non-cooperative reconstruction of a major cooperative theory for characteristic function games, the stable sets of von Neumann and Morgenstern proposed in their book on the theory of games and economic behavior (1944). Harsanyi's bargaining model generates a sequence of payoff vectors for a zero-normalized characteristic function with transferable utility. The first member of the sequence is the zero vector. The transition from one payoff vector to the next is as follows. A chairman, who is motivated to keep the game going, selects a coalition. The members of this coalition simultaneously propose imputations. If all these imputations agree and if the jointly proposed imputation dominates the preceding payoff vector with respect to the coalition, the proposed imputation becomes the next payoff vector. Otherwise the game ends with payoffs according to the last payoff vector.

For a class of characteristic function games including all three-person games and all simple games, Harsanyi established a close connection between stable sets and certain families of stationary equilibrium points of his bargaining model satisfying some additional selection criteria. For characteristic function games outside this class he obtained a similar result for a modified stable set concept involving a notion of indirect dominance. If one looks at Harsanyi's arguments in detail, one obtains the impression that it is very reasonable not to ignore indirect dominance in the definition of internal and external stability. Maybe only in this way the intuitive ideas behind the stable set concept can find a satisfactory expression.

Harsanyi's paper on the non-cooperative reconstruction of the stable set concept is a path-breaking contribution to the investigation of non-cooperative models of coalition formation. Even if later work in this area was not directly inspired by his efforts, it is indebted to Harsanyi's early insight into the necessity of modelling coalition formation as a non-cooperative extensive game.

6. Equilibrium selection

Already in his early work Harsanyi was guided by the idea that it is the task of game theory to produce unique rational recommendations. The equilibrium point concept is a necessary condition for a theory which recommends a unique strategy to every player in a non-cooperative game; everything else would be a self-destructing prophecy. Therefore, a general theory of unique rational strategy recommendations for non-cooperative games must be a theory which selects one equilibrium point out of many in games with a multiplity of equilibrium points.

John C. Harsanyi and the author decided to work on the construction of an equilibrium selection theory of this kind around 1970 after the completion of the paper on bargaining under incomplete information (1972). Originally, we planned to extend this paper to the variable threat case, but then we decided to subordinate this task to an attack on the general problem of equilibrium selection. We worked very hard on this problem for many years. The final result was our book "a general theory of equilibrium selection in games" (1988).

On the way to our final theory we ended up discarding two antecedent theories based on different concepts of risk dominance. We heatedly discussed many ideas most of which we had to reject in the light of general principles and specific examples. Unfortunately, it is not possible in the framework of this paper to explain in much more detail why it took us so long to agree on our final theory.

A crucial turning point towards our third approach to equilibrium selection occured with Harsanyi's discovery of the tracing procedure (1975). This procedure describes a continuous process of deliberation which starts with a given prior distribution for a normal form game. The prior distribution can be any strategy combination. The process works in a continuum of auxiliary games, characterized by a parameter in the unit interval. This tracing parameter t expresses the degree of trust put into the preliminary result of the reasoning process. The payoff to a player in the auxiliary game is defined as a weighted average of two payoffs in the original game: the payoffs against the prior distribution and the payoffs against the strategies actually played by the other players. t is the weight of the payoffs resulting from the strategies actually played.

The process runs through a path of equilibrium points of the auxiliary games until finally an equilibrium point of the original game is reached for $t=1$. The path may have to backtrack, which means that some values of t are met several times but finally it ends with t=1. What has been sketched up to now is the linear tracing procedure which only generically yields a unique result. A modified version, the logarithmic tracing procedure, which involves an additional small logarithmic payoff term, always yields a unique result.

In our equilibrium selection theory, the tracing procedure has a very important place. It is used in several ways, e.g. in the definition of risk dominance and in the determination of a substitute for a set of solution candidates among which risk dominance comparisons fail to discriminate. Of course, the tracing procedure has to be combined with other ideas of equal importance which cannot be discussed in the limited framework of this paper.

Our book also contains several applications of our theory to non–cooperative bargaining models. Harsanyi always insisted on the importance of judging theoretical ideas in the light of applications. We do not think that our theory is the final solution of the equilibrium selection problem, but at least it can be said that it performs well in the analysis of specific problems.

7. Basic problems of social science

Even if more could be said on Harsanyi's achievements as a game theorist – e.g. his paper on the oddness of the number of equilibrium points (1973) has not been mentioned – we shall now turn our attention to his significance as a social scientist. In his paper "rational choice models of political behavior vs. functionalist and conformist theories" (1969), Harsanyi presents a clear and well argued exposition of his views on the foundations of social theory. At that time, the functionalist view was prevailing in sociology. The functionalist approach to social theory explains social institutions by their function for society as a whole. This is based on the idea that evolutionary pressures force social institutions to perform a function for the maintenance of society.

According to the functionalists, individual behavior is determined by conventional roles and social values internalized by the process of socialization. Harsanyi introduces the term "conformist" for this type of explanation of individual behavior. He argues that functionalism overstates the degree of consensus and integration found in real–world societies. Moreover, he points out that functionalist explanations are purely static and cannot account for the phenomenon of social change.

Harsanyi proposes a rational choice approach to social theory firmly based on methodological individualism. He maintains that, unlike functionalist theories, rational choice models are capable of providing explanations of dynamic social phenomena.

Harsanyi's methodological individualism does not presuppose that behavior is motivated by strictly selfish goals. He postulates that people are willing to act in the public interest, if the cost of doing so is sufficiently low. Moreover, they adopt commitments to their families, to their friends and to social institutions. These commitments are made rationally and then constrain the selfishness of behavior.

In two papers in Behavioral Science (1962), Harsanyi made a very important contribution to the understanding of the concept of power. Following Dahl (1957) he looks at power as the capability to induce others to do something which they would not do otherwise. In the first paper he investigates bilateral power in the framework of Zeuthen–Nash theory and in the second paper he analyses multilateral power with the help of his general bargaining model for n–person games (see section 2). In this work, Harsanyi identified several new dimensions of the concept of power, and imbedded the problem of its measurement into a general game theoretical context. In doing so, he dramatically generalized earlier measures of power such as those of Dahl (1962) and of Shapley–Shubik (1954).

In a further paper in Behavioral Science (1966) Harsanyi presents a bargaining model for social status. In his view the advantage of a position of high status consists in the willingness of others to treat the bearer with special courtesy and cooperativeness. The model explains how status within a group is conferred in exchange for services rendered to the group. It makes use of Harsanyi's work on power and interprets social status as a relationship based on power.

Even if Harsanyi's game theoretical approach to power and status may not exhaust all aspects of these important concepts it has greatly deepened our understanding of the problems involved.

8. Utilitarian ethics

In the middle of the 20–th century, it was the prevailing opinion in economic theory that interpersonal utility comparisons are impossible. Utilitarianism seemed to be a closed chapter in the history of economic doctrines. The revival of utilitarian ethics is due to the work of John C. Harsanyi. He explained how interpersonal utility comparisons are possible and provided an axiomatic basis for social welfare measured by the sum (or the average) of all individual utilities. His basic contributions to utilitarian ethics appeared in two papers published in the Journal of Political Economy (1953, 1955). Chapter 4 of his book on rational behavior and bargaining equilibrium in games and social situations (1977) contains a systematic exposition of his views on the subject matter. Since then he has further elaborated his position in a number of later papers which can be identified by their titles in the list of his publications. He is now writing a book on utilitarianism which will be a comprehensive statement of his ethical system.

Interpersonal utility comparisons require that one put oneself in the place of others. For example, one must imagine what it is like to be 10 centimeters taller or shorter. This is not more difficult than to imagine what it is like to become blind. Intrapersonal comparisons involving blindness must be made in the face of risks of accidents which may be taken or avoided.

Utility may be looked upon as a function of a person's position A in society and a system R of all parameters such as body size which determine one's tastes. Accordingly, utility can be written as a function

$$U = F(A,R)$$

which is the same for every individual. The function F expresses a natural law which determines the cardinal utility of a position A as a function of genetic and environmental influences summarized by R. If this function is known, social welfare can be computed as the sum (or average) of individual utilities. Of course, since F is not completely known, each moral decision maker must try to base his welfare comparisons on his best judgment. Therefore in practice different individuals will have different social welfare functions, but in principle such differences of opinion can be resolved by gathering relevant factual information.

The computation of social welfare as the sum of all individual utilities is based on the idea that somebody who wants to make an impartial welfare judgment must imagine to be put in the place of each individual with equal probability. In this way, Harsanyi connects welfare judgments to Bayesian decision making under risk.

A distinction must be made between two kinds of utilitarianism. Whereas act utilitarianism requires the maximization of social welfare with respect to every single action, rule utilitarianism commands adherence to a set of socially optimal rules. Harsanyi convincingly argues in favor of rule utilitarianism.

The problem of rule utilitarianism versus act utilitarianism is only one of a number of questions which must be answered by a fully elaborated utilitarian ethics. The "boundary problem" concerns the question whose utilities should be considered in the computation of social welfare. Should unborn members of future generations be included? The "purification problem" consists in the treatment of other-directed motivations like sadism or benevolence towards relatives. Utilities must be "purified" of such motivations before the summation is performed. Harsanyi has commented on these and other issues arising in the elaboration of utilitarian ethics and undoubtedly will discuss them extensively in the book he is writing.

Of course, Harsanyi's utilitarianism is not the only formal theory of ethics proposed in the literature. However, it is the only one in full agreement with Bayesian decision theory. As Harsanyi has pointed out in his critique of Rawls in the Journal of Political Economy (1975) and in his paper on non-linear social welfare functions in Theory and Decisions (1978), other ways of deriving social preferences from individual preferences violate Bayesian principles. More than any other theory of ethics, Harsanyi's utilitarianism fits into the framework of present day economics.

Harsanyi's paper "ethics in terms of hypothetical imperatives" (1958) does not concern the contents of ethical theory but the logical structure of ethical statements. He maintains that ethical imperatives are necessarily hypothetical in the sense that they are addressed only to those who want to conform to certain ethical criteria. He argues that Kant did not succeed in proving the non–hypothetical character of the Categorical Imperative.

9. Philosophy of science

Harsanyi's philosophy of science develops the idea of Bayesian epistemological rationality. Science as the rational pursuit of the accumulation and systematization of knowledge should be guided by Bayesian principles. Bayesian rationality is primarily a normative requirement, but Harsanyi also wants to argue that good scientific practice observed in reality is compatible with Bayesianism.

In his paper "Popper's improbability criterion for the choice of scientific hypotheses" (1960), Harsanyi comments on Popper's methodological rule that among several alternative hypotheses in agreement with the facts, the a priori least probable one should be tentatively accepted as true, since it has the greatest chance to be refuted by further research. For Popper the a priori least probable hypothesis is the simplest one and scientists in fact prefer the simpler hypothesis in case of doubt.

Harsanyi following Jeffreys (1931) argues that the simplest hypothesis should actually be considered to be the a priori most probable one; moreover that hypothesis should be judged to be the simplest one which involves the smallest number of independent assumptions. Accordingly, good scientific practice is in agreement with the selection of the most probable hypothesis if a priori probabilities are defined in the right way.

A further important contribution to the philosophy of science is Harsanyi's paper "acceptance of empirical statements: a Bayesian theory without cognitive utilities" (1985). In this paper Harsanyi describes scientific procedures as approximately Bayesian. For the purpose of simplifying the decision process, a hypothesis may be accepted as practically certain. This means that in decision making conditional probabilities based on this hypothesis will be used instead of true probabilities. Accepting a hypothesis as practically certain is justified if one can expect that doing this will not entail significant mistakes. This does not necessarily require that the probability of the hypothesis is very high. Harsanyi formulates a principle of small disparate risks which may justify the acceptance of hypotheses as practically certain whose probability is actually quite low. The ways in which a hypothesis of this kind may be wrong are so diverse, that the different risks involved in accepting it as true will largely cancel out.

In a paper entitled "mathematics, the empirical facts, and logical necessity" (1983), Harsanyi has developed deep thoughts on the nature of mathematical statements. He maintains that mathematical statements are partially empirical and therefore synthetic a posteriori. He argues that our evidence for the correctness of simple arithmetic statements like 2+3=5 is empirical, since it is based on counting real or imagined physical objects. Even the use of the axiomatic method does not completely eliminate the empirical contents of mathematical statements. In order to prove the consistency of an axiom system one must exhibit a model whose consistency is asserted as an empirical fact. However, the use of the axiomatic method reduces the empirical contents to the consistency assertion. Therefore statements derived as logical consequences of axiomatic systems have a special status; Harsanyi calls them "structure-analytic" in order to indicate that they are very much like analytic statements but with the consistency assertion as synthetic component.

In the same paper, one also finds an interesting discussion of the literature concerning axiomizations of arithmetics and set theory. Unfortunately, Harsanyi's condensed and clear exposition cannot be easily summarized within this relatively short appraisal of his work.

10. Biographical remarks

The preceding sections have tried to give an impression of the scope, the depth, and the originality of the thinking of John C. Harsanyi as far as this is possible within a necessarily superficial short summary of his admirable achievements. The reader may now be curious about his life history. What will be said in the following is partly based on personal conversations. Maybe not everything is correctly remembered. I hope I shall succeed to avoid major distortions.

John C. Harsanyi was born in Budapest, Hungary, on May 29, 1920. His father was a pharmacist. In 1937, John Harsanyi won the First Prize in Mathematics in the National Competition of Hungarian High School Students. As the winner of this prize he received a ring which he is still wearing today. At this time Hungarian high school education in mathematics was known for its quality. A success in the national competition was an extraordinary achievement. Nevertheless, John C. Harsanyi did not choose to study mathematics when he entered the university.

John Harsanyi once told me that he had never been attracted to the study of mathematical proofs even if he was strongly interested in the results of mathematics in view of their philosophical significance. He studied pharmacy in the expectation to inherit his father's pharmacy. He also studied philosohpy and in 1947, he received a doctorate in this discipline at the University of Budapest. As far as I know his dissertation was on philosophical errors (he does not mention it in his list of publications). He also studied botany for some time, but he did not develop a lasting interest in the subject matter.

He must have been a very serious student of philosophy. He even heard lectures on Aristoteles given in Latin by a monk. In the years 1947–48 he held the position of a University Assistant at the University of Budapest. One of the students who visited his lectures was a lady who often came late. Once he approached her after the course and reprimanded her for her annoying behavior. Nevertheless, she later became his wife.

John Harsanyi successfully finished his study of pharmacy and eventually took over his father's pharmacy. However, it was very difficult to run a privately owned pharmacy after the war in Hungary. This was one of the reasons why in 1950, he and his wife decided to evade the communist oppression. They illegally crossed the border to Austria and stayed there for a short time before they emigrated to Australia.

In Australia, John Harsanyi worked in factories for some time. However, he was not successful in his career as a factory worker, since he was unable to learn simple tasks like operating an electrical knife without cutting the wire. He therefore tried to find a clerical job and eventually became an employee of the association of the coal mining industry. As far as I know it was his task to prepare statistical overviews.

Presumably in order to improve his career opportunities he studied economics. He succeeded to publish scientific papers on international trade in which he discussed complicated elasticities. (These papers are not included in the list of his publications prepared by himself. I do not know where they were published.) On the basis of this work, he was appointed to the position of a Lecturer of Economics at the University of Queensland, Brisbane, Australia, in 1954.

In the academic year 1956/57 John Harsanyi had the opportunity to visit Stanford on a Rockefeller Fellowship. In the preface to the first collection of Harsanyi's papers (1976), Kenneth J. Arrow writes:

> When John Harsanyi came to Stanford University as a candidate for the Ph.D., I asked him why he was bothering since it was most unlikely that he had anything to learn from us. He was already a known scholar; in addition to some papers in economics, the first two papers in his volume had already been published and had dazzled me by their originality and their combination of philosophical insight and technical competence".

Arrow refers to the papers of 1953 and 1955 in the Journal of political economy (see section 9).

After his visit to Stanford, John Harsanyi spent a year as a Ressearch Assistant at the Cowles Commission. He then came back to Stanford as a Visiting Assistant Professor. He received his Ph.D. from Stanford University in 1959. He then returned to Australia and spent two years as a Senior Fellow at the National University at Canberra. However, he felt that he was too far away from the leading centers of scientific research. Therefore, in 1961, he accepted a position as a

Professor of Economics at Wayne State University where he stayed until he moved to Berkeley in 1964. For many years he held the position of a Full Professor at the Schools of Business Administration of the University of California, Berkeley. In 1983, this institution honored him by an endowed chair; he became Flood Research Professor in Business Administration. He retired in 1990.

During his time at Berkeley, he spent several years elsewhere. He was a Fellow of the Institute for Advanced Study in the Behavioral Sciences at Stanford in the academic year 1964/65. He visited the University of Bielefeld, Germany, for the years 1973/74 and 1978/79. In this time, I was teaching there. He also visited several other universities for shorter periods. The Harsanyis have always liked to travel. Their son, Tom Harsanyi, had already seen quite a number of countries before he went to school.

John Harsanyi is a Fellow of the Econometric Society since 1968 and a Fellow of the American Academy of Arts and Sciences since 1984. Two years ago, in 1989, the importance of his work was also recognized by a honorary degree from Northwestern University, Evanstone, Illinois.

As nearly everybody of advanced age, John Harsanyi has some minor health problems, but they do not seem to be serious ones. We can expect that he will continue to do path–breaking work for many years to come.

References

Aumann, Robert J., 1959, Acceptable Points in General n–Person Games, in: Tucher and Luce (eds.), Contributions to the Theory of Games, Vol. IV, Princeton, N.J., pp. 287–324

Arrow, K. J., 1976, Foreword to John C. Harsanyi, Essays on Ethics, Social Behavior, and Scientific Explanation, Reidel, Dordrecht, VII–VIII

Dahl, R.A., 1957, The Concept of Power, *Behavioral Science*, 2, pp. 201–215

Jeffreys, H., 1931, Scientific Inference, Cambridge University Press, Cambridge, England

Nash, John F., 1950, The Bargaining Problem, *Econometrica* 18, pp. 155–162

Nash, John F., 1953, Two Person Cooperative Games, *Econometrica* 21, pp. 128–140

von Neumann, J., and Oskar Morgenstern, 1944, Theory of Games and Economic Behavior, Princeton University Press, Princeton, N.J.

Selten, R., 1990, Evolution, Learning, and Economic Behavior, 1989 Nancy L. Schwartz Memorial Lecture, Kellogg Graduate School, Northwestern University

Shapley, L.S., and M. Shubik (1954), A method for evaluating the distribution of power in a committee system, American Political Science Review, 48, p. 787–792

For references to the work of John C. Harsanyi, see the list of his publications.

List of Publications of John C. Harsanyi

May 1991

A. Books

A_1. Essays on Ethics, Social Behavior, and Scientific Explanation. With Foreword by Kenneth J. Arrow. Dordrecht, Holland: D. Reidel, 1976. xvi + 2262 pp.

A_2 Rational Behavior and Bargaining Equilibrium in Games and Social Situations. Cambridge, England: Cambridge University Press, 1977. x + 314 pp.
Paperback edition in 1986.

A_2' Comportamento Razionale e Equilibrio di Contrattazione. Milano, Italy: Il Saggiatore, 1985. xxx + 475 pp. (Italian translation of A_2 by Simona Morini.)

A_3 Papers in Game Theory. Dordrecht, Holland: D. Reidel, 1982. xii + 258 pp.

A_4 A General Theory of Equilibrium Selection in Games (joint work with Reinhard Selten). With Foreword by Robert Aumann. Cambridge, MA: MIT Press, 1988. xiv + 378 pp.

A_5 L'utilitarianismo (Utilitarianism). Italian translation by Simona Morini. Milano, Italy: Il Saggiatore, 1988. xxix + 169 pp.

B. Articles

1. "Cardinal Utility in Welfare Economics and in the Theory of Risk-taking," Journal of Political Economy, 61 (1953), 434–435. Reprinted in A_1.
2. "The Research Policy of the Firm," Economic Record, 30 (1954), 48–60.
3. "Welfare Economics of Variable Tastes," Review of Economic Studies, 21 (1953–54), 204–213.
4. Cardinal Welfare, Individualistic Ethics, and Interpersonal Comparisons of Utility," Journal of Political Economy, 63 (1955), 309–321. Reprinted in A_1.
5. "Approaches to the Bargaining Problem Before and After the Theory of Games: A Critical Discussion of Zeuthen's, Hichs's and Nash's Theories," Econometrica, 24 (1956), 144–157. Reprinted in A_3.
6. "On Incentives to Forecasters and to Decision Makers Under Uncertainty," Cowles Foundation Discussion Paper, 48 (1958).
7. "Ethics in Terms of Hypothetical Imperatives," Mind, 67 (1958), 305–316. Reprinted in A_1.
8. "A Bargaining Model for the Cooperative n-Person Game," Ch. 17 in A.W. Tucker and R.D. Luce (eds.), Contributions to the Theory of Games, 4. Princeton, NJ: Princeton University Press, 1959. 325–355.

9. "Explanation and Comparative Dynamics in Social Science," Behavioral Science, 5 (1960), 136–145. Reprinted in A_1.

10. "Popper's Improbability Criterion for the Choice of Scientific Hypotheses," Philosophy, 25 (1960), 332–340. Reprinted in A_1.

11. "Theoretical Analysis in Social Science and the Model of Rational Behavior," Australian Journal of Politics and History, 7 (1961), 60–74

12. "On the Rationality Postulates Underlying the Theory of Cooperative Games," Journal of Conflict Resolution, 5 (1961), 179–196. Reprinted in A_3.

13. "Measurement of Social Power, Opportunity Costs, and the Theory of Two–Person Bargaining Games," Behavioral Science, 7 (1962), 67–80. Reprinted in A_1.

14. "Measurement of Social Power in n–Person Reciprocal Power Situations," Behavioral Science, 7 (1962), 81–91. Reprinted in A_1.

15. "Models for the Analysis of the Balance of Power in Society." In the Proceedings of the 1961 International Congress for Logic, Methodology and Philosophy of Science. Stanford, CA: Stanford University Press, 1962, 442–462.

16. "Bargaining in Ignorance of the Opponent's Utility Function," Journal of Conflict Resolution, 6 (1962), 29–38.

17. "Rationality Postulates for Bargaining Solutions in Cooperative and in Non–cooperative Games," Management Science, 9 (1962), 141–153.

18. "Mathematical Models for the Genesis of War" (a review article on Lewis F. Richardson's work), World Politics, 14 (1962), 687–699

19. "A Simplified Bargaining Model for the n–Person Cooperative Game," International Economic Review, 4 (1963), 194–220. Reprinted in A_3.

20. "A General Solution for Finite Non–cooperative Games, Based on Risk–Dominance,". Ch. 29, in M. Dresher, L.S. Shapley, and A.W. Tucker (eds.), Advances in Game Theory. Princeton, NJ: Princeton University Press, 1964, 651–679.

21. "Bargaining and Conflict Situations in the Light of a New Approach to Game Theory," American Economic Review, 55 (1965), 447–457.

22. "Some Social–Science Implications of a New Approach to Game Theory," in Kathleen Archibald (ed.), Strategic Interaction and Conflict. Berkeley, CA: University of California, Institute of International Studies, 1966, 1–18.

23. "Game Theory and the Analysis of International Conflicts," Australian Journal of Politics and History, 11 (19965), 292–304. Reprinted in A_1.

24. "A General Theory of Rational Behavior in Game Situations," Econometrica, 34, (1966), 613–634.

25. "A Bargaining Model for Social Status in Informal Groups and Formal Organizations," Behavioral Science, 11 (1966), 357–369, Reprinted in A_1.

26. "Games with Incomplete Information Played by 'Bayesian' Players. Part I: The Basic Model," Management Science, 14 (November 1967), 159–182. Reprinted in A_3.

27. "Games with Incomplete Information Played by 'Bayesian' Players. Part II: Bayesian Equilibrium Points," Management Science, 14 (January 1968), 320–334. Reprinted in A_3.

28. "Games with Incomplete Information Played by 'Bayesian' Players. Part III: The Basic Probability Distribution of the Game," Management Science, 14 (March 1968), 486–502. Reprinted in A_3.

29. "Individualistic versus Functionalistic Explanations in the Light of Game Theory". In I. Lakatos (ed.), Problems in the Philosophy of Science. Amsterdam: North–Holland, 1968, 305–321 and 327–348.

30. "Rational–Choice Models of Political Behavior vs. Functionalist and Conformist Theories," World Politics, 21 (19699), 513–538. Reprinted in A_1.

31. "A Generalized Nash Solution for Two–person Bargaining Games with Incomplete Information" (joint paper with Reinhard Selten), Management Science, 18 (January 1972), Part II), 80–106.

32. "A Simple Probabilistic Model of Nuclear Multipolarity." Ch. 6 in R. Rosecrance (ed.), The Future of the International Strategic System. San Francisco, CA: Chandler, 1972, 93–102.

33. "Notes on the So–called Incompleteness Problem," Theory and Decision, 2 (1972), 342–352.

34. "Games with Randomly Disturbed Payoffs: A New Rationale for Mixed Strategy Equilibrium Points," International Journal of Game Theory, 2 (1973), 1–23. Reprinted in A_3.

35. "Oddness of the Number of Equilibrium Points: A New Proof," International Journal of Game Theory, 2 (1973), 2335–250. Reprinted in A_3.

36. "An Equilibrium–point Interpretation of Stable Sets and a Proposed Alternative Definition," Management Science, 20 (1974), 1472–1495. Reprinted in A_3.

37. "Can the Maximin Principle Serve as a Basis for Morality? A Critique of John Rawls's Theory," American Political Science Review, 69 (1975), 594–606. Reprinted in A_1.

38. "Nonlinear Social Welfare Functions: Do Welfare Economists Have a Special Exemption from Bayesian Rationality?" Theory and Decision, 6 (1975), 311–332. Reprinted in A_1.

39. "The Tracing Procedure: A Bayesian Approach to Defining a Solution for n–Person Noncooperative Games," International Journal of Game Theory, 4 (1975), 61–94.

40. "A Solution Concept for n–Person Noncooperative Games," International Journal of Game Theory, 5 (1977), 211–225.

41. "Rule Utilitarianism and Decision Theory," Erkenntnis, 11 (1977), 25–53.

42. "Liberty Under Socialism and the New Socialist Man: Comments on Dr. Cohen's Paper," Erkenntnis, 11 (1977), 427–428.

43. "The Problem Solving Ability of the Rule Utilitarian Approach Should Not Be Underestimated: Comments on Professor Scanlon's Paper," Erkenntnis, 11 (1977), 435–438.

44. "Morality and the Prisoner's Dilemma Problem: Comments on Professor Baier's Paper," Erkenntnis, 11 (1977), 441–446,

45. "Time and the Flow of Information in Noncooperative Games." In H. Albach et al. (eds.), Quantitative Wirtschaftsforschung (Festschrift for Wilhelm Krelle). Tübingen, West Germany: J.C.B. Mohr, 1977, 255–267.

46. "Advances in Understanding Rational Behavior." In R.E. Butts and J. Hintikka (eds.), Proceedings of the Fifth International Congress of Logic, Methodology and Philosophy of Science, Part II. Dordrecht, Holland: D. Reidel, 1977, 315–343. Reprinted in A_1.

47. "On the Rationale of the Bayesian Approach: Comments on Professor Watkin's Paper," in the same volume as item 46, 381–392.

48. "Nonlinear Social Welfare Functions: A Rejoinder to Professor Sen," in the same volume as items 46 and 47, 293–296.

49. "Morality and the Theory of Rational Behavior," Social Research, 44 (1977), 623–656.

50. "Sneed on Rawls's Theory of Social Institutions: Some Comments," Erkenntnis, 13 (1978), 225–230.

51. "Bayesian Decision Theory and Utilitarian Ethics," American Economic Review, 68 (1978), 223–228.

52. "A Solution Theory for Noncooperative Games and Its Applications to Cooperative Games." In P.C. Ordeshook (ed.), Game Theory and Political Science. New York: New York University Press, 1978, 39–95.

53. "A New General Solution Concept for Both Cooperative and Noncooperative Games," in Papers of the Rhineland–Westphalian Academy of Sciences, 287. Opladen, West Germany: Westdeutscher Verlag, 1979, 7–28. Reprinted in A_3.

54. "Rationality, Reasons, Hypothetical Imperatives, and Morality," in Hal Berghel et al., Wittgenstein, the Vienna Circle, and Critical Rationalism. Vienna, Austria: Verlag Hoelder–Pichler–Tempsky, 1979, 463–475.

55. "Bayesian Decision Theory, Rule Utilitarianism, and Arrow's Impossibility Theorem," Theory and Decision, 11 (1979), 289–317.

56. "Uses of Bayesian Probability Models in Game Theory." In D.H. Milnor (ed.), Science, Belief and Behaviour: Essays in Honour of R.B. Braithwaite. Cambridge, England: Cambridge University Press, 1980, 189–210. Reprinted in A_3.

57. "Rule Utilitarianism, Rights, Obligations, and the Theory of Rational Behavior," Theory and Decision, 12 (1980), 115–133. Reprinted in A_3.

58. "Analysis of a Family of Two–person Bargaining Games with Incomplete Information," International Journal of Game Theory, 9 (1979), 65–89.

59. "The Shapley Value and the Risk–Dominance Solutions of Two Bargaining Models for Characteristic–Function Games." In V. Böhm (ed.), Essays in Game Theory and Mathematical Economics in Honor of Oskar Morgenstern. Mannheim, West Germany: B.I. Wissenschaftsverlag, 1981, 43–68.

60. "Noncooperative Bargaining Models." In Deistler et al. (eds.), Games, Economic Dynamics, and Time Series Analysis. Vienna, Austria: Physica–Verlag, 1982, 49–58.

61. "Subjective Probability and the Theory of Games" and "Rejoinder to Professors Kadane and Larkey," Management Science, 28 (1982), 120–124 and 124–125.

62. "Some Epistemological Advantages of the Rule Utilitarian Position in Ethics." In P.E. French (ed.), Midwest Studies in Philosophy, VII. Minneapolis, MN: University of Minnesota Press, 1982, 389–402.

63. "Solutions for Some Bargaining Games under the Harsanyi–Selten Solution Theory," Parts I–II, Mathematical Social Sciences, 3 (1982), 179–191 and 259–279.

64. "Mathematics, the Empirical Facts, and Logical Necessity," Erkenntnis, 19 (1983), 167–192.

65. "Use of Subjective Probabilities in Game Theory," in B.P. Stigum and F. Wenstop (eds.), Foundations of Utility and Risk Theory with Applications. Dordrecht, Holland: D. Reidel, 1983, 297–310.

66. "Basic Moral Decisions and Alternative Concepts of Rationality," Social Theory and Practice, 9 (1983), 231–244.

67. "Bayesian Decision Theory, Subjective and Objective Probabilites, and Acceptance of Empirical Hypotheses," Synthese, 57 (1983), 341–365.

68. "Bargaining and Game Theory," Advances in Management Studies, 2 (1983), 47–73.

69. "Acceptance of Empirical Statements: A Bayesian Theory without Cognitive Utilities," Theory and Decision, 18 (1985), 1–30.

70. "Rule Utilitarianism, Equality, and Justice," Social Philosophy and Policy, 2 (1985), 115–127.

71. "Does Reason Tell Us What Moral Code to Follow, and, Indeed, to Follow Any Moral Code at All?" Ethics, 96 (1985), 42–55.

72. "On Preferences, Promises, and the Coordination Problem: Reply to Regan," Ethics, 96 (1985), 68–73.

73. "Individual Utilities and Utilitarian Ethics." In A. Diekmann et al. (eds.), Paradoxical Effects of Social Behavior: Essays in Honor of A. Rapoport. Vienna and Heidelberg: Physica–Verlag, 1986, pp. 1–12.

74. "Utilitarian Morality in a World of Very Half-hearted Altruists." In W.P. Heller et al. (eds.), Social Choice and Public Decision Making: Essays in Honor of K.J. Arrow, Vol. I. Cambridge, England: Cambridge University Press, 1986, pp. 57–73.

75. "Von Neumann–Morgenstern Utilities, Risk Taking, and Welfare." In G.R. Feiwel (ed.), Arrow and the Ascent of Modern Economic Theory. New York: New York University Press, 1987, pp. 545–558.

76. "Some Recent Developments in Game Theory." In G. Eberlein et al. (eds.), Theory and Decision. Dordrecht, Holland: D. Reidel, 1987, pp. 235–260.

77. to 79. "Bargaining." "Interpersonal Utility Comparisons." "Value Judgments," in The New Palgrave (an economic dictionary). Oxford, England: Blackwell, 1987

80. "The Tracing Procedure: A Self–Correcting Reasoning Procedure," Theory and Decision, 22 (1987), 7–23.

81. "Morals by Agreement." A review article on David Gauthier's book of the same title. Economics and Philosophy, 3 (1987), 339–373.

82. "Assessing Other People's Utilities," in B.E. Munier (ed.), Risk, Decision, and Rationality. Dordrecht, Holland: D. Reidel, 1988, pp. 127–138.
83. "Problems with Act Utilitariianism and Malevolent Preferences." A critique of Richard Hare's theory. In D. Seanor and N. Fotion (eds.), Hare and Critics. Oxford, England: Clarendon Press, 1988, pp. 89–99.
84. "Democracy, Equality, and Popular Consent." In I. Shapiro and G. Reeher (eds.), Power, Equality, and Politics. Boulder, CO: The Vestview Press, 1988, pp. 276–283.

GPSR Compliance

The European Union's (EU) General Product Safety Regulation (GPSR) is a set of rules that requires consumer products to be safe and our obligations to ensure this.

If you have any concerns about our products, you can contact us on ProductSafety@springernature.com

In case Publisher is established outside the EU, the EU authorized representative is:

Springer Nature Customer Service Center GmbH
Europaplatz 3
69115 Heidelberg, Germany

Batch number: 10848150

Printed by Printforce, the Netherlands